CELLULAR AND MOLECULAR BIOLOGY OF NEURONAL DEVELOPMENT

CELLULAR AND MOLECULAR BIOLOGY OF NEURONAL DEVELOPMENT

Edited by

Ira B. Black

Cornell University Medical College
New York, New York

PLENUM PRESS • NEW YORK AND LONDON

Library of Congress Cataloging in Publication Data

Main entry under title:

Cellular and molecular biology of neuronal development.

"Proceedings of a Symposium on Cellular and Molecular Biology of Neuronal Development held on St. Thomas, U.S. Virgin Islands, March 2–5, 1983" — Verso t.p.
Bibliography: p.
Includes index.
1. Developmental neurology — Congresses. 2. Cytology — Congresses. 3. Molecular biology — Congresses. I. Black, Ira B. II. Symposium on Cellular and Molecular Biology of Neuronal Development (1983: St. Thomas, V.I.) [DNLM: 1. Nervous system — Embryology — Congresses. 2. Neurons — Physiology — Congresses. 3. Nerve growth factors — Congresses. 4. Molecular biology — Congresses. WL 102.5 S9885c 1983]

QP363.3.C45 1984	591.1'88	83-26899

ISBN-13: 978-1-4612-9686-7   e-ISBN-13: 978-1-4613-2717-2
DOI: 10.1007/978-1-4613-2717-2

Proceedings of a symposium on Cellular and Molecular Biology of Neuronal Development held on St. Thomas, U.S. Virgin Islands, March 2–5, 1983

© 1984 Plenum Press, New York
Softcover reprint of the hardcover 1st edition 1984
A Division of Plenum Publishing Corporation
233 Spring Street, New York, N.Y. 10013

Contributors

Joshua E. Adler, Division of Developmental Neurology, Cornell Medical College, New York, New York 10021

Felicia B. Axelrod, Department of Pediatrics, New York University Medical Center, New York, New York 10016

Yves-Alain Barde, Department of Neurochemistry, Max-Planck-Institute for Psychiatry, D-8033 Martinsried, Federal Republic of Germany

Michael J. Bastiani, Department of Biological Sciences, Stanford University, Stanford, California 94305

Makonnen Belew, Department of Biochemistry (Faculty of Science), Uppsala University, S-751 22 Uppsala, Sweden

Cara Berman, Department of Molecular Biology, Genentech, Inc., South San Francisco, California 94080

Ira B. Black, Division of Developmental Neurology, Cornell Medical College, New York, New York 10021

Martha C. Bohn, Division of Developmental Neurology, Cornell Medical College, New York, New York 10021

Xandra O. Breakefield, Department of Human Genetics, Yale University School of Medicine, New Haven, Connecticut 06510

David E. Burstein, Department of Pharmacology, New York University School of Medicine, New York, New York 10016

Carmela M. Castiglione, Department of Human Genetics, Yale University School of Medicine, New Haven, Connecticut 06510

Gary Ciment, Department of Biology, University of Oregon, Eugene, Oregon 97403

Frank Collins, Department of Anatomy, University of Utah School of Medicine, Salt Lake City, Utah 84132

Michael Comb, Department of Chemistry, University of Oregon, Eugene, Oregon 97403

James L. Connolly, Department of Pathology, Harvard Medical School, Boston, Massachusetts 02115

Lisa M. Coussens, Department of Molecular Biology, Genentech, Inc., South San Francisco, California 94080

Stanley M. Crain, Department of Neuroscience and Physiology, and the Rose F. Kennedy Center for Research in Mental Retardation and Human Development, Albert Einstein College of Medicine, Bronx, New York 10461

Thomas J. Dull, Department of Molecular Biology, Genentech, Inc., South San Francisco, California 94080

Ted Ebendal, Department of Zoology, Uppsala University, S-751 22 Uppsala, Sweden

David Edgar, Department of Neurochemistry, Max-Planck-Institute for Psychiatry, D-8033 Martinsried, Federal Republic of Germany

Gerald D. Fischbach, Department of Anatomy and Neurobiology, Washington University School of Medicine, St. Louis, Missouri 63110

John Girdlestone, Department of Biology, University of Oregon, Eugene, Oregon 97403

Corey S. Goodman, Department of Biological Sciences, Stanford University, Stanford, California 94305

Alane Gray, Department of Molecular Biology, Genentech, Inc., South San Francisco, California 94080

Steven H. Green, Department of Pharmacology, New York University School of Medicine, New York, New York 10016

Lloyd A. Greene, Department of Pharmacology, New York University School of Medicine, New York, New York 10016

Edward Hawrot, Department of Pharmacology, Yale University School of Medicine, New Haven, Connecticut 06510

Edward Herbert, Department of Chemistry, University of Oregon, Eugene, Oregon 97403

Markus Hosang, Department of Neurobiology, Stanford University School of Medicine, Stanford, California 94305

Richard I. Hume, Department of Anatomy and Neurobiology, Washington University School of Medicine, St. Louis, Missouri 63110

Eugene M. Johnson, Jr., Department of Pharmacology, Washington University School of Medicine, St. Louis, Missouri 63110

G. Miller Jonakait, Division of Developmental Neurology, Cornell Medical College, New York, New York 10021

John A. Kessler, Division of Developmental Neurology, Cornell Medical College, New York, New York 10021

Nicole M. Le Douarin, Institut d'Embryologie du CNRS et du Collège de France, 94130 Nogent-sur-Marne, France

Pamela Toy Manning, Department of Pharmacology, Washington University School of Medicine, St. Louis, Missouri 63110

Marston Manthorpe, Department of Biology, School of Medicine, University of California, San Diego, La Jolla, California 92093

Keith A. Markey, Division of Departmental Neurology, Cornell Medical College, New York, New York 10021

Gerard Martens, Department of Chemistry, University of Oregon, Eugene, Oregon 97403

Lars Olson, Department of Histology, Karolinska Institute, Stockholm, Sweden

Paul H. Patterson, Biology Division, California Institute of Technology, Pasadena, California 91125

John Pearson, New York University Medical Center, New York, New York 10016

Edith R. Peterson, Departments of Neuroscience and Physiology, and the Rose F. Kennedy Center for Research in Mental Retardation and Human Development, Albert Einstein College of Medicine, Bronx, New York 10461

Pasko Rakic, Section of Neuroanatomy, Yale University School of Medicine, New Haven, Connecticut 06510

Lorna W. Role, Department of Anatomy and Neurobiology, Washington University School of Medicine, St. Louis, Missouri 63110

Haim Rosen, Department of Chemistry, University of Oregon, Eugene, Oregon 97403

P. John Seeley, Department of Pharmacology, New York University School of Medicine, New York, New York 10016

Åke Seiger, Department of Histology, Karolinska Institute, Stockholm, Sweden

Michael L. Shelanski, Department of Pharmacology, New York University School of Medicine, New York, New York 10016

Eric M. Shooter, Department of Neurobiology, Stanford University School of Medicine, Stanford, California 94305

Nicholas C. Spitzer, Department of Biology, University of California, San Diego, La Jolla, California 92093

Arne Sutter, Department of Neurobiology, Stanford University School of Medicine, Stanford, California 94305; present address: Freie Universität für Pharmakologie, Universitätsklinikum Charlottenburg, D-1000, Berlin 33, Federal Republic of Germany

Hans Thoenen, Department of Neurochemistry, Max-Planck-Institute for Psychiatry, D-8033 Martinsried, Federal Republic of Germany

Axel Ullrich, Department of Molecular Biology, Genentech, Inc., South San Francisco, California 94080

Ronald D. Vale, Department of Neurobiology, Stanford University School of Medicine, Stanford, California 94305

Silvio Varon, Department of Biology, School of Medicine, University of California, San Diego, La Jolla, California 92093

James A. Weston, Department of Biology, University of Oregon, Eugene, Oregon 97403

Introduction

A central problem in neurobiology concerns mechanisms that generate the profound diversity and specificity of the nervous system. What is the substance of diversification and specificity at the molecular, cellular, and systems levels? How, for example, do 10^{11} neurons each form approximately 10^4 interconnections, allowing normal physiological function? How does disruption of these processes result in human disease? These proceedings represent the efforts of molecular biologists, embryologists, neurobiologists, and clinicians to approach these issues.

The chapters in this volume are grouped by subject to present the varieties of methods used to approach each individual area. Section I deals with embryogenesis and morphogenesis of the nervous system. In Chapter 3, Weston and co-workers describe the use of monoclonal antibodies that recognize specific neuronal epitopes (including specific gangliosides) for the purpose of defining heterogeneity in the neural crest, an important model system. Immunocytochemical analysis reveals the existence of distinct subpopulations within the crest at extremely early stages; cells express neuronal or glial binding patterns at the time of migration. Consequently, interactions with the environment may select for predetermined populations. Le Douarin reaches similar conclusions in Chapter 1 by analyzing migratory pathways and developmental potentials in crest of quail–chick chimeras. Her work demonstrates that virtually all axial levels of crest are capable of expressing the entire spectrum of sensory and autonomic phenotypes on transplantation, but that only restricted phenotypes are expressed during normal development. Thus, selection occurs during normal ontogeny. Le Douarin concludes that the crest is composed not of a population of homogeneous, multipotential cells, but rather of separate pools of precursors for sensory and autonomic lineages. Within each line, precursors may give rise either to neurons or to nonneuronal, support cells. In Chapter 2, Rakic reaches remarkable analogous conclusions by studying the neocortex, cerebellum, and hippocampus of the rhesus monkey. Glial and neuronal cell lineages are distinguishable before most neurons are postmitotic. Moreover, glial phenotypic expression occurs prior to terminal mitosis. In summary, these studies suggest that cellular diversity is expressed far earlier in ontogeny than had been recognized formerly.

Experiments with identified neurons in the grasshopper, described by Bastiani and Goodman in Chapter 4, indicate that complex interactions between growth cones and the embryonic microenvironment determine pathway selection. In particular, growth cones interact through filopodial interpenetration. Specific and selective adhesivity of growth cones and their filopodia to "landmark" cells and structures may define pathway construction in the nervous system. This hypothesis would also suggest that cell-surface determinants change throughout ontogeny, a contention supported by the use of monoclonal antibodies. It need not be stressed that a variety of specific, selective ontogenetic processes may be mediated through such interactions, which, when deranged, result in disease.

What are some of the specific phenotypic characters that emerge during neuronal development? Can we begin to delineate underlying mechanisms? Section II is devoted to these issues. In Chapter 6, Spitzer describes the differentiation of spinal neurons *in vivo* and *in vitro* and draws a number of general conclusions. Conversion from Ca^{2+} to an Na^{+} action potential appears to be a general phenomenon, involving a wide variety of neuronal types. Further, stage-specific uncoupling of neurons also appears to be generalized. Finally, transient sensitivity to neurotransmitters is apparent during ontogeny. For example, transient glycine responsiveness occurs in Rohon–Beard neurons. Although the foregoing characters do change during ontogeny, the individual types of ion channels themselves are apparently stable in their constitution. In a series of perturbation experiments, Spitzer demonstrates that inhibition of the Ca^{2+} action potential does not affect neurite outgrowth or generation of Na^{+} components of the action potential. Since many of these characteristics are dependent on discrete gene products, examination in the future may be facilitated by gene-cloning techniques (see below). Black and colleagues describe in Chapter 8 the remarkable plasticity of transmitter phenotypic characters during development *in vivo,* a finding that parallels the work of Spitzer. Moreover, Black and colleagues indicate that dramatic plasticity persists into adulthood and perhaps for the life of the neuron. During embryogenesis, a population of presumptive neurons in the gut transiently expresses a number of catecholaminergic phenotypic characters, including the synthetic enzymes tyrosine hydroxylase (TH) and dopamine-β-hydroxylase, as well as endogenous catecholamines and the high-affinity uptake system for norepinephrine. However, after 3 days, by embryonic day 14.5, all these characters, save for uptake, are undetectable. The trophic protein nerve growth factor (NGF), or maternal stress, through the mediation of transplacental glucocorticoid hormones, causes persistence of these catecholaminergic characters. Consequently, a variety of environmental signals alter developmental phenotypic expression, in agreement with the work of LeDouarin, Bastiani and Goodman, and Weston and co-workers. Further, transient expression is not restricted to cells in the gut, but also occurs in cranial sensory ganglia of the rat embryo. Black and colleagues report that TH transiently appears in a rostrocaudal wave in the trigeminal, petrosal, nodose, and rostral dorsal-root sensory ganglia. It may be concluded that transient expression of phenotypic characters is generalized, as typified by transient transmitter responsiveness reported by Spitzer, transient epitope detectability cited by Bastiani and Goodman, and the work of Black and colleagues.

Work described by Black and colleagues further suggests that notable plasticity persists into adulthood. Neonatal and mature sympathetic neurons are capable of expressing the putative peptide transmitter substance P (SP). Transsynaptic impulses through the mediation of acetylcholine, postsynaptic depolarization, and transmembrane Na^+ flux normally suppress SP. However, conditions that reduce transsynaptic stimulation are associated with emergence of immunocytochemically detectable SP. It is not yet clear whether this represents the *de novo* appearance of SP. Persistence of plasticity into maturity may suggest that many "ontogenetic" mechanisms also govern function during adulthood and that common pathogenetic factors underlie diseases of development, maturity, and old age.

Patterson has been studying neuronal surface glycoproteins and glycolipids to identify mechanisms that may underlie specificity in synaptogenesis. In Chapter 5, he describes several potentially interesting molecules that may be related to transmitter expression and to endocytosis following transmitter release. This work may begin to elucidate the generation of specificity of neuronal networks and transmitter phenotype, and thus is closely related to issues described by Bastiani and Goodman.

Fischbach and colleagues describe early events in synaptogenesis at the neuromuscular junction in culture. Using a new assay approach, acetylcholine release by the advancing growth cone has been detected. The potential ontogenetic role of transmitter release prior to synapse formation is presently under investigation. These authors have also been studying the mechanisms underlying acetylcholine receptor accumulation on muscle membrane at the site of synapse formation. Evidence suggests that new receptors are inserted into the membrane at the synapse; lateral movement of already-inserted receptors in the membrane contributes little, if at all, to junctional receptor accumulation. Finally, a nerve-derived factor that elicits muscle receptor accumulation is presently being purified. These approaches are beginning to indicate how specific cellular interactions elicit functional differentiation, and underly the process of synaptogenesis itself.

NGF has been the most intensively studied neuronal growth factor, and Section III is entirely devoted to recent advances. In Chapter 9, Greene and colleagues present a model for the action of NGF on neurite outgrowth that consists of two separate and distinct components. The first involves rapid, local transcription-independent modulation of growth-cone shape, motility, and locomotion. The second involves a delayed, transcription-dependent stabilization of neurites, which may be mediated by alteration in the properties of microtubules. This particular model is also consistent with the previously presented "priming model" of NGF action on PC12 cells.

In Chapter 13, Sutter and colleagues describe studies on NGF receptors and identify low-affinity ($K_d = 10^{-9}$ M) and high-affinity ($K_d = 10^{-11}$ M) forms. Dose–response curves in several *in vitro* systems suggests that the high-affinity receptor mediates neurite outgrowth. Evidence suggests that NGF binding converts the low- to the high-affinity form. Cross-linking studies are now under way, in efforts to isolate the receptor molecules. Hawrot (Chapter 10) has used a culture system of primary dissociated sympathetic neurons to study the NGF receptor. Using this model, he has been able to characterize the retrograde transport of NGF *in vitro*, and he finds no evidence for nuclear accumulation of the protein. Moreover, using other cell types, he has found that mammalian nodose neurons, presumably derived from epibranchial placodes, appear to bind NGF. Finally, some human melanoma cell lines also bind NGF.

In Chapter 12, Crain and Peterson introduce a potentially valuable new tool for the analysis of NGF action and neurite outgrowth. The plant alkaloid taxol increases and prolongs the dependence of dorsal root ganglion neurons on NGF. This may provide a potentially valuable tool for the analysis of NGF action and may help in the understanding of diseases in which NGF is, in some manner, deficient. The role of NGF in diseases of neuronal development has been receiving increasing attention.

Johnson and Manning present in Chapter 11 a fascinating model that they use to analyze the effects of NGF. Guanethidine-induced destruction of sympathetic neurons appears to be a cell-mediated autoimmune disorder that may be prevented by NGF. This model has implications for the emerging field of neuroimmunology, in general, and for an analysis of autoimmune diseases of the nervous system and their relationship to NGF, in particular. Ullrich and colleagues present a major breakthrough with the cloning of the NGF gene. This work is described in greater detail in the discussion of Section V below.

The insights gained from the study of NGF have fostered the realization that a variety of different factors regulate neuronal ontogeny. Accordingly, Section IV describes some of these new factors. Thoenen and colleagues delineate in Chapter 16 the strategies required to define new neuronal growth and trophic factors. They specifically describe culture systems that are appropriate for characterizing new factors, for studying age-dependent changes that may alter detectability of factors, and for demonstrating that multiple factors may interact at the level of the neuron or the level of the culture substratum to confound results. Thoenen and colleagues outline potential approaches to be used in the purification and characterization of new neurotrophic factors and describe a new neurotrophic factor purified 1.4-million-fold from the pig brain. In Chapter 17, Varon and Manthorpe outline their research strategies for examining "neuronotrophic" and "neurite-promoting" factors. The authors have focused on peripheral and central cholinergic neurons, and they provide evidence that a number of factors may foster central as well as peripheral survival and neurite elongation. In addition, Varon and Manthorpe raise the intriguing possibility that there may also exist inhibitory factors that serve to regulate neurite elaboration and, possibly, the formation of neural connections.

Collins's studies, presented in Chapter 14, suggest that a variety of extrinsic factors may be responsible for appropriate directionality and rate of neurite outgrowth. These factors may include substrate conditioning factors, growth factors, and factors intrinsic to the neuron. In the last category are included stage-specific changes in neuronal responsiveness, as indicated by the example of ciliary-ganglion neurons. Collins also presents startling, persuasive evidence that NGF itself is capable of enchancing ciliary-neurite outgrowth at appropriate stages. This constitutes the first example of parasympathetic responses to NGF. In Chapter 15, Ebendal and co-workers provide evidence, using a newly evolved two-site radioimmunoassay in conjunction with a bioassay, that chick-embryo extract may contain multiple growth factors, one of which exhibits characteristics similar to those of classic NGF. It is evident that we have just begun to explore the area of neuronal growth factors, an area that promises to define new ontogenetic mechanisms and, perhaps, new disease mechanisms as well.

The generation of neuronal diversity and specificity involves interaction of genomic information with epigenetic factors. It is apparent, consequently, that molecular biology as a field has immense contributions to make to the understanding of neuronal development. In Section V, a number of contributions to this volume indicate that his promise is already being realized. Herbert and co-workers discuss in Chapter 18 the molecular biology of polyproteins, translation products that contain a series of biologically active molecules. The authors note that many of these polyproteins contain a series of peptides that serve to coordinate complex series of behaviors. Moreover, through differential posttranslational processing of the polyproteins, remarkable diversity and specificity may be acheived by the neuron, in particular, and the cell, in general. Herbert and colleagues describe features common to many of the polyproteins and their genes. For example, single exons code for all the biologically active peptides derived from pro-opiomelanocortin and proenkephalin. Moreover, in the genome, the sequence C_pG clusters at the $5'$ and $3'$ untranslated regions, and the methylation of cytosine residues has been implicated in gene expression during growth and development. Finally, these authors use this information to provide insights into the phylogeny of biologically active peptides and their genes. In aggregate, these observations serve to emphasize the power of recombinant DNA approaches in providing insights into mechanisms of neuronal growth and development. These studies suggest that many diseases of neural development may result from abnormalities at the level of the genome.

In Chapter 19, Ullrich and colleagues report the cloning of the NGF gene, which is of immense practical and theoretical importance. This pioneering work may allow us to understand the processing of the NGF molecule and abnormalities of NGF regulation. Moreover, employed in conjunction with restriction endonucleases, this work may make it possible to determine whether NGF is closely linked to abnormalities in familial dysautonomia as well as other developmental diseases. Additionally, cloning of the NGF gene may allow the production of NGF on an extensive basis. The potential clinical significance of this landmark achievement certainly need not be stressed. Breakefield and co-workers discuss in Chapter 20 the use of linkage analysis to determine whether the gene for the β subunit of NGF may be the locus of abnormality in the genetic autonomic-sensory disease familial dysautonomia. Use of the gene cloned by Ullrich and colleagues has made this work possible. The prospect of these studies exemplifies the synergistic nature of molecular biology, neurobiology, and clinical neurology and pediatrics. This may be called the beginning of a new era in our understanding of diseases of neural development.

Despite this optimistic note, the formidable problems facing the neuro-biologist and clinician are clearly delineated in Section VI by the contributions of Pearson and Axelrod. In Chapter 22, Pearson describes familial dysautonomia as well as congenital sensory neuropathies and outlines outstanding problems of pathogenesis. In Chapter 21, Axelrod describes the protean manifestations of familial dysautonomia, which must somehow be fathomed by the clinician and neurobiologist. Moreover, the apparent heterogeneity of other congenital sensory neuropathies is described. It seems safe to conclude that only the interdisciplinary inquiry represented by cellular and molecular neurobiology will be able to approach these problems.

It is with deep gratitude that all the participants in this symposium acknowledge the unwavering support of the National Foundation for Jewish Genetic Diseases, Inc., and the Dysautonomia Foundation. In particular, the guidance, advice, and generosity of George Crohn Jr. are most enthusiastically acknowledged. The efficiency and unstinting efforts of Joan Samsen Crohn are also most gratefully appreciated. The extraordinary talents, organizational capabilities, and unfailing cheerfulness of Elise Grossman are deeply appreciated.

Ira B. Black
Cornell University Medical College

Contents

Chapter 3

Heterogeneity in Neural Crest Cell Populations

James A Weston, John Girdlestone, and Gary Ciment

Chapter 4

The First Growth Cones in the Central Nervous System of the Grasshopper Embryo

Michael J. Bastiani and Corey S. Goodman

II. DEVELOPMENTAL EXPRESSION OF NEURONAL PHENOTYPIC CHARACTERS

Session Chairman: Ira B. Black

Chapter 5

Surface-Bound and Released Neuronal Glycoconjugates

Paul H. Patterson

Chapter 6

The Differentiation of Membrane Properties of Spinal Neurons

Nicholas C. Spitzer

Chapter 7

The Accumulation of Acetylcholine Receptors at Nerve-Muscle Synapses in Culture

Gerald D. Fischbach, Lorna R. Role, and Richard I. Hume

Chapter 8

Transmitter Phenotypic Plasticity in Developing and Mature Neurons in Vivo

Ira B. Black, Joshua E. Adler, Martha C. Bohn, G. Miller Jonakait, John A. Kessler, and Keith A. Markey

III.   NERVE GROWTH FACTOR AS A MODEL GROWTH FACTOR
Session Chairman: Eric M. Shooter

Chapter 9

Mechanisms of the Promotion of Neurite Outgrowth by Nerve Growth Factor

Lloyd A. Greene, David E. Burstein, James L. Connolly, Steven H. Green, P. John Seeley, and Michael L. Shelanski

Chapter 10

Cultured Sympathetic Neurons in the Study of Nerve Growth Factor Action

Edward Hawrot

Chapter 11

Guanethidine-Induced Destruction of Sympathetic Neurons: An Autoimmune "Disease" Prevented by Nerve Growth Factor

Eugene M. Johnson, Jr., and Pamela Toy Manning

Chapter 12

Enhanced Dependence of Fetal Mouse Neurons on Trophic Factors after Taxol Exposure in Organotypic Cultures

Stanley M. Crain and Edith R. Peterson

Chapter 13

The Interaction of Nerve Growth Factor with Its Specific Receptors

Arne Sutter, Markus Hosang, Ronald D. Vale, and Eric M. Shooter

IV.   NEW NEURONAL GROWTH FACTORS
Session Chairman: Hans Thoenen

Chapter 14

Multiple Sites for the Regulation of Neurite Outgrowth

Frank Collins

Chapter 15

Nerve Growth Factors in Chick and Rat Tissues

Ted Ebendal, Lars Olson, Åke Seiger, and Makonnen Belew

Chapter 16

Macromolecular Factors Involved in the Regulation of the Survival and
Differentiation of Peripheral Sensory and Sympathetic Neurons

Hans Thoenen, Yves-Alain Barde, and David Edgar

Chapter 17

Trophic and Neurite-Promoting Factors for Cholinergic Neurons

Silvio Varon and Marston Manthorpe

V. MOLECULAR BIOLOGY OF NEURAL DEVELOPMENT AND FUNCTION
Session Chairman: Xandra O. Breakefield

Chapter 18

Expression of Opioid Peptide Genes in Different Species

Edward Herbert, Michael Comb, Haim Rosen, and Gerard Martens

Chapter 19

Isolation and Characterization of DNA Sequences Coding for Mouse and Human β-Nerve Growth Factor

Axel Ullrich, Alane Gray, Cara Berman, and Thomas J. Dull

Chapter 20

Linkage Analysis in Familial Dysautonomia Using Variations in DNA Sequence in the β Nerve Growth Factor Gene Region: A Beginning

Xandra O. Breakefield, Carmela M. Castiglione, Lisa M. Coussens, Felicia B. Axelrod, and Axel Ullrich

VI.   DISEASES OF DEVELOPMENT
Session Chairman: Ira B. Black

Chapter 21

Familial Dysautonomia and Other Congenital Sensory and Autonomic Neuropathies

Felicia B. Axelrod

Chapter 22

Developmental Neurobiology of Human Diseases: Familial Dysautonomia and Related Disorders

John Pearson

I

Embryogenesis and Morphogenesis of the Nervous System

Session Chairman: James A. Weston

A Model for Cell-Line Divergence in the Ontogeny of the Peripheral Nervous System

Nicole M. Le Douarin

1. ORIGIN OF GANGLION CELLS IN THE PERIPHERAL NERVOUS SYSTEM

The neural crest gives rise to all neuronal and glial cells of the autonomic and spinal ganglia. Certain sensory ganglia of the cranial nerves are also derived entirely from the neural crest (proximal ganglia of nerves IX and X), while others have a more complex origin in that the epibranchial placodes contribute either to all (geniculate, petrosal, and nodose ganglia) or to some (trigeminal) of their neurons.

With the use of a cell-marking technique based on stable structural differences between nuclei of quail and chick cells (Le Douarin, 1969, 1973), the site of origin along the neural axis of the different peripheral ganglia was determined. By microsurgery, defined regions of the neural primordium were removed from a chick (or quail) embryo and replaced by the equivalent primordium from a quail (or chick) at the same developmental stage. Analysis of the resulting chimeras allowed a "fate map" of crest-cell derivatives to be constructed (Le Douarin and Teillet, 1973; for a review, see Le Douarin, 1982).

In the development of the autonomic nervous system (ANS), a regionalization of the neural crest could be recognized (Fig. 1): the sympathetic chain derives from the entire length of the neural crest, from the level of the 6th somite

NICOLE M. LE DOUARIN ● Institut d'Embryologie du CNRS et du Collège de France, 94130 Nogent-sur-Marne, France.

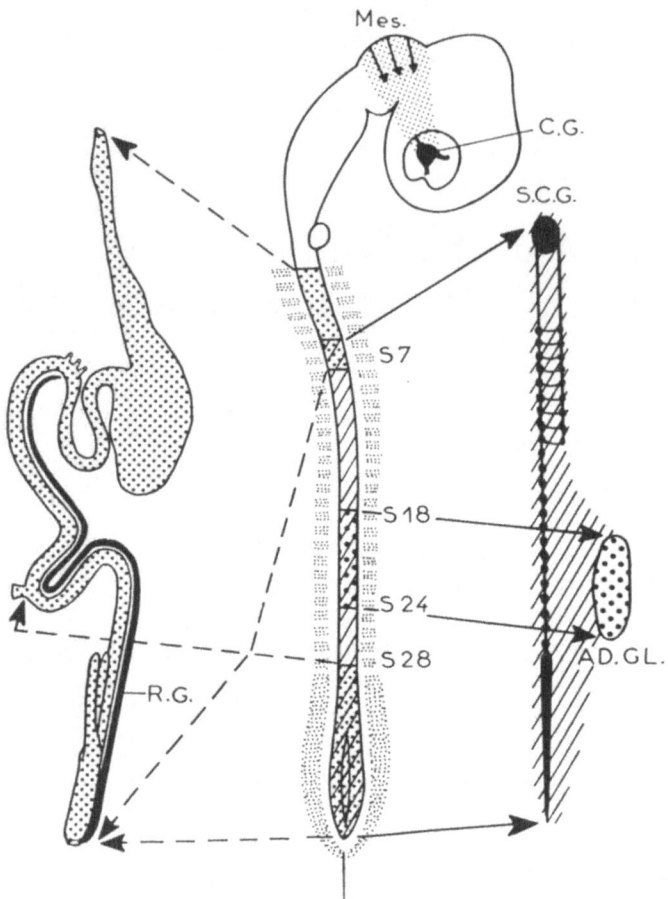

Figure 1. Levels of origin of the sympathetic ganglia and plexuses and the enteric and ciliary ganglia in the neural crest. The vagal level of the neural crest [somites (S)1–7] provides all the enteric ganglia of the preumbilical gut and contributes to the innervation of the postumbilical gut. The lumbosacral level of the neural crest gives rise to the ganglion of Remak (R.G.) and to some ganglion cells of the postumbilical gut. The ciliary ganglion (C.G.) arises from the mesencephalic crest (Mes.). The sympathetic chain and plexuses are derived from the level of the neural crest posterior to the S5, and the adrenomedullary cells originate from the level of S18–24. (AD.GL.) Suprarenal gland; (S.C.G.) superior cervical ganglion.

caudad, with the chromaffin cells of the adrenal medulla originating specifically from the level of somites 18–24. The great majority of enteric ganglia arise from "vagal" neural crest, opposite somites 1–7. Neural-crest cells from this region start migrating at around stage 8–10 somites in a ventral direction and become localized in the area of the branchial arches (Le Lièvre and Le Douarin, 1975;

Thiery *et al.*, 1982). The precursors of the enteric ganglia become incorporated into the developing wall of the foregut, which is of mesodermal origin. Thereafter, they migrate caudally along the gut, which they colonize down to the cloacal end, giving rise to the myenteric and submucosal plexuses. An additional, although minor, contribution to these structures in the postumbilical gut is made by the lumbosacral level of the crest, which gives rise essentially to the parasympathetic ganglion of Remak (Teillet, 1978).

It is evident from Fig. 1 that the cervicodorsal crest, located between somites 7 and 28, does not provide the developing gut with ganglionic cells. The migration of crest cells from this area is limited to the dorsal trunk structures, and, apart from the Schwann cells lining the nerves, these crest cells do not penetrate the dorsal mesentery.

The origin of the dorsal-root ganglia (DRGs) at the cervicotruncal level of the neural crest can be traced easily in quail–chick chimeras: the ganglia arise

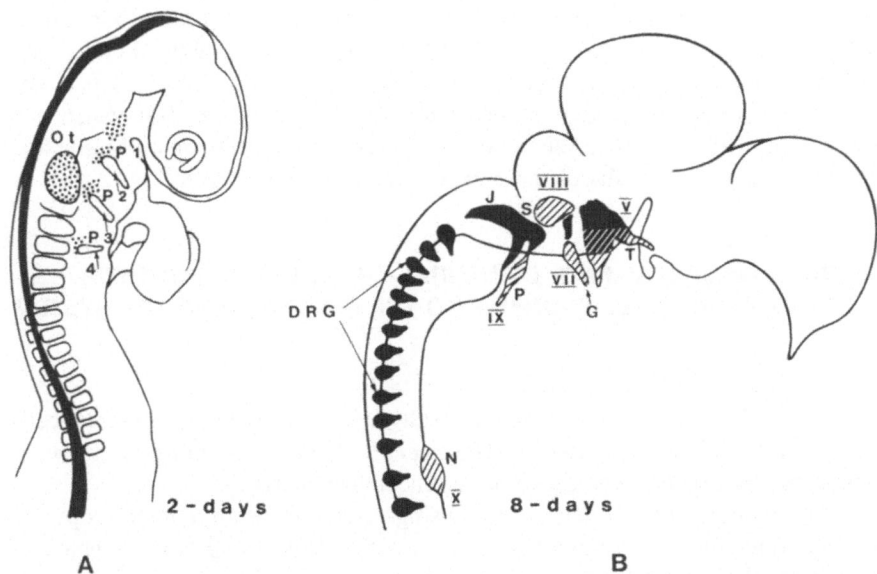

Figure 2. Diagram showing the cellular origin of the cranial sensory ganglia. (A) Localization of the ectodermal placodes located either in the region of the presumptive inner ear [otic placode (Ot)] or at the levels of branchial arches 1–4 [ectobranchial placode (P)]. (B) The spinal sensory ganglia (DRG) and the proximal ganglia of cranial nerves VII, IX [superior (S)], and X [jugular (J)] are entirely of neural crest origin. The distal cranial ganglia of nerves VII [geniculate (G)], IX [petrosal (P)], and X [nodose (N)] have a mixed crest and placodal origin, with the neurons originating from the placode and the satellite cells from the crest. The trigeminal (T) ganglion of nerve V is a composite ganglion in which the neurons originate from both the crest and a placode and the glia originate from the crest. VIII otic ganglion.

at the level of each somite from the corresponding transverse region of the neural crest from somite 6 downward.

Our knowledge of the ontogeny of the sensory ganglia located along the cranial nerves came first from the pioneering work of embryologists at the turn of the century and later who noticed that cells break off from the cephalic neurogenic placodes and form ganglia in conjunction with neural-crest cells. Descriptions of this sort were made in the embryos of fish (Von Kuppfer, 1894; Landacre, 1910, 1912), amphibians (Landacre and McLellan, 1912; Coghill, 1916), birds (His, 1868; Goronowitsch, 1893; Goldby, 1928), and, somewhat later, mammals (Adelmann, 1925; da Costa, 1931; Halley, 1955; Verwoerd and van Oostrom, 1979). Since cells of placodal and crest origin rapidly become indistinguishable from each other during ontogenesis, the conclusions drawn from the simple histological observations, although careful and detailed, were contradictory.

The second type of approach to this problem consisted of either labeling or removing the possible sources of ganglion cells, i.e., the placodes and the neural crest (Van Campenhout, 1937, 1940; Yntema, 1942, 1943, 1944; Hamburger, 1961; Johnston and Hazelton, 1972; Noden, 1973, 1975). Finally, use of the quail–chick chimera system enabled the respective roles of the placodal and crest cells to be defined more accurately (Noden, 1978; Narayanan and Narayanan, 1980; Ayer-Le Lièvre and Le Douarin, 1982; D'Amico-Martel and Noden, 1983). The results of these experiments are indicated in Fig. 2.

2. DEVELOPMENTAL FATE OF NEURAL-CREST CELLS AND THEIR TOPOGRAPHIC RELATIONSHIP TO THE CENTRAL NERVOUS SYSTEM

2.1. Level of the Trunk

A detailed study of the migratory pathways taken by the neural-crest cells at the trunk level has been done by Thiery *et al*. (1982), who followed migration within the fibronectin network of the extracellular matrix.

Observations of Tosney (1978), using scanning electron microscopy, as well as those from our laboratory, have revealed that during the first phase of migration, the crest-cell population expands laterally and covers the dorsal half of the neural tube. Then the cells stop for a while, as though the migration pathways were too narrow for them to proceed farther. Thereafter, three pathways become successively available for crest-cell migration. The first is located between two consecutive somites; the second and third consist of the spaces between the somite and the neural tube, on one hand, and the somite and the dorsal ectoderm, on the other (Fig. 3). For example, at the level of somite 15 in the chick embryo, lateral expansion of the neural folds starts from about the 18-somite stage. The cells opposite the bulk of the somite accumulate in the junction

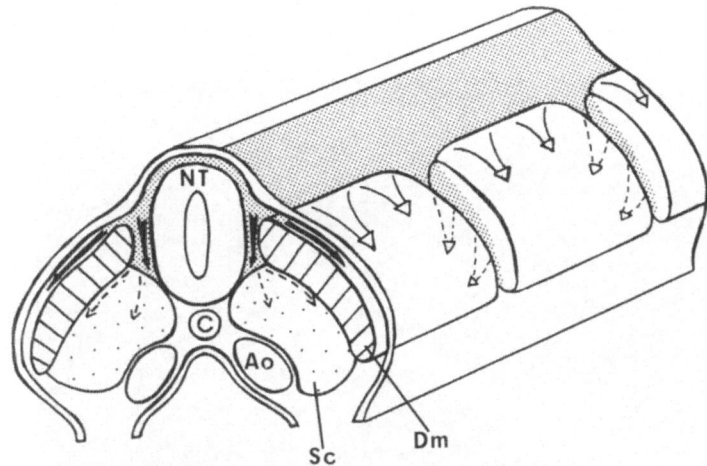

Figure 3. Diagram showing the three migratory pathways taken by the crest cells in the trunk region: the intersomitic pathway, the pathway located between the bulk of the somite and the neural tube (NT), and the superficial pathway underneath the ectoderm. (Ao) Aorta; (C) notochord; (Sc) sclerotome; (Dm) dermomyotome.

of the somite, the neural tube, and the superficial ectoderm, where they remain for more than 7 hr (Fig. 4). The dorsoventral pathway between somite and neural tube then becomes available, and cell migration can take place ventrally. This coincides with the dissociation of the somite into dermomyotome and sclerotome. The lateromedial expansion of sclerotomal cells, which will surround the axial organs (notochord and neural tube) to form the vertebrae, prevents the crest cells from reaching the paranotochordal region. As a result, they accumulate between the neural tube and the somitic mesenchyme and form the DRGs (Fig. 5). In contrast, the neural-crest cells located opposite the intersomitic space do not encounter this obstacle and utilize the narrow pathway formed between the basement membranes of the two adjacent somites (Fig. 6). This pathway, the first to appear, allows crest cells to reach the paranotochordal region rapidly, where they form the sympathetic chains, the aortic plexus, and, at the level of somites 18–24, the adrenal medulla (Fig. 7). The pathway located between the superficial ectoderm and the dermomyotome is taken by presumptive melanocytes, which later colonize the skin.

2.2. Rhombencephalic Level of the Neural Crest

These levels have been investigated in great detail, since they are the sites of origin of the sensory ganglia of cranial nerves VII, IX, and X and also of

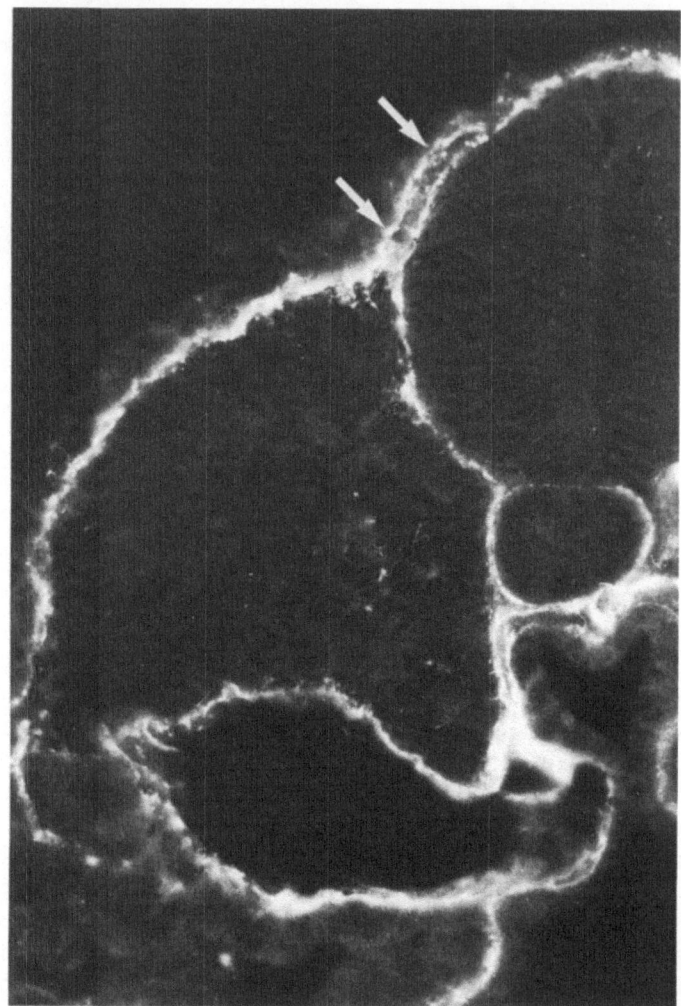

Figure 4. Level of the 15th somite at the 20-somite stage. Immunofluorescence labeling for fibronectin (FN). Crest cells (⇧) form a single layer between the dorsal ectoderm and the neural tube. The different tissues retain their compact epithelial structure. No space is available in the FN-rich basement membranes ahead of the front of migration. ×470. From Thiery *et al.*, (1982) with permission.

most of the enteric ganglia (Duband and Thiery, 1982; Thiery *et al.*, 1982; Ayer-Le Lièvre and Le Douarin, 1982; Teillet and Le Douarin, 1983). In this region, the crest cells migrate massively underneath the superficial ectoderm. Especially at the level of the first 5 somites, the dorsoventral pathway between somites and neural tube does not open up, and most crest cells take the superficial route, between dermomyotome and ectoderm, to reach the endodermal wall of the pharynx (Fig. 8).

It is striking that no DRGs are formed at the level of the first 5 somites, suggesting that their development is directly related to the availability of the dorsoventral pathway.

A detailed study has been carried out on the early steps of the formation of sensory ganglia of cranial nerves VII, IX, and X in quail–chick chimeras (Ayer-Le Lièvre and Le Douarin, 1982). At the cephalic and cervical levels of the body, growth proceeds more rapidly in the dorsal than in the ventral areas (see Le Douarin, 1964). This process, plus the onset of cephalic flexions, results in a segregation of the migrating crest cells into dorsal and ventral flows.

The dorsal group of crest cells remains close to the hindbrain and contributes to the proximal sensory ganglia of cranial nerves IX and X (i.e., superior and jugular ganglia), in which both neurons and glia are of crest origin.

In contrast, the ventral group of migrating cells reaches the ventrolateral region of the branchial arches. Most give rise to mesectodermal cells (see Le Lièvre and Le Douarin, 1975), but some remain in contact with the epibranchial placodes, and together they form the trunk (distal) ganglia of cranial nerves VII, IX, and X (geniculate, petrosal, and nodose), the neurons of which are derived from the placodes. In these ganglia, neuronal potentials of the neural crest are therefore not expressed, since the crest derivatives give rise to satellite cells exclusively.

The following conclusion can be drawn from this study: *sensory neurons develop from the neural crest only in ganglia located in close proximity to the brain and the spinal cord.*

3. DEVELOPMENTAL POTENTIALS OF THE NEURAL CREST

It has long been known that the developmental fate of a given anlage does not necessarily reflect the totality of developmental potentials of component cells. Isolation *in vitro* or heterotopic transplantation *in vivo* has often revealed broader developmental potentials than those that are actually expressed in normal development. This means that certain potentials are repressed in one way or another during the normal course of ontogeny. Whether they are repressed through an active inhibitory mechanism or because certain favorable conditions or factors are lacking is unknown.

In the case of the neural crest, marked regionalization has been demonstrated in normal development, particularly with respect to the precursors of the ANS (see Fig. 1). However, our experiments have clearly demonstrated that the developmental potentials of the various crest-cell populations (e.g., vagal, cervicodorsal, or lumbosacral) are not restricted to those expressed during normal ontogeny. In fact, heterotopic transplantations of pieces of the neural primordium, e.g., transfer of the vagal primordium to the adrenomedullary region of the neuraxis and vice versa, demonstrated that any level of the crest can give rise to the whole spectrum of sensory and autonomic structures (Le Douarin and Teillet, 1974). This was confirmed more recently by another approach in which pieces of neural crest from various levels (including the prosencephalon) of the neural axis of a quail were transplanted into the crest migratory pathway of a chick embryo at the adrenomedullary level (Fig. 9). This supernumerary crest dissociates and migrates rapidly, finally stopping in the normal sites of arrest of the trunk crest cells of the host, i.e., in the DRGs, the sympathetic ganglion chains and plexuses, and the adrenal medulla. In each of these sites, the grafted crest cells differentiated according to their final location, regardless of axial origin in the donor embryo. In one striking example, the prosencephalic crest, from which no peripheral nervous system (PNS) ganglia are known to originate, gave rise to DRG neurons as well as sympathetic and adrenomedullary cells. Other levels of donor crest behaved similarly. Only the relative proportion of neurons and glia arising from the various crest populations, initially similar in size, were found to be variable (Le Lièvre *et al.*, 1980; Schweizer, 1980).

---➤

Figure 5. Trunk region; level of the 15th somite. Immunofluorescence labeling for fibronectin (FN). (a) 29-Somite stage. As soon as the somite disrupts into dermomyotome (dmt) and sclerotome (sc), ventral penetration of crest cells is very rapid. Note that crest cells remain segregated from the somitic cells. (b) 30-Somite stage. At 3 hr after somite dissociation, pioneer crest cells are found near the notochord (n). The dermomyotome (dmt) ventral basement membrane assembles progressively. Note that the ectoderm (e) and the dermomyotome basement membranes are still fused. No crest cells penetrate this area, even though it contains more FN than the ventral pathway. (c) 32-Somite stage. Crest cells accumulate in an FN-rich area between the dermomyotome (dmt), the sclerotome (sc), and the neural tube (nt). They occupy more space laterally, but do not progress further ventrally toward the notochord, since sclerotomal cells have invaded that space. (d) 35-Somite stage. Crest cells continue to accumulate along the side of the neural tube (nt). The precise boundary between crest cells and sclerotomal cells was determined using chick–quail chimeras. The dermomyotome (dmt) is now completely surrounded by a continuous basement membrane. Note that FN disappears along the lateral side of the neural tube, whereas it is still present more ventrally and dorsally.(♦) Location of crest cells; (▷) front of migration. (a–d) ×420. From Thiery *et al.*, (1982) with permission.

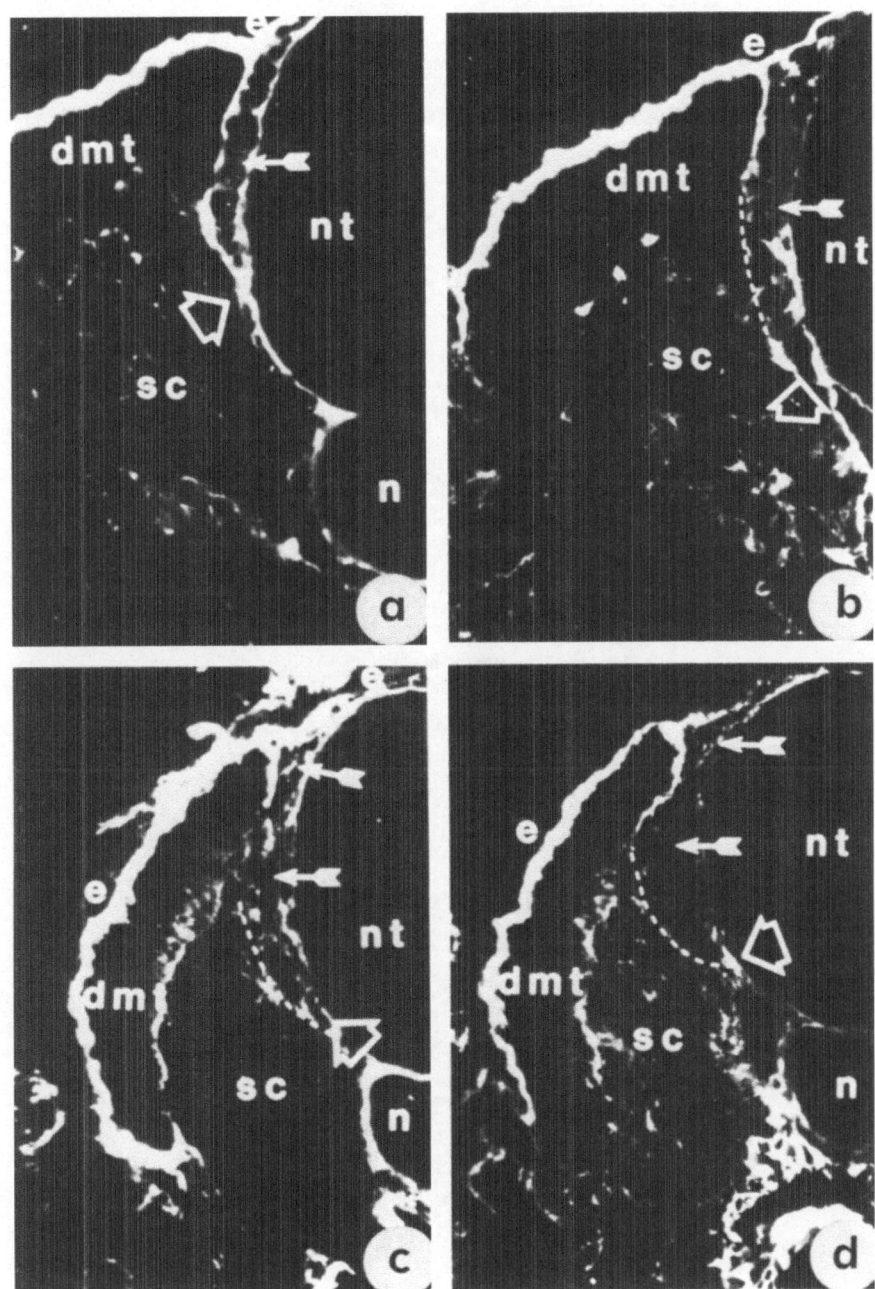

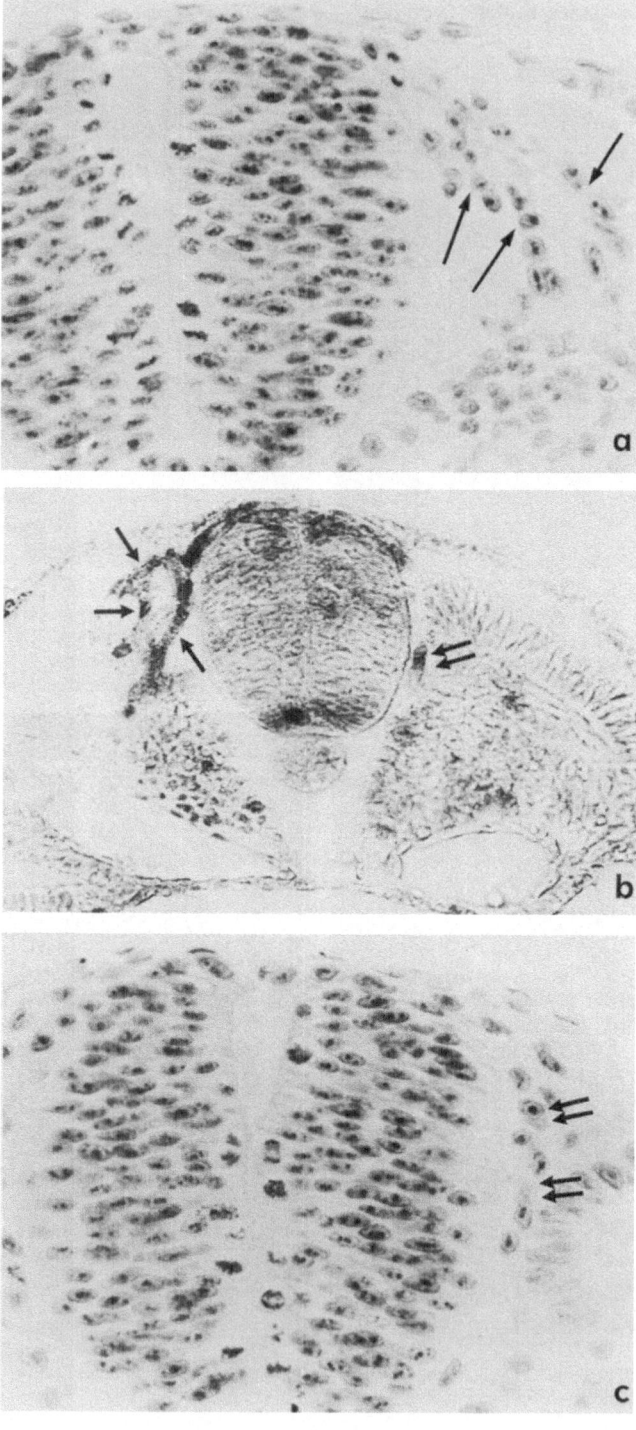

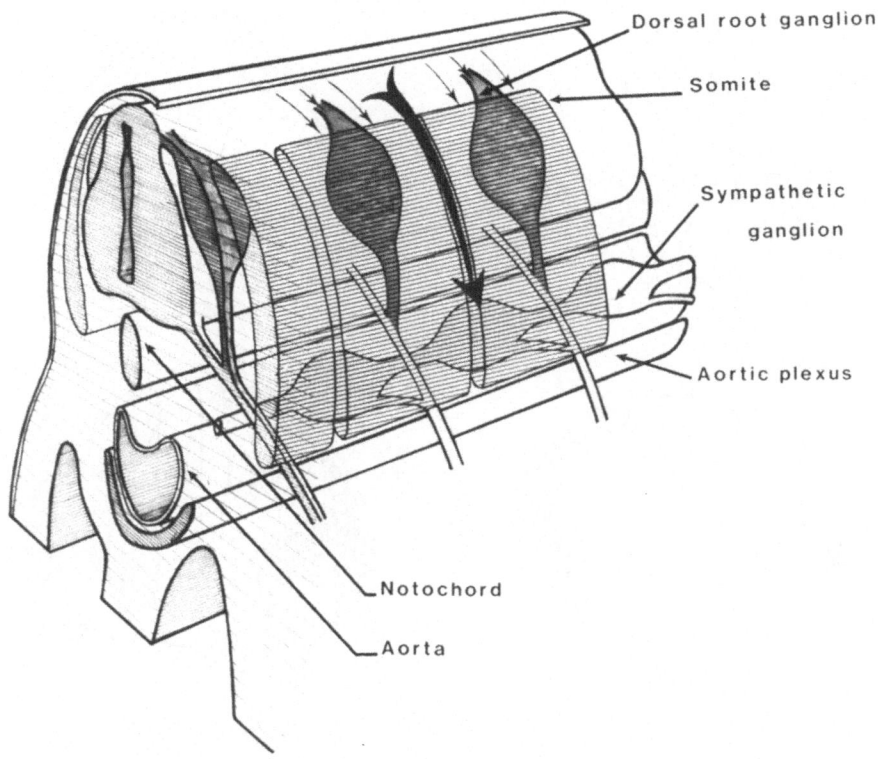

Dorsal root ganglion

Somite

Sympathetic
ganglion

Aortic plexus

Notochord

Aorta

Figure 7. Diagram showing the final distribution of the trunk neural-crest derivatives with respect to the pathway of migration taken by neural crest cells. The single heavy arrow denotes the inter-somitic pathway through which most of the cells that give rise to the sympathetic-chain ganglia and plexuses migrate; the small arrows denote the migratory pathway located between somite and neural tube through which the cells that form the DRGs migrate.

It is clear that although developmental potentials regarding PNS cell components are distributed throughout the neural crest (with only quantitative differences), some kind of selection elicits expression of only some capabilities during ontogeny. One outstanding question is whether the variety of presumptive phenotypes are contained in a multipotential cell or in several partly or completely committed cells.

←

Figure 6. Neural crest in the process of migration at the level of the trunk. The crest cells are labeled either through the quail–chick chimera system (a, c) or through their natural marker acetylcholinesterase (b). [a; b (left)] The pathway located between two consecutive somites is illustrated (→); [b (right); c] the pathway between somite and neural tube is illustrated (⇐).

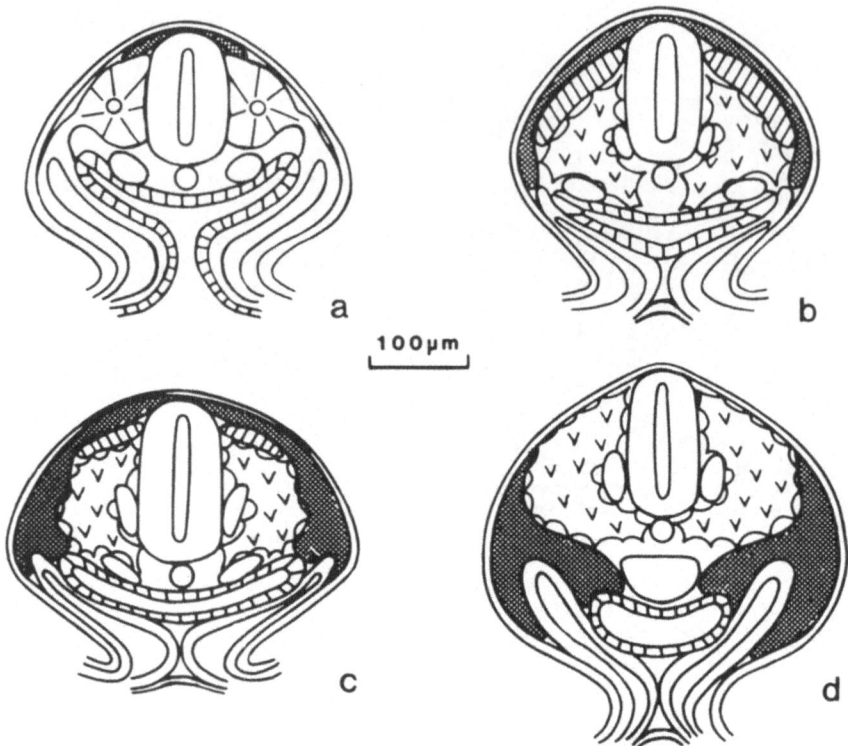

Figure 8. Crest-cell pathway in the vagal region. (a) 11-Somite stage. Crest cells initiate their migration under the ectoderm. (b) 14-Somite stage. Within 3 hr, the crest-cell population occupies a narrow space between the dissociated somite and ectoderm. The apparent rate of migration is close to 100 μm/hr. (c) 18-Somite stage. Crest cells have reached the apex of the gut and become subdivided into two streams. (d) 28-Somite stage. The last crest cells are separated from the neural tube by the expanding sclerotome. The number of crest cells increases rapidly, while they penetrate between the splanchnopleural mesoderm and the endoderm (the two components of the gut) and between the somatopleural mesoderm and ectoderm. (▨) Neural-crest cells. From Thiery *et al.*, (1982) with permission.

4. DEVELOPMENTAL POTENTIALS IN GANGLIA OF NEURAL-CREST ORIGIN

To approach this question, we determined whether nonexpressed differentiative capacities could be elicited in developing PNS ganglia. A variety of quail cellular populations were introduced into the crest migratory pathway of a 2-day chick host.

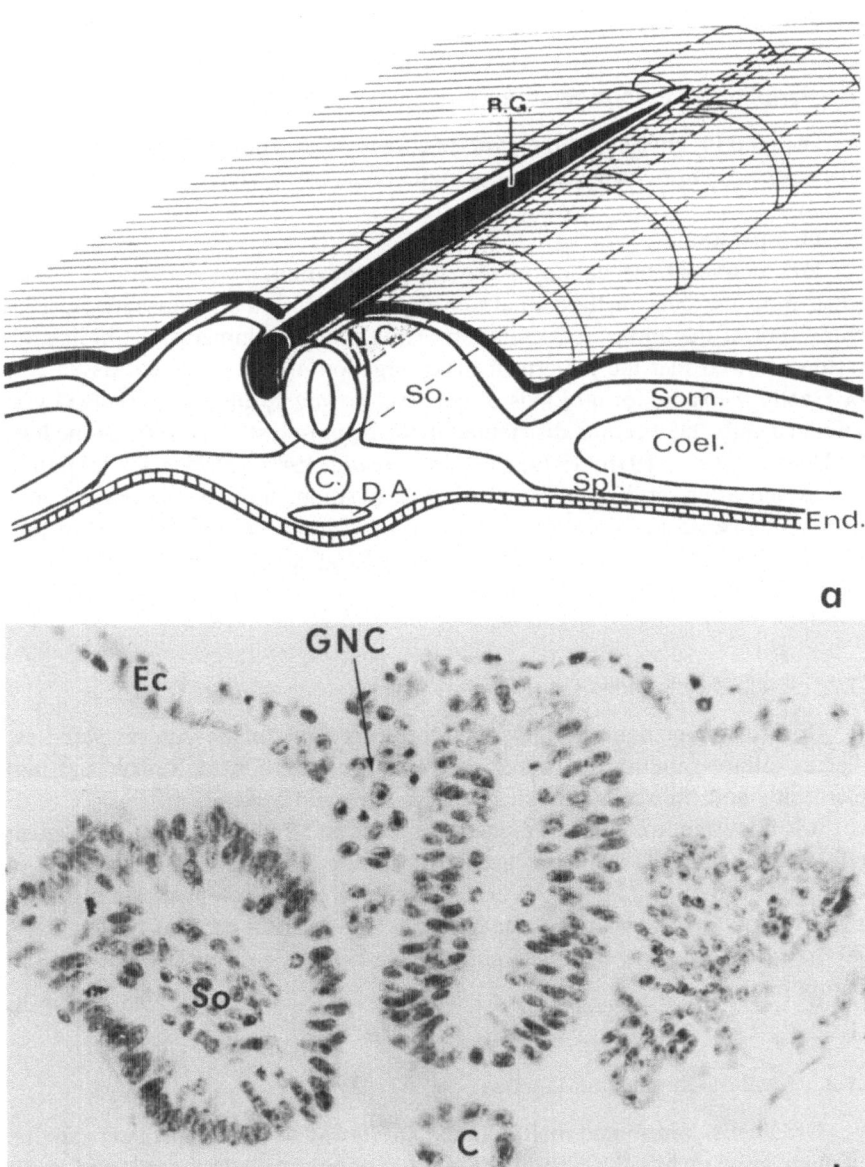

Figure 9. (a) Diagramatic representation of the grafting experiments into the neural-crest migration pathway of a 2-day-old chick embryo. (C.) Notochord; (D.A.) dorsal aorta; (Coel.) coelomic cavity; (End.) endoderm; (N.C.) neural crest; (Gr.) graft; (So.) somite; (Som.) somatopleure; (Spl.) splanchnopleure. (b) Cross section of a 19-somite chick embryo 2 hr after the graft, at the level of somites 18–22, of a fragment of the neural crest from a quail embryo. (GNC) Grafted neural-crest cells; (Ec) host ectoderm; (So) Somite; (C) notochord. ×340.

4.1. Peripheral Nervous System Ganglia of Neural-Crest Origin

The ganglia chosen for the experiments were autonomic (ciliary and sympathetic chain) and sensory (DRG), ganglia taken from quail donors after 4.5–6 days of incubation. The evolution of the graft was followed at various times after the operation.

Irrespective of the nature of the grafted ganglion, its peripheral cells detached from the bulk of the explant on the side facing the neural tube and the ventral root of the aorta. Sequential observation of the chimeras 12–48 hr after grafting showed that the pattern of disaggregation of the implanted tissue was not exactly the same for all kinds of ganglia. However, all finally resolved into individual cells that became distributed in the neural-crest derivatives of the host (Le Douarin *et al.*, 1978, 1979; Le Lièvre *et al.*, 1980). During the migration process and after settling in their definitive position, most of the grafted cells underwent intense proliferation (Dupin, 1982; Dupin and Le Douarin, in preperation). From 6 days onward, the definitive localization of the implanted cells was established for all types of grafts.

4.1.1. Graft of Autonomic Ganglia

Grafted sympathetic ganglia did not contribute to the enteric plexuses, whereas ciliary-ganglion cells migrated into the ganglion of Remak and into Auerbach's and Meissner's plexuses of the mid- and hindgut.

It is important to emphasize that during the first 2 postoperative days, quail cells often localized where the host DRGs were forming. However, later in development, they were not apparent in these structures, except in rare instances in which they formed scattered satellite cells among chick neurons. In any case, they never showed the neuronal phenotype characterized by large nuclei and neurofibrils.

4.1.2. Graft of Dorsal-Root Ganglia

DRG cells contributed mainly to the DRGs and to the sympathetic ganglia, although some quail cells were also found in adrenomedullary cords and in the adrenal and aortic plexuses. In the DRGs, the majority of quail cells appeared as large neurons and supporting cells. In none of the cases was the host DRG predominantly composed of quail cells, as it was when a piece of trunk crest was implanted (Fig. 10).

Associated formol-induced fluorescence and Feulgen–Rossenbeck techniques applied to the host embryos showed fluorescent quail cells in the sympathetic ganglia and plexuses and in the adrenal medulla, whether the graft was of sensory or autonomic origin. Electron microscopy of the chimeric suprarenal

24 hr

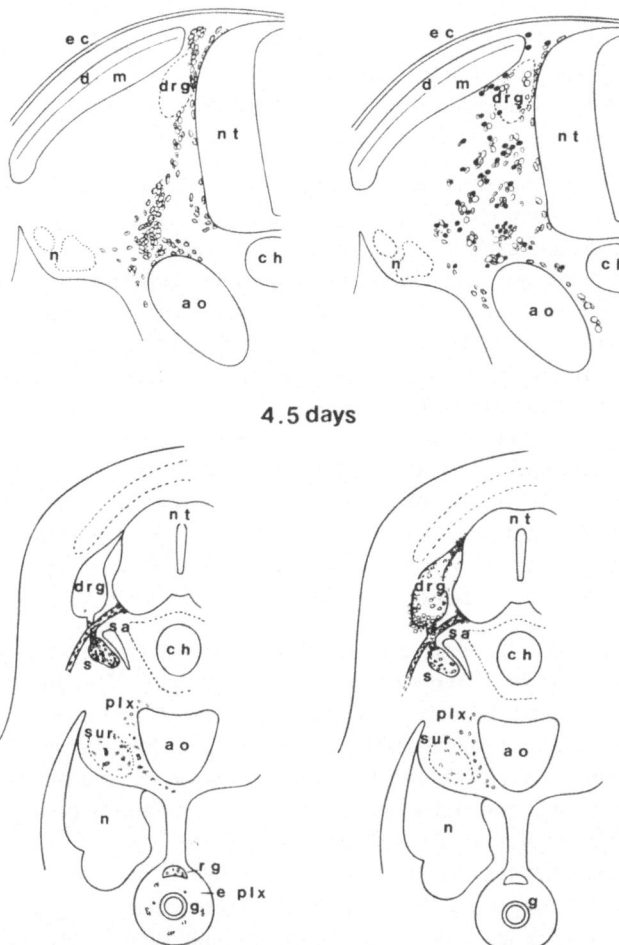

4.5 days

Figure 10. Diagram summarizing the observations of several grafts of quail autonomic ganglia and DRGs at the adrenomedullary level of chick embryos 24 hr and 4.5 days after transplantation. At 24 hr after the graft, cells of the autonomic ganglia are partially dissociated and dispersed in the host sclerotomal mesenchyme. Notice quail cells located along the neural tube (nt) and the dorsal aorta (ao) and also in the host DRG region (drg). In case of DRG grafts, the cells are generally dispersed in the sclerotomal mesenchyme, forming small clumps with many necrotic cells (●) that are not seen in the autonomic grafts. At 4.5 days after the graft, the differences between the two types of grafts are more conspicuous, namely because DRG grafts result in the colonization of the host DRG region. This does not occur in the autonomic grafts, in which the quail cells are located in the region of the ventral root of the rachidian nerve, the area where the sympathetic ganglion (s) of the host is in the process of formation, and more ventrally in the area where the aortic plexus (plx) and the adrenal medulla will be localized. Notice the quail cells are also present at these levels after grafting a DRG. In addition, the contribution of quail cells to the ganglion of Remak (rg) and to the enteric plexuses (e plx) is fully apparent in the case of ciliary ganglion grafts. This, however, has not been observed after grafting of sympathetic-chain ganglia. (ch) Notochord; (d m) dermo-myotome; (ec) ectoderm; (g) gut; (n) nephros; (sa) intersegmentary artery; (sur) suprarenal gland. From Le Lièvre *et al.* (1980) with permission.

gland revealed cells with both the typical DNA-rich quail nucleolus and the characteristic catecholamine-secretory granules.

These results led to the following conclusions:

1. The observed distribution of grafted ganglion cells in the host is the result of differential affinities of autonomic and sensory ganglion cells for the various target sites of the neural-crest cells. Autonomic ganglionic neurons and satellite cells become localized in the autonomic structures of the host (i.e., the autonomic ganglionic cells become localized in the sympathetic ganglia and plexuses, adrenal medulla, and in some cases enteric ganglia). In contrast, the grafted sensory neurons (of the DRG) colonize the host DRG in addition to the sympathetic and adrenome-dullary structures of the host.

2. At 4.5–6 days of incubation, the developmental capabilities of DRGs and autonomic ganglia are different: those of the DRG cells are broader, since, in the back-transplantation system, they differentiate along both "sensory" and "autonomic" cell lines. The capacity of DRG cells to give rise to adrenergic structures was also found in another type of experiment by Newgreen and Jones (1975).

It has long been recognized that during ontogeny, two types of neurons can be distinguished in the DRGs of birds: a lateroventral (LV) region, where large neurons differentiate relatively early in embryogenesis, and a mediodorsal (MD) region, where neurons are smaller (Hamburger and Levi-Montalcini, 1949; Pannese, 1974). Birth-date studies demonstrated that most of the neurons become post-mitotic between 4.5 and 6.5 days in the LV regions and between 4.5 and 7.5 days in the MD regions (Carr and Simpson, 1978).

To see whether the capacity to produce sensory neurons after back-trans-plantation into younger hosts is due to the presence in the graft of neuronal precursors that are not yet postmitotic, we explanted the LV or MD regions of 5- to 10-day quail DRGs separately (Schweizer *et al.*, 1983). The results showed that the ability to provide the host embryo with autonomic neurons and para-ganglia was maintained in both regions of the DRG through this whole devel-opmental period. In contrast, sensory neurons arose in the host only from LV regions taken from a 5-day donor, and not from donors aged 6 days or older. Consequently, at this later stage, neuroblasts have become postmitotic and are therefore committed neurons. In the MD regions, the capacity to produce sensory neurons is maintained until 7 days, but disappears thereafter, when all the neurons are postmitotic (Schweizer *et al.*, 1983).

We conclude that only undifferentiated, committed stem cells in any given DRG possess the potential to give rise to sensory neurons prior to their terminal mitoses. Since, after back-transplantation, no sensory neurons arise from the graft, all the transplanted postmitotic neurons probably die. We (Le Lièvre *et al.*, 1980) have actually observed cell death during the first 2 or 3 postoperative days.

If postmitotic sensory neurons do not survive under the conditions of back-transplantation, one may ask whether postmitotic autonomic neurons exhibit similar behavior. To investigate this point, Dupin (1982), in this laboratory, has labeled embryonic quail ciliary neurons during their last mitosis with tritiated thymidine injection at 3 and 4 days of incubation. The ciliary ganglion was back-transplanted when the donor had reached 9–11 days of age. At this time, the neurons alone were still heavily labeled, while in the dividing satellite-cell population, the isotope had become diluted. Through this radioactive labeling, the quail neurons remained distinguishable from the quail glial cells when transplanted into the chick embryo. It quickly became apparent that the neurons did not survive in grafts more than 2–3 days. Therefore, the transplanted cells that exhibited the quail nuclear marker were all derived from the nonneuronal-cell population.

Our observations may be applicable to transplantation experiments involving all types of peripheral ganglia. In other words, ganglionic nonneuronal cells retain the potential to generate autonomic neurons, even after the ganglion neurons are postmitotic. In spinal sensory as well as autonomic ganglia, only ANS precursors remain after the neurons are postmitotic.

4.2. Cranial Sensory Ganglia of Mixed Placodal and Crest Origin

These ganglia, such as the nodose, were of particular interest, since their neurons are normally placodal and not crest derivatives [at least in the two avian species considered (see Ayer-Le Lièvre and Le Douarin, 1982)]. We constructed chimeric nodose ganglia in which either the neurons or the nonneuronal cells selectively exhibited the quail nuclear marker. It was therefore possible to ask the following questions:

1. Do the postmitotic neurons of the nodose ganglion die after back-transplantation within the crest migratory pathway of the chick embryo?
2. Do neuronal (sensory, autonomic, or both) potentials exist in the non-neuronal population of the ganglion that can be revealed in back-transplantation experiments?

In fact, after transplantation of quail nodose–ganglion neurons into chick embryos, no quail neurons were detectable several days postgrafting. As for the second question, no quail sensory neurons developed in the host DRGs. However, besides Schwann and satellite cells located in the expected structures of the host (Figs. 11 and 12), many autonomic neurons developed from the graft. Neurons were distributed not only in sympathetic structures and adrenergic paraganglia but also in the intramural enteric ganglia (Fig. 13).

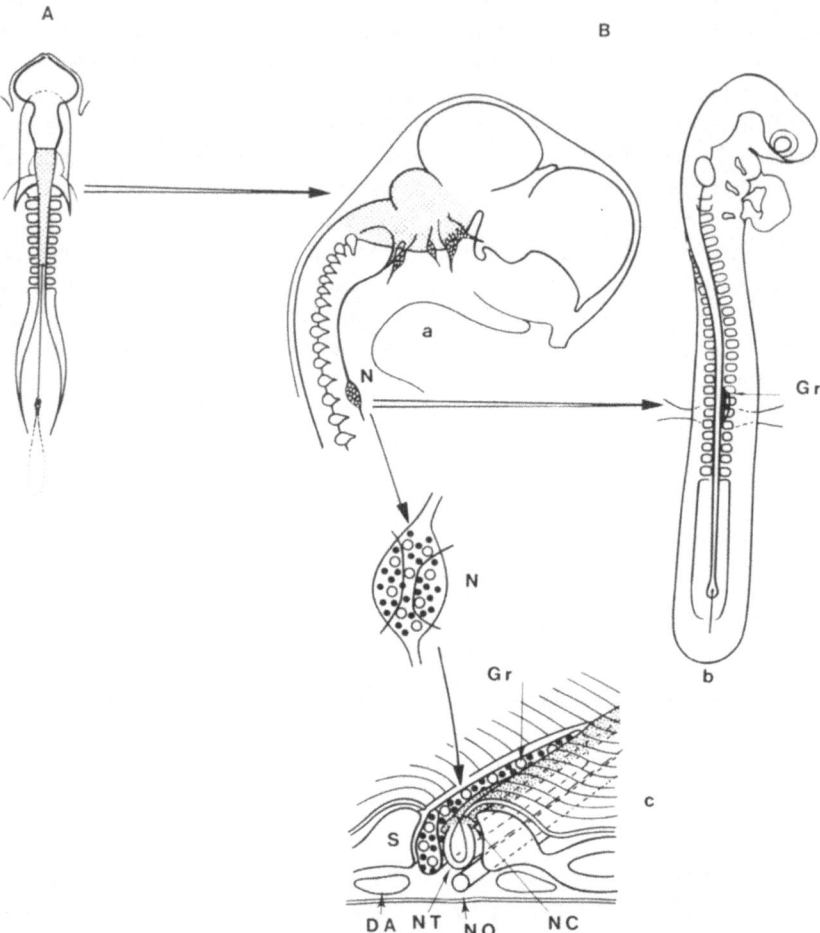

Figure 11. Construction of a chimeric nodose ganglion and its transplantation into a 2-day chick host. (A) A quail–chick (or chick–quail) chimera is made by replacing the rhombencephalovagal neural primordium of the host by the same piece of neural primordium from the donor (▨). The graft is made in embryos at stage 6–11 somites. (B) Grafting of pieces of the chimeric nodose (N) ganglion from 5- to 9-day chimeras (a) between neural tube (NT) and somites (S) at the brachial level of a 2-day chick host embryo (b). (c) Transverse section of the host at the graft level. (DA) Dorsal aorta; (Gr) graft; (NC) host neural crest; (NO) notochord. From Ayer-Le Lièvre and Le Douarin (1982) with permission.

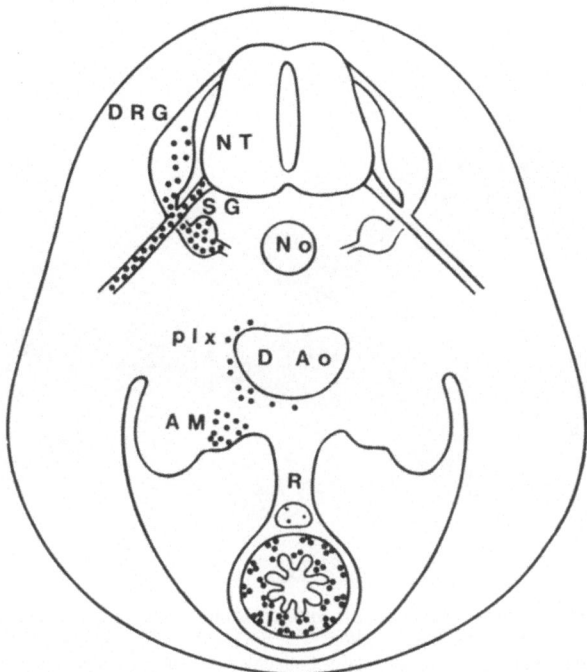

Figure 12. Diagram showing the results of the experiment represented in Fig. 11. Section through an 8-day host embryo indicating the localization of the quail cells derived from the grafted quail–chick chimeric nodose ganglia: quail cells are found in the dorsal root ganglion (DRG), in the rachidian nerves, in the sympathetic ganglia (SG) and plexuses (plx), in the adrenal medulla (AM), in the ganglia of the enteric plexuses (I), and in the Remak's ganglion (R). In the DRG, nerves, and Remak's ganglion, the quail cells are of the glial types. In the other structures, they also form neurons and paraganglion cells. (D.Ao.) Dorsal aorta; (No) notochord; (NT) neural tube. From Ayer-Le Lièvre and Le Douarin (1982) with permission.

5. GENERAL CONCLUSIONS AND FUTURE PERSPECTIVES

In view of the experiments described herein and of the observations concerning the normal development of PNS ganglia, one can formulate the following hypotheses (Fig. 14):

1. The neural crest is not composed of a homogeneous population of multipotential cells from which any of the phenotypes encountered in crest derivatives can arise. As far as the PNS is concerned, it seems that there exist in the neural crest separate pools of precursor cells for the sensory and the autonomic cell lines.

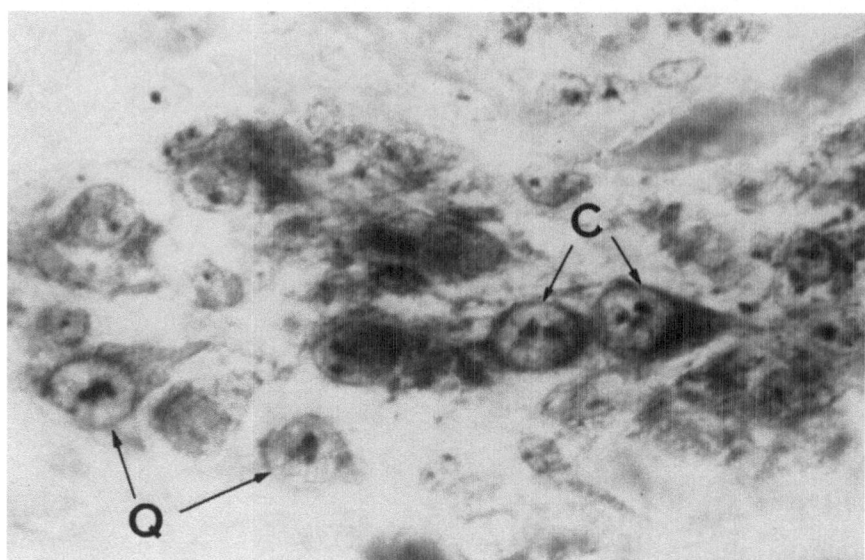

Figure 13. Silver staining of enteric ganglia differentiating in the chick host gut after the graft of a 7-day quail–chick chimeric nodose ganglion. The host was 14 days old at the time of sacrifice. The enteric ganglion is chimeric; i.e., it contains both chick (C) and quail (Q) neurons distinguishable through the mass of heterochromatin, which in quail nuclei appears strongly stained but in chick nuclei exists only as small dispersed chromocenters. × 1200. From Ayer-Le Lièvre and Le Douarin (1982) with permission.

2. Sensory neurons develop only in ganglia located in the vicinity of the central nervous system (CNS). If neural-crest-cell migration leads them to ganglia located ventrally, in an area remote from the CNS, such precursor cells die, and only autonomic cells develop. In other words, the cells of the "sensory line" would have strict survival requirements that would be fulfilled *in vivo* only if they develop close to the CNS. One observes that DRG cells extend neurites rapidly toward the CNS. That the neural tube produces some specific growth factor that would ensure the survival and growth of the neurites is an attractive hypothesis. A possible candidate for such a role has recently been described by Barde *et al.* (1983).

In addition, an interesting observation has to be mentioned in this respect: while neural-crest cells from any part of the neural axis transplanted into the crest migration pathway develop into DRG sensory neurons (Le Douarin *et al.*, 1979; Le Lièvre *et al.*, 1980; Schweizer, 1980), the same crest, when cultured *in vitro* for 2 days, gives rise to autonomic, but not sensory, neurons (Erickson *et al.*, 1980). This suggests strongly that the presumptive sensory-neuron pre-

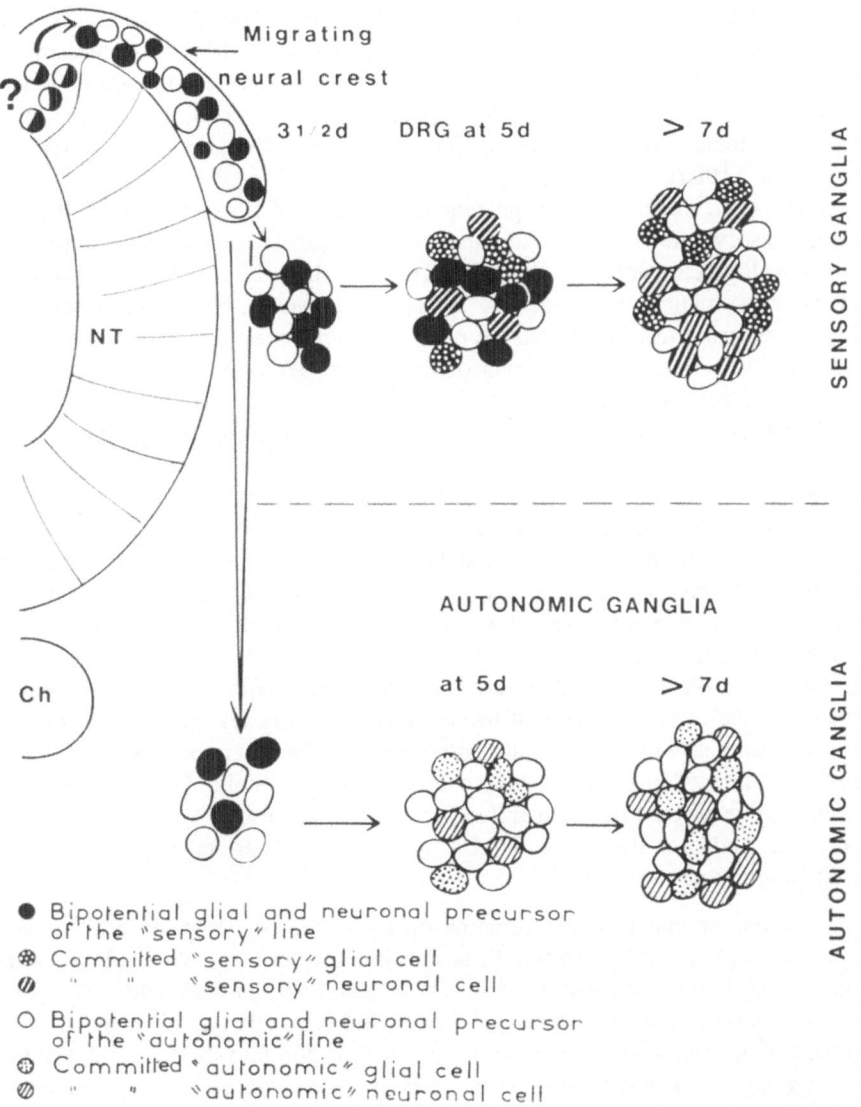

Figure 14. Hypothetical model for segregation of "sensory" and "autonomic" cell lines in the neural crest and its derivatives at the trunk levels. Sensory precursors survive only in ganglia located in the vicinity of the neural tube (NT). Both sensory and autonomic precursors exist in DRG and autonomic ganglia. (Ch) Notochord. See the text for further explanation.

cursor cells die in culture, while the autonomic neuronal precursors survive and develop further in the graft.

3. The autonomic precursors have a large range of developmental options with respect to the transmitter that they will synthesize [e.g., catecholamine (CA), acetylcholine (ACh), neuropeptides], and these options are largely under the control of environmental factors produced by nonneuronal cells of the embryonic rudiment in which the cells differentiate (see Le Douarin, 1982; Fontaine-Pérus et al., 1982).

Autonomic precursors are present in all types of PNS ganglia, including cranial sensory ganglia such as the nodose. They survive there at least until day 10 of development, although they never express their differentiating capacities in the sensory-neuron environment. These can, however, be elicited if sensory neurons are back-transplanted into the neural-crest migration pathway of a younger host (Fig. 15), since various kinds of autonomic derivatives arise in the host from grafted sensory ganglia.

Whether there exists a single autonomic precursor from which various neuronal phenotypes (i.e., adrenergic, cholinergic, peptidergic) can arise or whether several specialized precursors are precommitted early remains a question.

However, the fact that both ACh and CA can be produced simultaneously by the same autonomic neuron (Furshpan et al., 1976; Landis, 1976; Potter et al., 1981) supports the view that within a given committed autonomic neuroblast, both cholinergic and adrenergic developmental options coexist. It could even be proposed that the autonomic neurons are basically bifunctional with respect to their ability to synthesize ACh and CA, the essential effect of environmental cues being the increase of one of these metabolic pathways (adrenergic or cholinergic) coupled with the decrease of the other (cholinergic or adrenergic) in a flip-flop manner (Potter et al., 1981; for a discussion see Smith, 1983). Additional studies are required to demonstrate, for example, the activity of both sets of genes that code for the CA and ACh synthetic enzymes, respectively, in the developing autonomic neurons.

Another crucial problem remains open, and it concerns the relationship between the glial and the neuronal lines during development. As can be seen in Fig. 14, my hypothesis postulates that both glial and neuronal cells arise from the initial sensory and autonomic precursors, indicating that in this respect, such precursors are bipotential. This view is based on the fact that following autonomic-ganglia transplantation, no (or very few) satellite cells are encountered populating the host DRGs, suggesting that satellite cells as well as neuronal cells are specified as either sensory or autonomic. However, it must be emphasized that such evidence is indirect; it relies only on the specificity of the cell–cell recognition process that presumably occurs during dispersion and homing of the grafted cells after back-transplantation into the neural-crest migratory pathway

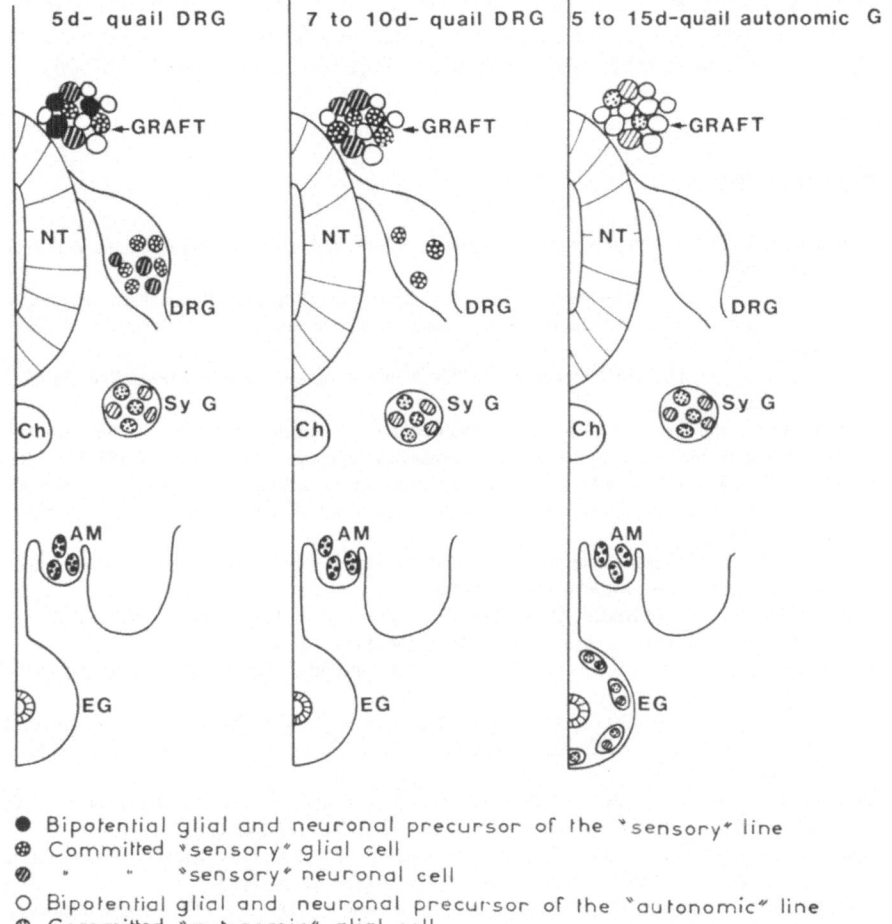

| 5d- quail DRG | 7 to 10d- quail DRG | 5 to 15d-quail autonomic G |

- ● Bipotential glial and neuronal precursor of the "sensory" line
- ⊕ Committed "sensory" glial cell
- ⊘ " " "sensory" neuronal cell
- O Bipotential glial and neuronal precursor of the "autonomic" line
- ◉ Committed "autonomic" glial cell
- ⊘ " " "autonomic" neuronal cell

Figure 15. Diagram summarizing the results of back-transplantation experiments involving sensory and autonomic ganglia at various ages. Sensory neurons of graft origin are found in the host only when the transplanted ganglion contains sensory precursor cells, i.e., in 5-day quail DRG. (AM) Adrenal medulla; (Ch) notochord; (EG) enteric ganglion; (NT) neural tube; (Sy G) sympathetic ganglion.

of a younger host. The most desirable test in this instance would be the transplantation into selected regions of a chick host of single quail cells taken either from the neural-crest or ganglion-cell populations. The hypothesis presented above may be tested by determining whether glia plus neurons are derived from these precursor sources.

ACKNOWLEDGMENTS. This work was supported by the Centre National de la Recherche Scientifique, the Délégation Générale à la Recherche Scientifique et Technique (Contrat 80 E 0877), and the NIH (Grant 2 RO1 DEO 4257-06).

REFERENCES

Adelmann, H.B., 1925, The development of the neural folds and cranial ganglia of the rat, *J. Comp. Neurol.* **39**:19–171.

Ayer-Le Lièvre, C., and Le Douarin, N.M., 1982, The early development of cranial sensory ganglia and the potentialities of their component cells studied in quail–chick chimeras, *Dev. Biol.* **94**:291–310.

Barde, Y.A., Edgar, D., and Thoenen, H., 1983, New neurotrophic factors, *Annu. Rev. Physiol.* **45** (in press).

Carr, V.McM., and Simpson, S.B., 1978, Proliferative and degenerative events in the early development of chick dorsal root ganglia. I. Normal development, *J. Comp. Neurol.* **182**:727–740.

Coghill, G.E., 1916, Correlated anatomical and physiological studies of the growth of the nervous system of amphibia. II. The afferent system of the head of *Amblystoma*, *J. Comp. Neurol.* **26**:247–340.

Da Costa, A.C., 1931, Sur la constitution et le développement des ébauches ganglionnaires crâniennes chez les Mammifères, *Arch. Biol.* **42**:71–106.

D'Amico-Martel, A., and Noden, D.M., 1983, Contributions of placodal and neural crest cells to avian cranial peripheral ganglia, *Am. J. Anat.* **166**:445–468.

Duband, J.L., and Thiéry, J.P., 1982, Distribution of fibronectin in the early phase of avian cephalic neural crest cell migration, *Dev. Biol.* **93**:308–323.

Dupin, E., 1982, Différenciation et prolifération cellulaires au cours de l'ontogenèse du système nerveux autonome chez l'Oiseau: Etudes *in vivo* et *in vitro*, Thèse de 3e cycle, Université de Paris VI.

Erickson, C.A., Tosney, K.W., and Weston, J.A., 1980, Analysis of migratory behavior of neural crest and fibroblastic cells in embryonic tissues, *Dev. Biol.* **77**:142–156.

Fontaine-Pérus, J.C., Chanconie, M., and Le Douarin, N.M., 1982, Differentiation of peptidergic neurones in quail–chick chimaeric embryos, *Cell Differ.* **11**:183–193.

Furshpan, E.J., McLeish, P.R., O'Lague, P.H., and Potter, D.D., 1976, Chemical transmission between rat sympathetic neurons and cardiac myocytes developing in microcultures: Evidence for cholinergic, adrenergic and dual-function neurons, *Proc. Natl. Acad. Sci. U.S.A.* **73**: 4225–4229.

Goldby, F., 1928, On the presence of a series of ectodermal placodes in the head region of a sparrow embryo, *Am. J. Anat.* **62**:135–138.

Goronowitsch, N., 1893, Untersuchungen über die Entwicklung der sog. "Ganglienleisten" im Kopfe der Vögelembryonen, *Morphol. Jahrb.* **20**:187–259.

Halley, G., 1955, The placodal relations of the neural crest in the domestic cat, *J. Anat.* **89**:133–152.

Hamburger, V., 1961, Experimental analysis of the dual origin of the trigeminal ganglion in the chick embryo, *J. Exp. Zool.* **148**:91–124.

Hamburger, V., and Levi-Montalcini, R., 1949, Proliferation, differentiation and degeneration in the spinal ganglia of the chick embryo under normal and experimental conditions, *J. Exp. Zool.* **111**:457–502.

His, W., 1868, *Untersuchungen über die erste Anlage des Wirbeltierleibes: Die erste Entwicklung des Hühnchens im Ei*, (F.C.W. Vogel, ed.) Leipzig.

Johnston, M.C., and Hazelton, R.D., 1972, Embryonic origins of facial structures related to oral sensory and motor function, in: *Third Symposium on Oral Sensation and Perception: The Mouth of the Infant* (J.B. Bosma, ed.), Charles C Thomas, Springfield, Illinois, pp. 76–97.

Landacre, F.L., 1910, The origin of the cranial ganglia in *Ameiurus*, *J. Comp. Neurol.* **20**:309–411.

Landacre, F.L., 1912, The epibranchial placodes of *Lepidosteus osseus* and their relation to the cerebral ganglia, *J. Comp. Neurol.* **22**:1–55.

Landacre, F.L., and McLellan, M., 1912, The cerebral ganglia of the embryo of *Rana pipiens*, *J. Comp. Neurol.* **22**:461–486.

Landis, S.C., 1976, Rat sympathetic neurons and cardiac myocytes developing in microcultures: Correlation of the fine structure of endings with neurotransmitter function in single neurons, *Proc. Natl. Acad. Sci. U.S.A.* **73**:4220–4224.

Le Douarin, N., 1964, Etude expérimentale de l'organogenèse du tube digestif et du foie chez l'embryon de poulet, *Bull. Biol. Fr. Belg.* **98**:544–676.

Le Douarin, N., 1969, Particularités du noyau interphasique chez la caille japonaise *(Coturnix coturnix japonica):* Utilisation de ces particularités comme "marquage biologique" dans les recherches sur les interactions tissulaires et les migrations cellulaires au cours de l'ontogenèse, *Bull. Biol. Fr. Belg.* **103**:435–452.

Le Douarin, N.M., 1973, A biological cell labeling technique and its use in experimental embryology, *Dev. Biol.* **30**:217–222.

Le Douarin, N., 1982, *The Neural Crest,* Cambridge University Press, Cambridge, 259 p.

Le Douarin, N.M., and Teillet, M.A., 1973, The migration of neural crest cells to the wall of the digestive tract in avian embryo, *J. Embryol. Exp. Morphol.* **30**:31–48.

Le Douarin, N.M., and Teillet, M.A., 1974, Experimental analysis of the migration and differentiation of neuroblasts of the autonomic nervous system and of neurectodermal mesenchymal derivatives, using a biological cell marking technique, *Dev. Biol.* **41**:162–184.

Le Douarin, N.M., Teillet, M.A., Ziller, C., and Smith, J., 1978, Adrenergic differentiation of cells of the cholinergic ciliary and Remak ganglia in avian embryo after *in vivo* transplantation, *Proc. Natl. Acad. Sci. U.S.A.* **75**:2030–2034.

Le Douarin, N.M., Le Lièvre, C.S., Schweizer, G., and Ziller, C.M., 1979, An analysis of cell line segregation in the neural crest, in: *Cell Lineage, Stem Cells and Cell Determination* (N. Le Douarin, ed.), Elsevier/North-Holland, Amsterdam, pp. 353–365.

Le Lièvre, C., and Le Douarin, N.M., 1975, Mesenchymal derivatives of the neural crest: Analysis of chimaeric quail and chick embryos, *J. Embryol. Exp. Morphol.* **34**:125–154.

Le Lièvre, C.S., Schweizer, G.G., Zilller, C.M., and Le Douarin, N.M., 1980, Restrictions of developmental capabilities in neural crest cell derivatives as tested by *in vivo* transplantation experiments, *Dev. Biol.* **77**:362–378.

Narayanan, C.H., and Narayanan, Y., 1980, Neural crest and placodal contributions in the development of the glossopharyngeal–vagal complex in the chick, *Anat. Rec.* **196**:71–82.

Newgreen, D.F., and Jones, R.O., 1975, Differentiation *in vitro* of sympathetic cells from chick embryo sensory ganglia, *J. Embryol. Exp. Morphol.* **33**:43–56.

Noden, D.M., 1973, The migratory behavior of neural crest cells, in: *Fourth Symposium on Oral Sensation and Perception: Development in the Fetus and Infant* (J. Bosma, ed.), U.S. Department of Health, Education and Welfare, Bethesda, Maryland, pp. 9–33.

Noden, D.M., 1975, An analysis of the migratory behavior of avian cephalic neural crest cells, *Dev. Biol.* **42**:106–130.

Noden, D.M., 1978, The control of avian cephalic neural crest cytodifferentiation. II. Neural tissues, *Dev. Biol.* **67**:313–329.

Pannese, E., 1974, The histogenesis of the spinal ganglia, *Adv. Anat. Cell Biol.* **47**:1–97.

Potter, D.D., Landis, S.C., and Furshpan, E.J., 1981, Adrenergic–cholinergic dual function in cultured sympathetic neurons of the rat, *Ciba Found. Symp.* **83**:123–138.

Schweizer, G., 1980, Recherches sur la ségrégation des lignées cellulaires dans la crête neurale de l'embryon d'Oiseau, Thèse de 3e cycle, Université de Paris VI.

Schweizer, G., Ayer-Lelièvre, C., and Le Douarin, N.M., 1983, Restrictions of developmental capacities in the dorsal root ganglia during the course of development. *Cell Differentiation* (in press).

Smith, J., 1983, Early events in autonomic neuron development: The cholinergic/adrenergic choice, in: *Dale's Principle and Communication between Neurones* (N. Osborne, ed), Pergamon Press pp. 143–159.

Teillet, M.A., 1978, Evolution of the lumbo-sacral neural crest in the avian embryo: Origin and differentiation of the ganglionated nerve of Remak studied in interspecific quail–chick chimaeras, *Roux's Arch. Dev. Biol.* **184:**251–268.

Teillet, M.A., and Le Douarin, N.M., 1983, Consequences of neural tube and notochord excision on the development of the peripheral nervous system in the chick embryo, *Dev. Biol.* **98:**192–211.

Thiery, J.P., Duband, J.L., and Delouvée, A., 1982, Pathways and mechanisms of avian trunk neural crest cell migration and localization, *Dev. Biol.* **93:**324–343.

Tosney, K.W., 1978, The early migration of neural crest cells in the trunk region of the avian embryo: An electron microscopic study, *Dev. Biol.* **62:**317–333.

Van Campenhout, E., 1937, Le développement du système nerveux crânien chez le poulet, *Arch. Biol.* **48:**611–666.

Van Campenhout, E., 1940, Expériences concernant le développement des placodes épiblastiques des ganglions crâniens chez le poulet, *C. R. Seances Soc. Biol. Paris* **134:**112–113.

Verwoerd, C.D.A., and van Oostrom, C.G., 1979, Cephalic neural crest and placodes, *Adv. Anat. Embryol. Cell Biol.* **58:**1–71.

Von Kuppfer, C., 1894, Studien zur vergleichenden Entwicklungsgeschichte des Kopfes der Cranioten. II. Die Entwicklung des Kopfes von *Ammocoetes planeri,* Lehmann, Munich.

Yntema, C.L., 1942, Experiments on the origin of some of the sensory cranial ganglia in the chick, *Anat. Rec.* **82:**455.

Yntema, C.L., 1943, An experimental study of the origin of the sensory neurons and sheath cells of the IXth and Xth cranial nerves in *Amblystoma punctatum, J. Exp. Zool.* **92:**93–120.

Yntema, C.L., 1944, Experiments on the origin of the sensory ganglia of the facial nerve in the chick, *J. Comp. Neurol.* **81:**147–167.

Emergence of Neuronal and Glial Cell Lineages in Primate Brain

Pasko Rakic

1. INTRODUCTION

It has become increasingly evident that the interaction between glial and neuronal cells in the developing mammalian brain plays an important role in the migration and compartmentalization of young neurons and, later, in their maintenance and the regulation of their environment. Furthermore, it appears that genetically or environmentally induced disturbances of neuronal–glial interactions during development may lead to structural, biochemical, and functional abnormalities of the brain (e.g., Rakic, 1975a; Caviness and Rakic, 1978; Volpe, 1981). Since the majority of cells in the mammalian brain are of glial lineages, it is safe to say that each neuron is likely to be in direct contact with at least one glial cell, but probably with several glial cells. However, although it is well established that both neuronal and glial cells derive from the neuroepithelium of the primitive neural tube, it has proved difficult to determine the exact timing and cellular mechanisms that lead to divergence of these two cell lines.

In this chapter, I will describe a series of Golgi, electron-microscopic, [³H]thymidine ([³H]-TdR) autoradiographic, and immunocytochemical analyses designed to unravel the emergence of separate neuronal and glial cell lines in the developing primate brain. The emphasis will be on the cellular events in developing rhesus monkey embryos that have been a major object of research

PASKO RAKIC • Section of Neuroanatomy, Yale University School of Medicine, New Haven, Connecticut 06510.

in my laboratory during the past decade. This type of *in vivo* analysis, even when carried out with the most advanced neurobiological methods, has certain limitations. However, studies in monkeys have an important advantage in that they provide the essential link between results obtained from transplantation experiments or from various tissue-culture studies carried out on cells from nonprimate species and normal and pathological findings that can be observed in human postmortem tissue. Furthermore, similarities in the pattern of neuronal organization, timing, and sequence of developmental events in monkey and human embryos allow inferences about cellular events that could not be learned, or even suspected, from examination of human autopsy material. I will limit my description to the emergence of the astrocytic cell line in the cerebral cortex, cerebellar cortex, and dentate gyrus of the hippocampus. These three regions are selected arbitrarily, since radial glial cells are ubiquitous in the developing primate brain. Among the cells to be discussed are a transient population of telencephalic radial glial cells, cerebellar Bergmann glial cells, specialized glial cells of the dentate gyrus, and more common forms of protoplasmic and fibrillary astrocytes. By choice, and, perhaps more important, due to lack of information, I will not discuss the origin and genesis of oligodendroglial cells.

2. HISTORICAL PERSPECTIVE

The timing and place of and the mechanisms that underlie separation of neuronal and glial cell lines in mammalian brain have long been unsettled and controversial. In a classic series of studies published almost a hundred years ago, His (1889) proposed that the germinative epithelium, situated near the ventricular surface of the developing cerebral wall in the human embryo, consists of two classes of precursor cells: one that produces neuronal and another that produces glial cells. This was a fortuitous and prophetic idea that in time proved to be correct, even though the histology of the human autopsy material hardly justified his lucid conclusions. An opposing view was advanced in 1897 by Schaper (1897), who suggested that the germinal zone consists of a single indifferent precursor cell type that gives rise to both neurons and glia. This view seems to have prevailed because His erroneously considered, due to poor fixation of the tissue, which precluded distinction between cell boundaries, that the glial precursor forms a multinuclear syncytium.

The cellular homogeneity of the ventricular zone gained support from histological studies of F.C. Sauer (1935), who found that morphological and shape differences between ventricular cells can be explained by interkinetic, to-and-fro nuclear movement. He concluded that all proliferative cells belong to the same class. This interpretation has been further supported by a histochemical study using Feulgen's technique (M.E. Sauer and Chittenden, 1959) and later by electron-microscopic observations (H. Fujita and S. Fujita, 1963, 1964) and [^{3}H]-TdR autoradiographic analyses (S. Fujita, 1963; Sidman *et al.*, 1959). In addition, since studies of the time of cell origin revealed few labeled glial cells in adult animals that had been exposed to [^{3}H]-TdR as embryos, it was believed that glial cells are generated after all, or after the majority of, the neurons destined for a given structure have been formed (S. Fujita 1963, 1966). The sequential generation of neurons and glia assumes homogeneity of the proliferative ventricular zone that initially produces neuronal cells only and at later stages glial cells. This concept was generally accepted and adopted by standard textbooks and monographs dealing with developmental neurobiology and neuropathology (e.g., Jacobson, 1978; Lund, 1978; S. Fujita, 1980).

3. EVIDENCE FOR EARLY CELLULAR DIVERGENCE

For many years, however, the presence of radial glial cells in the fetal brain provided at least one line of evidence that glial and neuronal cell lines coexist during embryonic development. This transient cell class was first revealed and classified as nonneuronal (glial or epithelial) in nature by observations made using the Golgi impregnation method at the turn of the century (Golgi, 1885; Lenhossek, 1891; Retzius, 1894; Ramón y Cajal, 1911). The use of electron microscopy in the early 1970s showed that radial glial cells are indeed present during the middle and late stages of primate corticogensis in both cerebellum (Rakic, 1971) and cerebrum (Rakic, 1972). The somata of the radial glial cells are usually located near the ventricular surface, and their elongated processes traverse the entire width of the cerebral or rhombencephalic wall and terminate with endfeet at the pial surface (Fig. 1). The glial nature of this cell class and their fibers can be rather securely established in the rhesus monkey at the midgestational period using higher-power examination of the Golgi-impregnated section (Fig. 2A) or in electron micrographs (Fig. 2B). Lamellate expansions are a distinguishing feature in both types of preparations. However, since neither Golgi silver impregnation nor electron microscopy could establish the identity of such cells at the early embryonic stages (Schmechel and Rakic, 1979b), we decided to use an antibody to glial fibrillary acidic protein (GFAP) that was shown to have a high specificity and affinity for the astroglial cell lineage (Bignami *et al.*, 1972; Eng and Bigbee 1978). The peroxidase–antiperoxidase (PAP) immunohistochemical method convincingly demonstrated the existence of radial glial cells during the first third of the 165-day gestational period in all major subdivisions of the embryonic monkey brain including cerebrum, cerebellum, and hippocampus (Levitt and Rakic, 1980; Levitt *et al.*, 1983; Eckenhoff and Rakic, 1982, 1983).

Use of the PAP immunohistochemical method at older stages, when cell

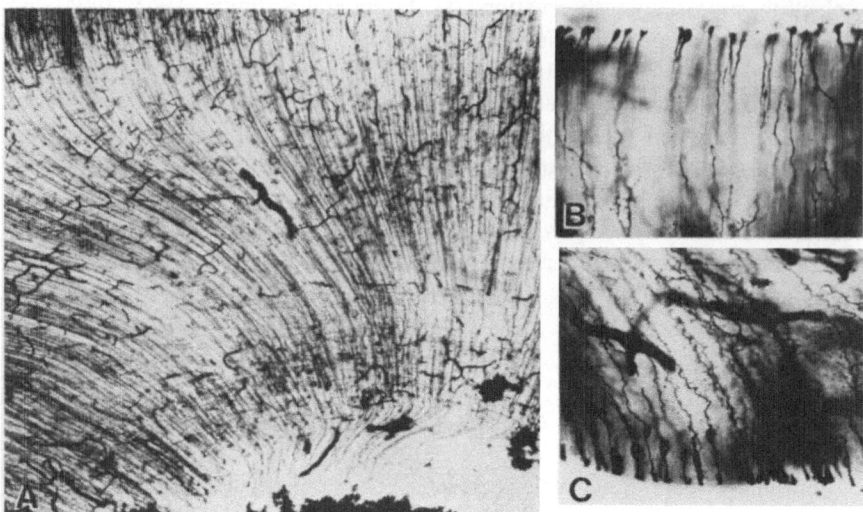

Figure 1. Photomicrographs of the Golgi-impregnated telencephalic wall of the rhesus monkey at midgestational age. (A) View of the portion of cerebral wall extending from the ventricular surface (bottom) to just beneath the cortical plate. The low-magnification photograph of the 125-μm-thick section displays a characteristic shape and distribution of elongated radial glial fibers as they traverse the intermediate zone of the occipital lobe. (B) Cell somas in the ventricular zone impregnated by the Golgi method. The only indication of their glial nature is the presence of elongated radial fibers with characteristic lamellate expansions. (C) Cone-shaped termination (endfeet) of radial glial fibers at the pial surface of the fetal cerebral wall.

phenotype can be verified by other morphological criteria, demonstrated that only cells that belong to the astrocytic class are immunoreactive and that no neuronal elements contain reaction product within their cytoplasm. The specificity of the antibodies directed against GFAP is particularly explicit in areas where radial glial fibers are apposed to the somata of neurons that are identified by counterstaining with the Nissl method. In such instances, the radial glial fibers are densely stained with GFAP, while the adjacent migrating or mature neurons remain devoid of reaction product (Levitt and Rakic, 1980; Eckenhoff and Rakic, 1983). The GFAP seems to be distributed throughout the entire shaft of the fetal radial glial fibers from the cell body to the endfeet at the pial surface, leaving usually lamellate expansions unstained (Figs. 3 and 4). As a consequence, these processes at the light-microscopic level appear rather smooth when stained by the PAP method for GFAP, in contrast to the elaborate "bushy" character of radial glial fibers in neocortical regions, cerebellum, or hippocampus stained with the Golgi method (Rakic 1971, 1972; Schmechel and Rakic, 1979b; Duffy and Rakic, 1983). PAP staining at the electron-microscopic level shows a clear distinction between GFAP-positive and GFAP-negative cells and fibers (Fig. 5).

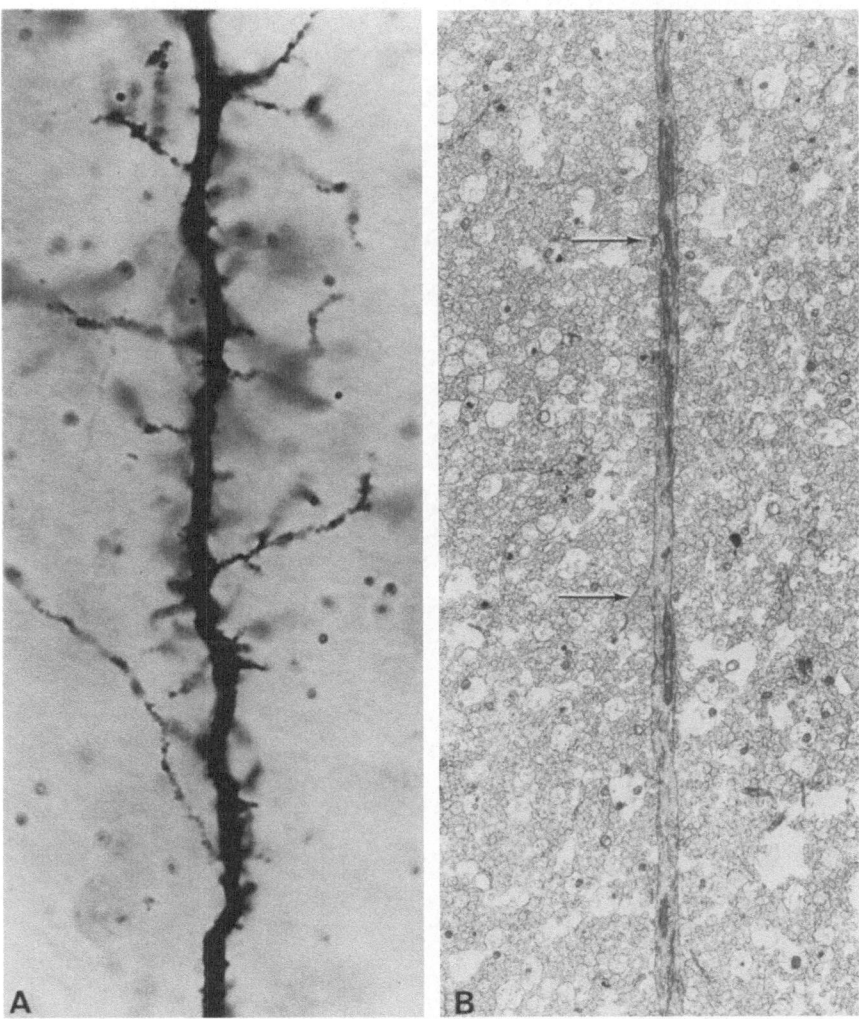

Figure 2. Fetal radial glial fibers crossing the intermediate zone of the occipital cerebral wall impregnated according to the Golgi method (A) and processed for electron microscopy (B). Both pictures are reproduced at approximately the same magnification. Numerous transversely cut round profiles in the electron micrograph are transversely cut afferent and efferent axons in the optic radiation. (→) Lamellate expansions emanating from the glial fiber.

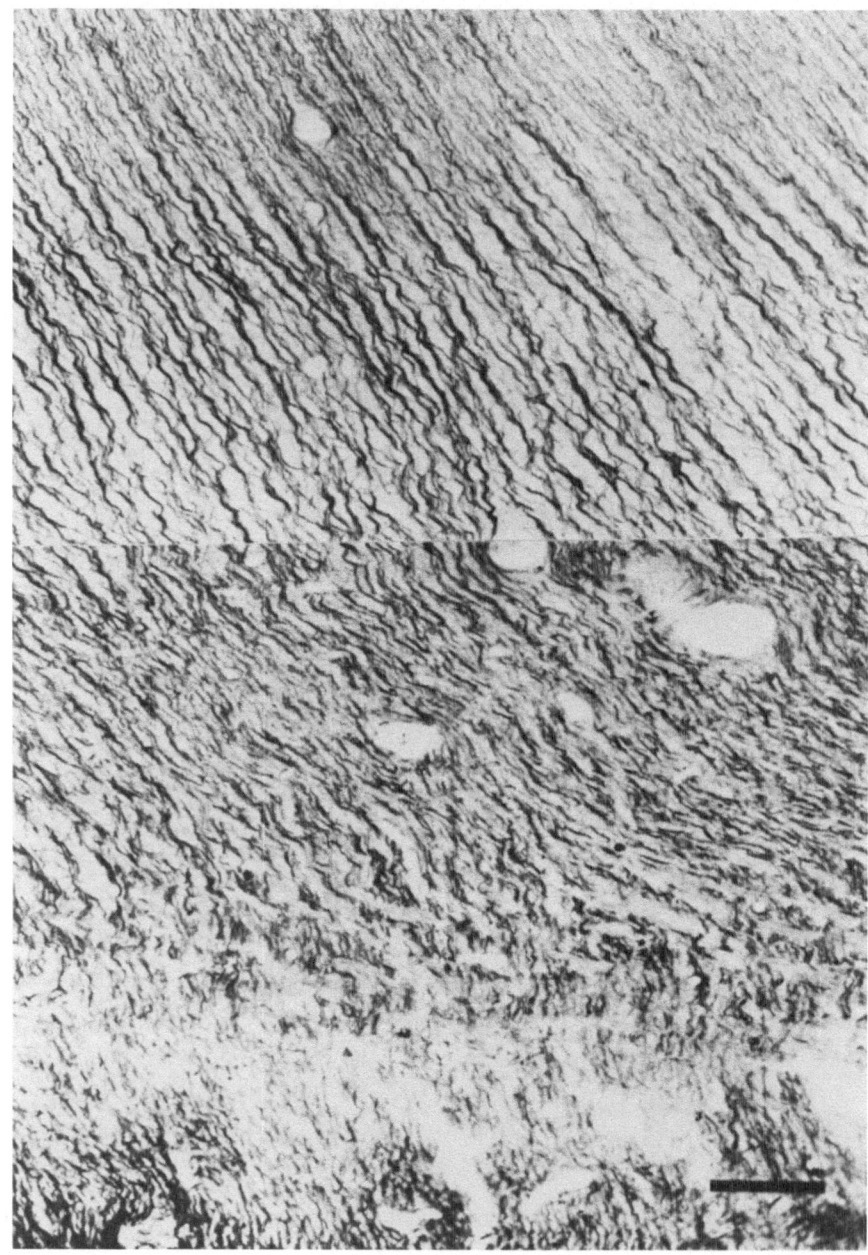

Figure 3. Photomontage of an 8-μm-thick section across the cerebral wall in a monkey fetus in midgestational period stained with the PAP immunohistochemical method using anti-GFAP primary antibody. This technique reveals a myriad of radial glial cells that in this 8-μm-thick section are denser than in the 125-micro-thick Golgi preparation (Fig. 1A), in which many fibers remained unstained. Scale bar: 75μm. From Levitt and Rakic (1980).

Both Golgi and immunohistochemical analyses show that the radial glial cells are the earliest class of astrocytes to appear in all regions of the primate brain examined. PAP staining, however, enables detection of glial cells at younger embryonic ages than they can be recognized by their morphology in Golgi impregnations (Levitt *et al.*, 1983). Although GFAP-positive radial glial cells were detected earliest, around the 40th embryonic day (E39), it is possible that a separate glial cell line exists even during earlier stages. Commitment to glial-cell lineage does not necessarily begin with the expression of GFAP. More sensitive antibodies or the *in situ* DNA-hybridization method may expose different enzymes specific for glial cells at an age before they produce GFAP in amounts detectable by the PAP immunohistochemical method. At any rate, our data at present indicate that cells with glial phenotype exist already at the onset of neurogenesis, at least in the primate occipital lobe. These findings contradict the commonly held view that glial cells are generated only toward the end of neuronal production in any given brain region.

4. HETEROGENEITY OF PROLIFERATIVE CELLS

Theoretically, heterogeneous cell populations can be generatd by proliferation of a single precursor pool that produces cells capable of expressing two phenotypes or by proliferation of different classes of precursor cells. The Golgi method cannot help in resolving this issue, since mitotic figures are resistant to silver impregnation. Nissl-stained sections or transmission-electron-microscopic (TEM) examination of the proliferative zones in monkey cerebrum fails to expose dissimilarities among cells in various phases of the mitotic cycle (Fig. 6). The ultrastructural similarity of dividing cells appears to support the concept that the ventricular zone consists of a cellularly homogeneous population (H. Fujita and S. Fujita, 1963, 1964). However, immunohistochemical examination shows the presence of two distinct precursor cell types in the ventricular zone of the rhesus monkey cerebrum during neurogenesis (Levitt *et al.*, 1981, 1983). One type of these mitotic figures contains GFAP, while the other does not (Fig. 7). Dividing cells that contain GFAP presumably give rise to radial glial cells at early ages and to astrocytes at later developmental stages. Thus, the unlabeled mitotic cells contribute to the postmitotic neuron population (Levitt *et al.*, 1981). The presence of immunoreactivity in very young embryos (E39 and E40) provides evidence that neuronal and glial cell lines diverge within the population of dividing precursor cells and that cells already committed to the glial line retain their capacity to divide.

Quantitative electron-microscopic analysis of immunohistochemically stained sections of the ventricular zone of the occipital lobe in a series of fetal monkeys (Levitt *et al.*, 1983) reveals a systematic change in the proportion of GFAP-

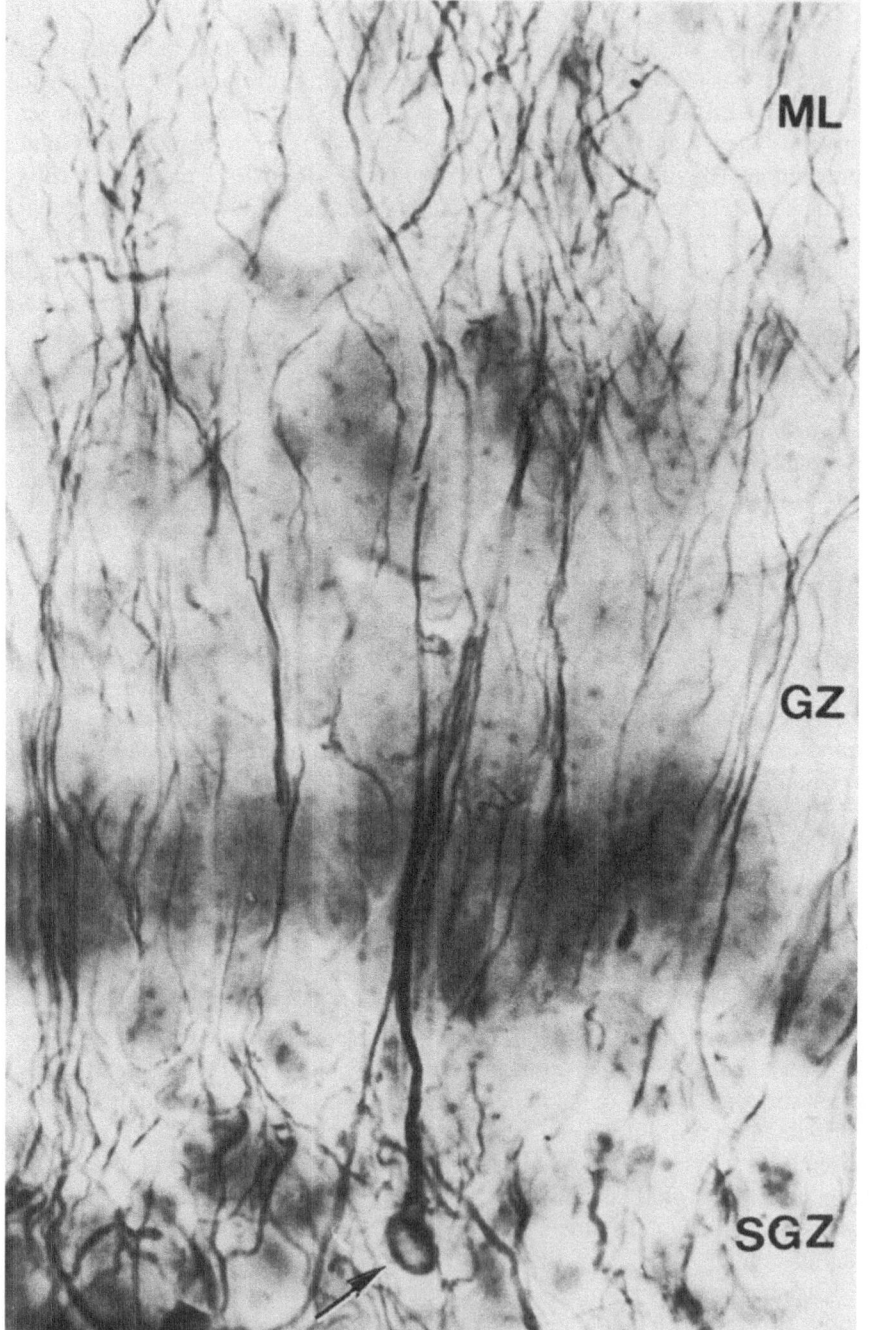

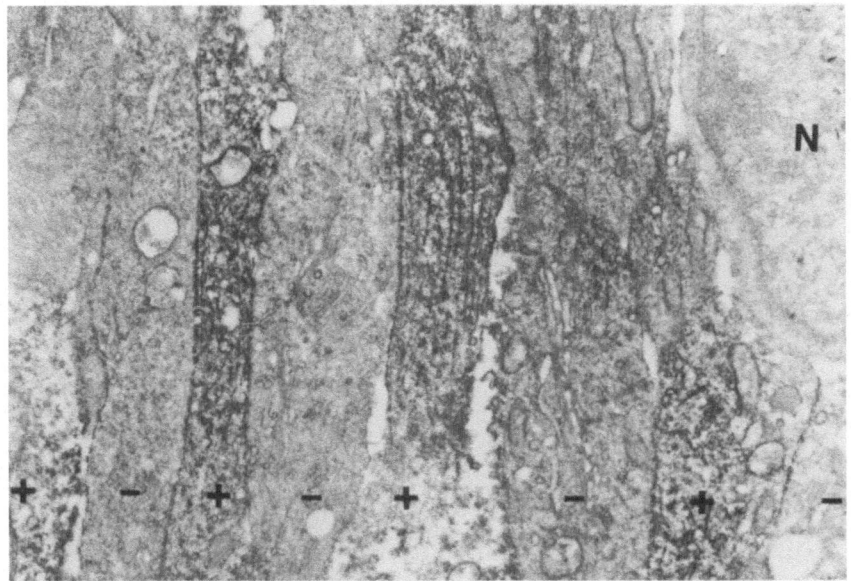

Figure 5. Electron micrographs of the outer portion of the ventricular zone in the occipital lobe of a 61-day-old monkey embryo processed with the PAP immunohistochemical method using anti-GFAP antibody. Labeled cells (+) are easily distinguished from unlabeled (−) in the uncounterstained sections. Satisfactory penetration of antibody is evident by close apposition of GFAP-positive cells (+) and GFAP-negative cells (−). (N) Nucleus. From Levitt *et al.* (1983).

positive to GFAP-negative mitotic cells (Fig. 8). These changes in the ratio of the two cell classes occur in harmony with the number of postmitotic neurons destined for the overlying cortical plate as calculated from the [^{3}H]-TdR autoradiographic data (Rakic, 1974, 1976). Thus, at early embryonic stages (E47), when only 3% of cortical neurons have been generated, about one third of all mitotic figures in the ventricular zone are GFAP-positive. The unstained proliferative cells presumably will produce future neurons, while the GFAP-positive mitoses produce radial glial cells. At somewhat later stages (E61) of corticogenesis, when about 15% of neurons in the occipital cortex have already been

←——————————————————————————————————————

Figure 4. Immunohistochemical localization of GFAP of the dentate plate at embryonic day 144. By this age, unique to the dentate gyrus, most of the radial glial cell bodies (→) have translocated from the ventricular zone to the subgranular zone (SGZ). Nevertheless, the radial organization of the glial fibers is maintained. Radially oriented glial processes are interposed between columns of cells in the granular zone (GZ) and then immediately branch on entering the molecular layer (ML) and finally terminate at the pial surface (above the micrograph). From Eckenhoff and Rakic (1984).

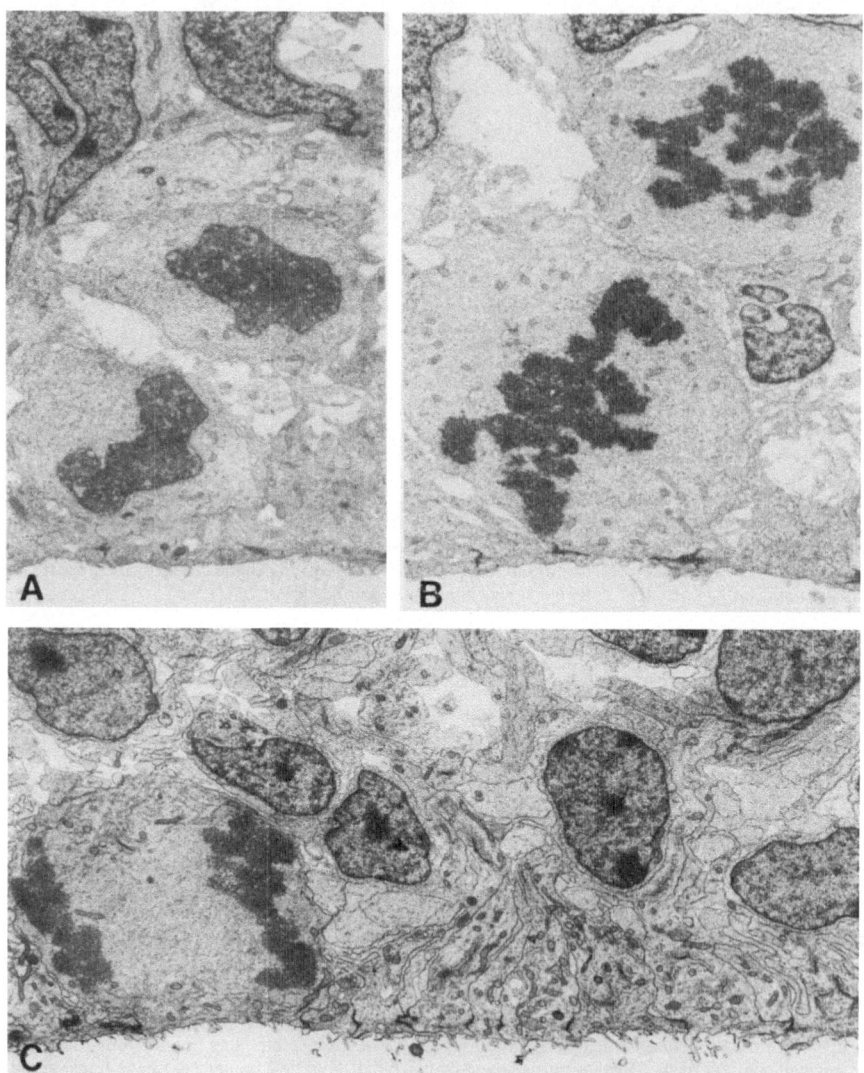

Figure 6. (A–C) Examples of mitotic figures in the ventricular zone of the occipital lobe in a monkey fetus at the midgestational period. Cytological characteristics shown by TEM provide no evidence for heterogeneity of proliferative cells.

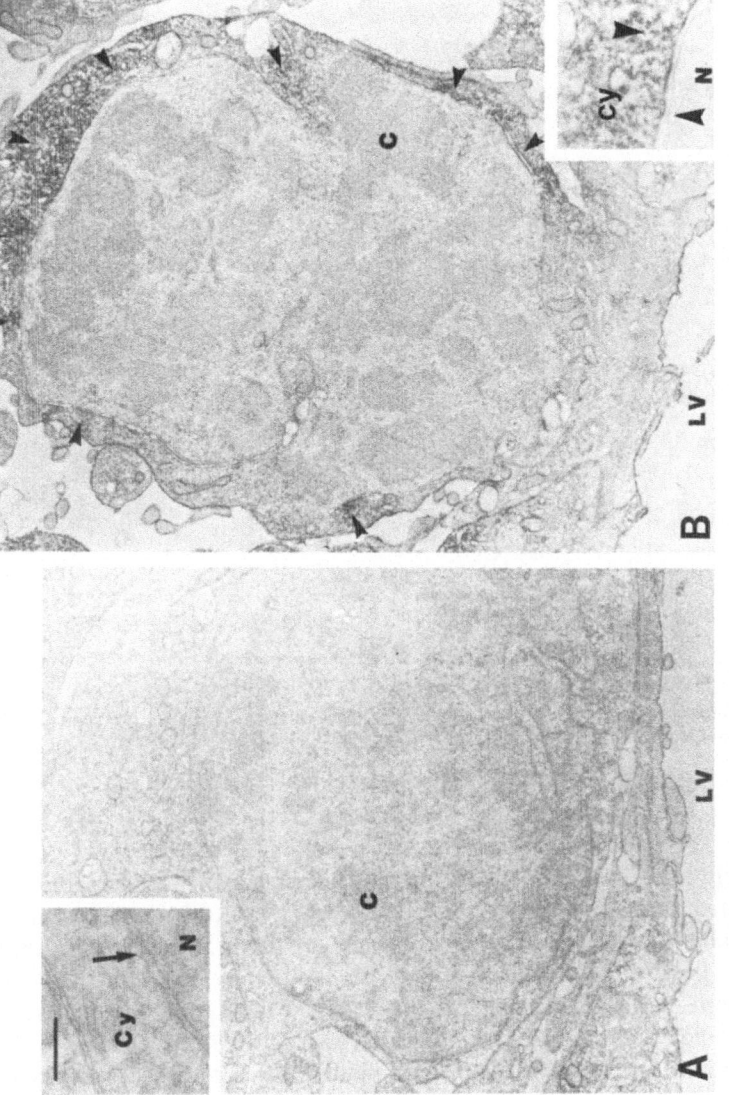

Figure 7. GFAP-negative (A) and GFAP-positive (B) mitotic figures situated within the ventricular zone of the monkey fetus in the midgestational period. (A) The absence of any immunoreactivity is confirmed at higher power *(inset)*; (B) the dense horseradish peroxidase reaction product characteristic of anti-GFAP PAP staining (A) is clearly visible in the cytoplasm (cy) and at higher power *(inset)*. (C) chromatine; (LV) lateral ventricle; (N) nucleus. From Levitt *et al.* (1981).

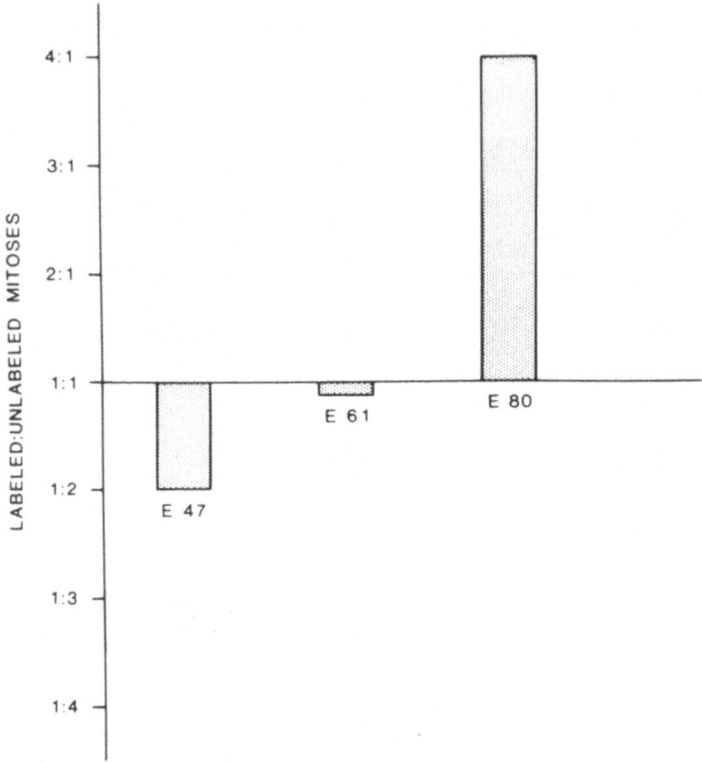

Figure 8. Proportion of GFAP-positive and GFAP-negative mitotic figures in the ventricular zone at embryonic (E) days 47, 61, and 80. From Levitt *et al.* (1983).

generated, almost half the mitotic cells in the subjacent ventricular zone are GFAP-positive (Fig. 8). At about this embryonic age, few morphologically differentiated astrocytes are visible in the cerebral wall of the rhesus monkey occipital lobe (Schmechel and Rakic, 1979b; Levitt and Rakic, 1980). Accordingly, it is possible that some astrocytes in addition to radial glial cells arise directly from the labeled precursors that might leave the mitotic cycle at this stage. By midgestation (E80), when about two thirds of the neurons destined for area 17 have been generated, fewer dividing cells in the ventricular zone are unstained. It is likely that these cells will produce the remaining neurons, while approximately 80% of the mitotic figures in the ventricular surface that immunoreact with antibody directed against GFAP (Fig. 8) will produce astroglial cell lines.

Thus, as the genesis and subsequent migration of prospective neurons proceed, there is a systematic change in the relative numbers of GFAP-positive and GFAP-negative cells. Unlabeled cells predominate at early stages of cortical neurogenesis, while most proliferative cells are GFAP-positive at the late stages. After E102, the ventricular zone is exhausted and gradually becomes replaced by an ependymal layer (Rakic, 1975a). In specimens older than E102, almost all proliferative cells are situated in the cerebral subependymal layer, and these cells are GFAP-positive (Levitt et al., 1983; Eckenhoff and Rakic, unpublished observation). The few unlabeled dividing cells at the ventricular surface may represent ependymal cells, which do not stain for GFAP in the adult (Eng and DeArmond, 1982). Eventually, the ependymal layer becomes totally clear of GFAP immunoreactivity (Levitt and Rakic, 1980).

Although the evidence and reasoning outlined above seem rather convincing that GFAP-positive and GFAP-negative precursor cell types produce glial and neuronal cell lines, respectively, there are at present several questions that need to be addressed. First, it is not possible to determine whether unlabeled mitotic figures produce only neurons. Since many cortical neurons of the deep cortical layers in the monkey occipital lobe have become postmitotic by E47 (Rakic, 1974), at least a portion of the unlabeled mitotic figures at this age do produce neurons. However, one cannot exclude the possibility that at this age some unlabeled cells are still pluripotential cells. Likewise, one cannot be totally sure that a portion of unlabeled mitotic cells have already been committed to the astroglial cell line, but do not as yet contain a sufficient amount of GFAP to be detectable with the PAP method. Second, one may argue that some precursor cells could express GFAP transiently or that they contain a larger amount of intermediate filaments that cross-react with antiserum directed against GFAP. If this unlikely biochemical event does indeed occur, some of the positively stained cell pool conceivably would not produce glia. There are, however, several additional lines of evidence supporting the notion that GFAP immunoreactivity in mitotic figures is indeed a reliable indicator of their commitment to glial lineages. Two classes of dividing cells (GFAP-positive and GFAP-negative) exist in the proliferative zone exclusively during the period of neurogenesis when both neuronal and glial phenotypes are being produced. As mentioned above, the proportions of GFAP-positive and GFAP-negative mitotic figures change systematically and in remarkable harmony with increments of postmitotic neurons destined for the neocortex. After all neurons of a given structure have been generated, virtually all proliferative cells become GFAP-positive, as shown in the subgranular zone of the dentate gyrus (Eckenhoff and Rakic, 1984; Levitt et al., 1983). Furthermore, the proliferative external granular layer of the developing cerebellar cortex, which produces granule cells and interneurons of the molecular layer (but not astrocytes), contains no labeled mitotic figures (Levitt et al., 1981).

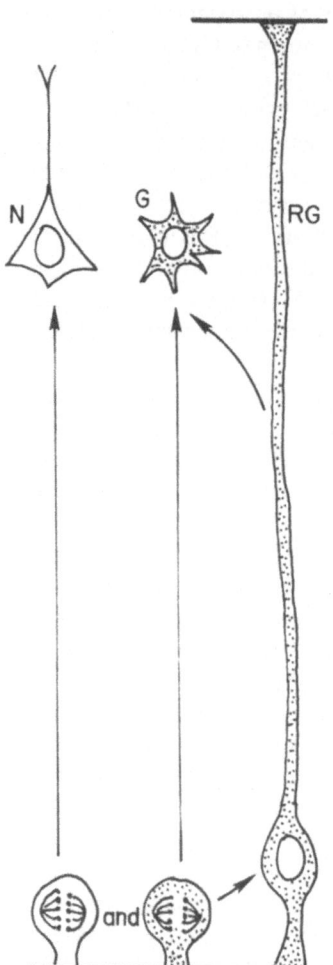

Figure 9. Diagram of dual cell-line and morphogenetic-pathway concept. On the basis of a series of immunocytochemical analyses of gliogenesis in the fetal monkey cerebrum (Levitt *et al.*, 1981, 1983; Eckenhoff and Rakic, 1983), it appears that both GFAP-positive (□) and GFAP-negative (⊞) dividing cells coexist throughout the period of neurogenesis. The GFAP-positive cells initially produce radial glial (RG) cells; later, either directly or indirectly (↑), they generate astrocytes (G) and various specialized astrocytelike cells. (N) Neuron. From Rakic (1981).

Finally, our results based on the presence of glial-specific marker are in agreement with demonstrations of neurofilament protein in some mitotically active cells in the ventricular zone of the embryonic chick spinal cord, which presumably produces neurons (Tapscott *et al.*, 1981). Thus, available data lead to the conclusion that neuronal and glial cells are committed to their future phenotype within the proliferative zone before their mitotic cycle has ended. The concept of two separate precursor cell lines and two alternative morphogenetic pathways for production of astrocytes is illustrated schematically in Fig. 9.

5. FATE OF FETAL RADIAL GLIAL CELLS

Initial studies on several forms of submammalian vertebrates (fishes, amphibians, and reptiles) led to the notion that the basic supporting framework of the nervous parenchyma in these species is provided by radial glial cells. These cells have been designated as ependymal or epithelial cells and fibers, and the astrocytes *sensu stricto* were considered to be absent in these species. Although subsequent research demonstrated the presence of astrocytes in these species (e.g., Kuhlenbeck, 1970), they are less numerous, and radial glial cells are the predominant glial form. A recent comparative immunohistochemical study using antibody against GFAP in various vertebrate species shows a distinct immunoreactivity of glial cells in cyclostomes, teleosts, amphibians, reptiles, birds, and all mammals examined (Onteniente *et al.*, 1983). In submammalian species, radial glial cells persist throughout life, while in mammals, these cells gradually disappear from most regions after reaching a peak during development. It is generally agreed that most of the radial glial cells transform into fibrillary astrocytes, protoplasmic astrocytes, or (in the cerebellum) into Bergmann glial cells. The transformation of radial glial cells into astrocytes was originally proposed by Ramón y Cajal (1911) on the basis of observations in developing spinal cord. Our analysis of transitional forms, and the timetable of the disappearance of radial glial cells in neocortex, hippocampus, and cerebellum of developing rhesus monkey, support this hypothesis (Rakic, 1971, 1975a; Schmechel and Rakic, 1979a,b; Eckenhoff and Rakic, 1984).

Figure 10 provides a semischematic sequence of morphogenetic transformation that leads to the formation of adult glial forms in the neocortex, cerebellar cortex, and hippocampus. In all three regions, initially the only glial elements are radial glial cells [Fig. 10A–C (a)]. In the telencephalon of the primates, radial fibers may attain enormous length, reaching several centimeters in the midterm human fetus. In the monkey, transitional forms between radial glial cells and astrocytes are first observed around E65 (Schmechel and Rakic, 1979b), when the somata of some radial glial cells detach from the ventricular surface and become displaced to the intermediate zone [Fig. 10A (b)]. Subsequently, cell somas of some radial glial cells may move at various depths of the cerebral wall and acquire bushy astrocytelike appendages (c). Some cells translocate their body close to the brain surface and form characteristic astrocytes of the molecular layer, retaining their endfeet at the pial membrane (d). Some radial glial cells, however, lose their attachment to both cerebral surfaces and form various types of astrocytes. Thus, protoplasmic astrocytes are first seen in the deep layers of the cortical plate (h), and only subsequently do they populate the entire width of the neocortex in an inside-out gradient of appearance (Rakic, 1975a; Schmechel and Rakic, 1979b). The fibrillary astrocytes populate subcortical strata that

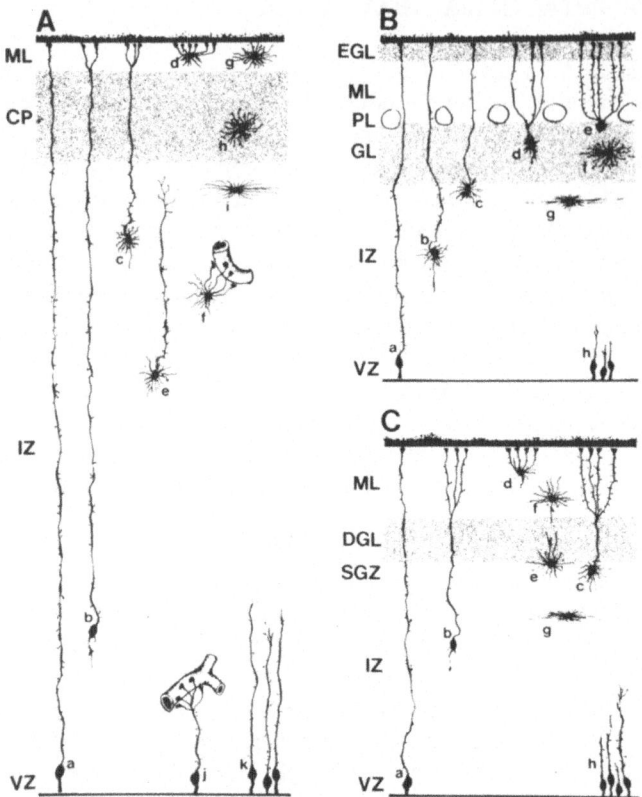

Figure 10. Semischematic diagram illustrating morphogenetic transformation of fetal radial glial cells into various astrocytic forms in the cerebral hemisphere (A), cerebellar hemisphere (B), and dentate gyrus of the hippocampus (C). (CP) Cerebral cortical plate; (DGL) dentate gyrus–granular layer of the hippocampal region; (EGL) external granular layer; (GL) granular layer (internal) of the cerebellum; (IZ) intermediate zone; (ML) molecular layer; (PL) Purkinje-cell layer; (SGZ) subgranular zone of the dentate gyrus; (VZ) ventricular zone. Further explanation and designation of various transitional cell forms (a–k) are provided in the text.

are prospective white matter (i). Many of the astrocytes become attached to the surface of blood vessels by their conical endfeet (f). However, some radial glial cells persist for a longer time, possibly continuing to transform into astrocytes, while others lose their outer contact with the pia and retain endfeet at the ventricular surface. These cells may attach their endfeet on the blood vessels (j) or have long, freely arborizing tips (k). Such elongated cells are frequently referred to, particularly in the hypothalamic area, as tanocytes.

In the cerebellar hemisphere, the early phases of gliogenesis are quite similar to those described above for the cerebral cortex, and the basic mechanism may be essentially the same (Rakic, 1971). Thus, the earliest recognizable glial element is radial glial cells that span the full thickness of the rhombencephalic wall [Fig. 10B (a)]. These cells begin to lose their attachment at the ventricular surface (b, c) and eventually transform into Bergmann glial cells through several transitional forms (d, e) or become protoplasmic astrocytes of the granular layer (f) and fibrillary astrocytes of the subcortical white matter (g). Some radial glial cells may, however, lose their pial attachment while keeping contact with the ventricular surface in a manner reminiscent of tanocytes in other brain regions [Fig. 10B (h)].

Gliogenesis in the dentate gyrus of the hippocampal region proceeds according to the same general principle (Eckenhoff and Rakic, 1982, 1984). Here, however, the cellular relationships are more complex due to considerable tectogenetic changes, rotation, and twisting of the dentate gyrus during development. The diagram in Fig. 10C provides a simplified version of cellular events; a more detailed description and the documentation of evidence for this schema should be sought in the recent study by Eckenhoff and Rakic (1984). However, the basic transformation of radial glial cells [Fig. 10C (a)] through phasic detachment from the ventricle (b) and subsequent transformation into astrocytes of the molecular layer (d, f), protoplasmic astrocytes in the subgranular zone (e), fibrillary astrocytes in subjacent white matter (g), or a specialized glial cell of the dentate gyrus (c) is essentially the same. It should be emphasized that the radial glial cell, the transitional forms, and the mature astrocytic classes described in these three regions are all GFAP-positive (Levitt and Rakic, 1980; Eckenhoff and Rakic, 1984).

Our finding that astrocytic cell lines appear early in the developmental history of the cerebral cortex, cerebellar cortex, and dentate gyrus of the rhesus monkey (Rakic, 1971; Schmechel and Rakic, 1979b; Levitt and Rakic, 1980; Eckenhoff and Rakic, 1984) is in general agreement with an early appearance of glial cells in the human fetus (Oksche, 1968; Sidman and Rakic, 1973; Kostovic et al., 1975; Antanitus et al., 1976; Choi and Lapham, 1978; Polak et al., 1982). Thus, the timing, sequence, and mode of cellular events described herein for the rhesus monkey apply rather accurately for the corresponding fetal ages in the human. Despite initially rapid formation of spatial gradients of astrocytic genesis, the transformation of radial glial cells into astrocytes takes place over a relatively prolonged period (Schmechel and Rakic, 1979b). For example, in the cerebrum, the main phase of transformation occurs from E60 to E120 for those radial glial cells that are attached to blood vessels in the cortical plate (roughly 2 months), from E75 to 1 or 2 months after birth for those radial glial cells that span the cerebral wall (roughly 5 months), and from E120 to several months after birth for those radial glial cells that transform into periventricular astrocytes (roughly 4 months). It should be emphasized that the transformation of radial glial cells into astrocytes is most prominent during the second half of gestation. However, our [³H]TdR autoradiographic evidence suggests that after E90, glial proliferation also occurs locally in the cortical plate (Rakic, 1974, 1975a). The number of astrocytes produced by local glial proliferation compared to the number of cells that are transformed from radial glial cells is unknown. The number of radial glial cells present in the cerebral wall at fetal and early postnatal ages is relatively small compared to the total population of astrocytes in the growing cerebral wall including the hippocampal region (Ramón y Cajal, 1911; Stensaas, 1967; Eckenhof and Rakic, 1984). However, the radial glial cells in the subventricular zone of the telencephalon or subgranular zone of the hippocampus may continue to divide, and even a slow rate of proliferation is adequate to produce a majority of the astrocytes in these brain regions.

Genetic mechanisms or factors that initiate translocation of the somas of radial glial cells from their initial position in the proliferative zones and subsequent transformation into astrocytes are unknown. The hypothesis that the increasing width of the cerebral wall induces displacement of radial glial cell somas outward, which in turn triggers their transformation into astrocytes (Oksche, 1968; Stensaas, 1967), seems unlikely in the light of our findings. Thus, we have not observed a significant difference in the persistence of radial glial cells that span the cerebral wall in expanded vs. nonexpanded regions of the convoluted primate telencephalon (Schmechel and Rakic, 1979b). Although typical astrocytes can be found in the first cortical layer when the cerebral wall is still relatively thin (around E73), some radial glial cells that span the distance between the ventricle and pia continue to be present even during early postnatal ages.

Autoradiographic [³H]-TdR analysis shows that many radial glial cells stop dividing during approximately a 2-month period (Schmechel and Rakic, 1979a). During this period, they remain attached to both cerebral surfaces, and the pause in proliferation may be essential to ensure a point-to-point contact between the proliferative ventricular zone and the expanding primate cortex that begins to fold (Rakic, 1982). In the second half of gestation after neuronal migration is completed, radial glial cells reenter the mitotic cycle, and many of them simultaneously transform into astrocytes (Schmechel and Rakic, 1979a). Nothing is known about the factors that may be responsible for this change in cell behavior. One possibility is that radial glial fibers may transmit from the cortical plate to their somas information that could delay their transformation into astrocytes until corticogenesis is completed. The possibility of an active retrograde flow of information, suggested by Oksche (1968), gained support by the demonstration of retrograde transport of horseradish peroxidase from the endfeet of radial glial cells to their somas in developing telencephalon (Ivy and Killackey, 1978). Thus, unlike neurons that withdraw from the cell cycle, radial glial cells retain their capacity to divide. Likewise, their progeny, the astrocytes, may be continuously turned over at a slow rate even under normal circumstances or may divide at an accelerated rate in response to various pathological agents (Strichartz *et al.*, 1982).

6. COMPARISON WITH THE PERIPHERAL NERVOUS SYSTEM

Since this conference is concerned mainly with mechanisms that regulate the development, growth, and function of the autonomic ganglia, it is perhaps appropriate to make some comparison between gliogenesis in the central nervous system (CNS) and peripheral nervous system (PNS). The comparison is somewhat artificial, since in the sympathetic and dorsal-root ganglia, the only cells that can possibly be considered roughly equivalent to central astrocytes are satellite cells (Pannese, 1974). The first and most obvious difference in development between the two systems is that unlike the situation in the brain, peripheral ganglia do not have stationary zones or proliferative centers with a specific function of producing cells. While in the mammalian brain, all neurons with the exception of the granule cells of the hipocampus begin migration only after their last cell division in the ventricular zone (Rakic, 1975a,b), crest cells continue to divide during the process of migration, and keep proliferation within the ganglion itself. Second, the astroglial cells in the developing CNS may play different roles, and the transient population of the radial glial cells that is so prominent in the fetal brain does not exist during genesis of the peripheral ganglia. Third, unlike an early determination of glial precursor cells in the primate CNS, neural-crest cell precursors in the chick PNS seem to have an extended period of pluripotentiality. Thus, as elegantly described by Nicole Le Douarin (Chapter 1), chimeric nodose ganglia containing quail neural-crest cells, which normally produce exclusively satellite glial cells within the ganglion, can yield neurons with the quail nuclear marker when transplanted to the trunk region of the chick embryo (LeLièvre and Le Douarin, 1982).

The phenotypic plasticity of developing neural-crest cells is well documented (see Le Douarin, 1980; Black, 1982) (see also Chapter 8), and it is reasonable to assume that precursors of the glial cells in the PNS may also behave differently from central, neuroectodermal cells. However, one cannot exclude the possibility that some of the GFAP-negative cells in our material are pluripotential and that under certain conditions they could be transformed into either cell class (Levitt *et al.*, 1981). The type of analysis performed does not provide information either on cellular mechanisms or on environmental or positional cues responsible for initiation of phenotypic expression. It simply establishes stages at which the fate of cells under normal circumstances becomes restricted. Our findings in the CNS indicate that once cells become committed to one developmental pathway, they do not give rise to, or behave as, other cell types. For example, astrocytes retain their capacity to divide under normal and pathological conditions, but neurons do not. Although conditions and experimental designs for the study of the PNS and CNS are difficult to compare, there nevertheless seem to be several differences that have to be considered in the interpretation of developmental disorders and reactive responses of immature glial cells to various pathogenic agents that affect the fetal nervous system.

7. SUMMARY

A series of recent studies using a variety of methods including Golgi, [³H]-TdR, electron microscopy, and immunocytochemistry lead to a reevaluation of the conventional view of proliferative zones as consisting of a homogeneous dividing population and dispute a dogma that implies sequential order with neurogenesis followed by gliogenesis. Data available at present for the neocortex, cerebellum, and hippocampal regions of the rhesus monkey indicate that (1) glial and neuronal cell lines are distinguishable and coexist in the primate brain at ages when few, if any, neurons have become postmitotic; (2) onset of glial phenotypic expression occurs prior to the last cell division; (3) some radial glial cells pass transiently through a latent amitotic stage; (4) after reentering the cell cycle, radial glial cells transform into protoplasmic and fibrillary astrocytes at a specific age for each of the three structures described; and (5) the proportions of glial and neuronal precursors and postmitotic cells change systematically during pre- and postnatal development.

REFERENCES

Antanitus, D.B., Choi, B.H., and Lapham, L.W., 1976, The demonstration of glial fibrillary acidic protein in the cerebrum of the human fetus by indirect immunofluorescence, *Brain Res.* **103**:613–616.

Bignami, A., Eng, L.F., Dahl, D., and Uyeda, C.T., 1972, Localization of the glial fibrillary acidic protein in astrocytes by immunofluorescence, *Brain Res.* **43**:429–435.

Black, I.B., 1982, Stages of neurotransmitter development in autonomic neurons, *Science* **215**:1198–2104.

Caviness, V.S., Jr., and Rakic, P., 1978, Mechanisms of cortical development: A view from mutations in mice, *Annu. Rev. Neurosci.* **1**:297–326.

Choi, B.H., and Lapham, L.W., 1978, Radial glia in the human fetal cerebrum: A combined Golgi, immunofluorescent and electron microscopic study, *Brain Res.* **148**:295–311.

Duffy, C.J., and Rakic, P., 1983, Differentiation of granule cells in the dentate gyrus of the rhesus monkey: A quantitative Golgi study, *J. Comp. Neurol.* **214**:224–337.

Eckenhoff, M.F., and Rakic, P., 1982, Golgi, immunocytochemical and electron microscopic evidence for the radial organization of the developing dentate gyrus in rhesus monkey, *Abstr. Soc. Neurosci.* **8**:821.

Eckenhoff, M.F., and Rakic, P., 1984, Radial organization of the hippocampal dentate gyrus: A Golgi, ultrastructural and immunohistochemical analysis in the developing rhesus monkey, *J. Comp. Neurol.* (in press).

Eng, L.F., and Bigbee, J.W., 1978, Immunochemistry of nervous system-specific proteins, in: *Advances in Neurochemistry*, Vol. 3 (B.W. Agranoff and M.H. Aprison, eds.) Plenum Press, New York, pp. 43–98.

Eng, L.F., and DeArmond, S.J., 1981, Immunocytochemical studies of astrocytes in normal development and disease, in: *Advances in Cellular Neurobiology*, Vol. 3 (S. Fedoroff and L. Hertz, eds.), Academic Press, New York, pp. 145–171.

Fujita, H., and Fujita, S., 1963, Electron microscopic studies on neuroblast differentiation in the central nervous system of domestic fowl, *Z. Zellforsch.* **60**:463–478.

Fujita, H., and Fujita, S., 1964, Electron microscopic studies on the differentiation of the ependymal cells and the glioblast in the spinal cord of domestic fowl, *Z. Zellforsch.* **64**:262–272.

Fujita, S., 1963, The matrix cell and cytogenesis in the developing central nervous system, *J. Comp. Neurol.* **120**:37–42.

Fujita, S., 1966, Application of light and electron microscopic autoradiography to the study of cytogenesis of the forebrain, in: *Evolution of the Forebrain* (R. Hassler and H. Stephan, eds.), Georg Thieme, Stuttgart, pp. 180–196.

Fujita, S., 1980, Cytogenesis and pathology of neuroglia and microglia, *Pathol. Res. Pract.* **168**:271–278.

Golgi, C., 1885, Sulla fina anatomia delgi organi centrali del sistema nervoso, republished in: *Opera Omnia,* Hoepli, Milan, pp. 397–536.

His, W., 1889, Die Neuroblasten und deren Entstehung im embryonalen Marke, *Abh. Math.-Phys. Kl. Saechs. Akad. Wiss.* **15**:313–372.

Ivy, G.O., and Killackey, H.P., 1978, Transient population of glial cells in developing rat telencephalon revealed by horseradish peroxidase, *Brain Res.* **158**:213–218.

Jacobson, M. (ed.), 1978, *Developmental Neurobiology,* 2nd ed., Plenum Press, New York, 562 pp.

Kostovic, I., Krmpotic-Nemanic, J., and Kelovic, A., 1975, The early development of glia in human neocortex, *Rad Jugosl. Akad. Znan. Umjet. Nat. Sci. Ser.* **17**:155–159.

Kuhlenbeck, H. (ed.), 1970, *Central Nervous System of Vertebrates,* Vol. 3, *Part I, Structural Elements: Biology of Nervous Tissue,* S. Karger, Basel, 818 pp.

Le Douarin, N.M., 1980, The ontogeny of the neural crest in avian chimaeras, *Nature (London)* **286**:633–669.

Le Lièvre, C.S., and Le Douarin, N.M. 1982, The early development of cranial sensory ganglia and the potentialities of their component cells studied in quail–chick chimeras, *Dev. Biol.* **94**:291–310.

Lenhossek, M.V., 1891, Zur Kenntnis der ersten Entstehung der Nervenzellen und Nervenfasern beim Vogelembryo, *Verh. X Int. Med. Congr. Berl. Abth.* **2**:115–124.

Levitt, P.R., and Rakic, P., 1980, Immunoperoxidase localization of glial fibrillary acid protein in radial glial cells and astrocytes of the developing rhesus monkey brain, *J. Comp. Neurol.* **193**:815–840.

Levitt, P., Cooper, M.L., and Rakic, P., 1981, Coexistence of neuronal and glial precursor cells in the cerebral ventricular zone of the fetal monkey: An ultrastructural immunoperoxidase analysis, *J. Neurosci.* **1**:27–39.

Levitt, P., Cooper, M.L., and Rakic, P., 1983, Early divergence and changing proportion of neuronal and glial precursor cells in the primate cerebral ventricular zone, *Dev. Biol.* **96**:427–484.

Lund, R.D., 1978, *Development and Plasticity of the Brain: An Introduction,* Oxford University Press, Oxford.

Oksche, A., 1968, Die prenatale und vergleichende Entwicklungs geschichte der Neuroglia, *Acta Neuropathol. Suppl.* **4**:4–19.

Onteniente, B., Kimura, H., and Maeda, T., 1983, Comparative study of the glial fibrillary protein in vertebrates by PAP immunohistochemistry, *J. Comp. Neurol.* **215**:427–436.

Pannese, E., 1974, The histogenesis of the spinal ganglia, *Adv. Anat. Embryol. Cell Biol.* **47**:1–97.

Polak, M., Haymaker, W., Johnson, J.E., and D'Amelio, F., 1982, Neuroglia and their reaction, in: *Histology and Histopathology of the Nervous System* (W. Haymaker and R.D. Adams, eds.), Charles C Thomas, Springfield, Illinois, pp. 363–480.

Rakic, P., 1971, Neuron–glia relationship during granule cell migration in developing cerebellar cortex: A Golgi and electron microscopic study in macaqus rhesus, *J. Comp. Neurol.* **141**:283–312.

Rakic, P., 1972, Mode of cell migration to the superficial layers of fetal monkey neocortex, *J. Comp. Neurol.* **145**:61–84.

Rakic, P., 1974, Neurons in rhesus monkey visual cortex: Systematic relation between time of origin and eventual disposition, *Science* **183**:425–427.

Rakic, P., 1975a, Timing of major ontogenetic events in the visual cortex of the rhesus monkey, in: *Brain Mechanisms in Mental Retardation* (N.A. Buchwald and M. Brazier, eds.), Academic Press, New York, pp. 3–40.

Rakic, P., 1975b, Cell migration and neuronal ectopias in the brain, in: *Birth Defects: Orig. Artic. Ser., Morphogenesis and Malformation of the Face and Brain* (D. Bergsman, ed.), Alan R. Liss, New York, **9**:95–129.

Rakic, P., 1976, Differences in the time of origin and in eventual distribution of neurons in areas 17 and 18 of visual cortex in rhesus monkey, *Exp. Brain. Res. Suppl.* **1**:244–248.

Rakic, P., 1981, Neuronal–glia interaction during brain development, *TINS* **4**:184–187.

Rakic, P., 1982, Early developmental events: Cell lineages, acquisition of neuronal positions, and areal and laminar development, *Neurosci. Res. Prog. Bull.* **20**:439–451.

Ramón y Cajal, S., 1911, *Histologie du Systeme Nerveux de l'Homme et des Vertebres,* Vol. 2, Paris, A. Maloine; reprinted by Consejo Superior de Investigaciones Cientificas, Instituto Ramón y Cajal, Madrid.

Retzius, G., 1894, Die Neuroglia des Gehirns beim Meuschen und bei Säugethieren, *Biol. Untersuchungen (New Ser.)* **6**:1–24.

Sauer, F.C., 1935, The cellular structure of the neural tube, *J. Comp. Neurol.* **63**:13–23.

Sauer, M.E., and Chittenden, A.C., 1959, Deoxyribonucleic acid content of cell nuclei in the neural tube of the chick embryo: Evidence for intermitotic migration of nuclei, *Exp. Cell Res.* **16**:1–6.

Schaper, A., 1897, The earliest differentiation in the central nervous system of vertebrates, *Science* **5**:430–431.

Schmechel, D.E., and Rakic, P., 1979a, Arrested proliferation of radial glial cells during midgestation in rhesus monkey, *Nature (London)* **277**:303–305.

Schmechel, D.E., and Rakic, P., 1979b, A Golgi study of radial glial cells in developing monkey telecephalon: Morphogenesis and transformation into astrocytes, *Anat. Embryol.* **156**:115–152.

Sidman, R.L., and Rakic, P., 1973, Neuronal migration, with special reference to developing human brain: A review, *Brain Res.* **62**:1–35.

Sidman, R.L., Maile, I.L., and Feder, N., 1959, Cell proliferation in the primitive ependymal zone: An autoradiographic study of histogenesis in the nervous system, *Exp. Neurol.* **1**:322–333.

Stensaas, L.J., 1967, The development of hippocampal and dorsolateral pallial regions of the cerebral hemisphere in fetal rabbits. I. Fifteen millimeter stage, spongioblast morphology, *J. Comp. Neurol.* **129**:59–70.

Strichartz, G.R., Aguayo, A.J., Cowan, M.W., Distel, H., Lim, L., McKhann, G.M., Mugnaini, E., Rakic, P., Rickmann, M.J., Spitzer, N.C., Webster, H.F., 1982, Ontogeny. State of the art in: *Neuronal-Glial Cell Interrelationship* (T.A. Sears, ed.) Springer–Verlag, Berlin, pp. 93–114.

Tapscott, S.J., Bennett, G.S., and Holtzer, H., 1981, Neuronal precursor cells in the chick neural tube express neurofilament proteins, *Nature (London)* **292**:836–838.

Volpe, J.J. (ed.), 1981 *Neurology of the Newborn,* W.B. Saunders, Philadelphia, 648 pp.

Chapter 3

Heterogeneity in Neural Crest Cell Populations

James A. Weston, John Girdlestone, and Gary Ciment

1. NEURAL CREST CELLS MIGRATE AND DIFFERENTIATE IN RESPONSE TO ENVIRONMENTAL CUES

The remarkable diversity of neural crest derivatives has been extensively documented (Noden, 1980; Weston, 1982; Le Douarin, 1982). During its development, the embryonic neural crest gives rise to pigment cells, connective tissue of the head and face, and various endocrine cells, as well as neuronal and glial cells of the peripheral nervous system.

Likewise, there is convincing evidence that the pattern of crest cell migration and differentiation is normally determined by local environmental stimuli encountered during migration. When crest cells are grafted heterotopically into host embryos, for example, cells of the graft differentiate according to their new location rather than their original source (Le Douarin and Teillet, 1974; Le Douarin et al., 1975; Nodel, 1975). Moreover, when cells from embryonic peripheral ganglia are introduced into the crest migratory spaces of young host embryos, remigration (Erickson et al., 1980) and expression of novel phenotypic markers (Le Lièvre et al., 1980) often occur. Finally, isolated avian crest cell populations can be influenced to express melanotic, cholinergic, or catecholaminergic characters under various culture conditions (Weston, 1982, 1983; Le Douarin, 1982). Thus, the range of developmental responsiveness of the neural crest, of nonneuronal cell populations in grafted spinal and cranial sensory ganglia, and of crest-derived cells in culture appears to be rather broad.

JAMES A. WESTON, JOHN GIRDLESTONE, and GARY CIMENT ● Department of Biology, University of Oregon, Eugene, Oregon 97403.

2. ENVIRONMENTAL REGULATION OF NEURAL CREST DEVELOPMENT HAS A NUMBER OF POSSIBLE EXPLANATIONS

The experimental results discussed above can be interpreted in at least two ways. One interpretation is that homogeneous, pluripotent neural crest cell populations are induced by specific environmental cues to express distinct phenotypic traits. It should be emphsized, however, that most of the reported experimentally induced metaplasias, which have led to the idea that crest cells respond to environmental cues, involve cell *populations* and therefore may not apply to conclusions about the developmental potentiality of individual crest cells. Hence, the alternative explanation of the experimental results, which cannot be excluded in most cases, is that local environmental factors differentially promote survival, proliferation, and modulation of phenotypic expression of specific, developmentally distinct, crest subpopulations.

2.1. Many of the Macromolecular Constituents Encountered by Migrating Crest Cells are Known

It would, of course, be of considerable general interest to define precisely how identified environmental cues regulate the development of individual crest-derived cells. Since it seems clear that local environmental conditions in the crest migratory spaces can influence differentiation, as well as cell migratory behavior, there have been a number of efforts to identify the macromolecular components in these spaces that might play a regulatory role. Many of the macromolecular constituents in the crest migratory spaces have been identified (see Weston *et al.*, 1978; Mayer *et al.*, 1981; Duband and Thiery, 1982; Weston, 1982), and the effects of some of these constituents on differentiation have been characterized (Sieber-Blum *et al.*, 1981; Loring *et al.*, 1982; Weston, 1982). To analyze the precise role of such cues in regulating phenotypic expression [i.e., selection vs. induction], however, we must first elucidate the range of developmental capabilities of the responding cells.

2.2. The Developmental State of the Cells Responding to Environmental Cues is Unclear

The premise of the first hypothesis, that individual crest cells are pluripotent, is operationally difficult to verify. Nevertheless, cloning experiments with crest cells provided some evidence for bipotentiality (Sieber-Blum and Cohen, 1980). In addition, the dual expression of different crest phenotypes in individual cultured crest-derived cells (Patterson *et al.*, 1975; Furshpan *et al.*, 1976; Loring *et al.*, 1982) also suggests that individual crest-derived cells retain some differentiative plasticity. On the other hand, there is also some suggestive evidence to support the assumption, implicit in the latter hypothesis, that developmentally restricted subpopulations exist in the crest. For example, although some neuronal derivatives (e.g., sympathetic neurons) arise in appropriate locations when cultured crest-cell populations are retransplanted into crest migratory spaces of younger host embryos, graft-derived dorsal root ganglion (DRG) neurons do not arise in locations known to support sensory gangliogenesis in the host embryo. Similar results have been obtained when either peripheral autonomic ganglia or specifically marked nonneuronal cells of cranial sensory ganglia were transplanted back into younger host embryos (Le Douarin et al., 1978; Le Lièvre et al., 1980; Erickson et al., 1980; Ayer-Le Lièvre and Le Douarin, 1982). Thus, some early developmental restrictions seem to have been imposed on some of the neurons and on the ability of migrating crest-derived cells to produce these neurons.

3. CELL-TYPE-SPECIFIC MARKERS, RECOGNIZING EARLY PHENOTYPIC EXPRESSION IN INDIVIDUAL CELLS, ARE NEEDED TO TEST ALTERNATIVE HYPOTHESES EXPLAINING NEURAL CREST DEVELOPMENT

If distinct subpopulations exist among migrating neural crest cells, as predicted by the latter hypothesis, they might be detectable with suitable cell-type-specific markers. Since the time and location of embryonic crest formation are precisely defined, the questions about when and under what conditions crest cells execute their differentiative transitions can, in principle, be readily related to the time that they begin to migrate. In practice, however, the small number of cells involved and our inability to monitor the early differentiative state of migrating cells has made this analysis difficult. Therefore, we have been attempting to apply novel cell-type-specific markers to determine whether developmentally distinct subpopulations exist among early migrating crest cells.

3.1. A Monoclonal Antibody (E/C8) Recognizes an Epitope Characteristic of Neuronal Cells and Can Be Used to Disclose a Previously Indistinguishable Crest Cell Subpopulation

A monoclonal antibody has been selected for its ability to bind to the neurons of cultured DRGs, but not to the glial and fibroblastic cells in the culture (Ciment and Weston, 1982) (Fig. 1). When the antibody is applied to cultured crest cells, moreover, some E/C8$^+$ cells are seen initially (Ciment and Weston, 1982), and some of these persist in older crest cultures that contain mostly melanocytes (J. Girdlestone, unpublished observation). When grown in a defined medium [N$_2$ plus nerve growth factor (Bottenstein and Sato, 1978)] that permits the survival

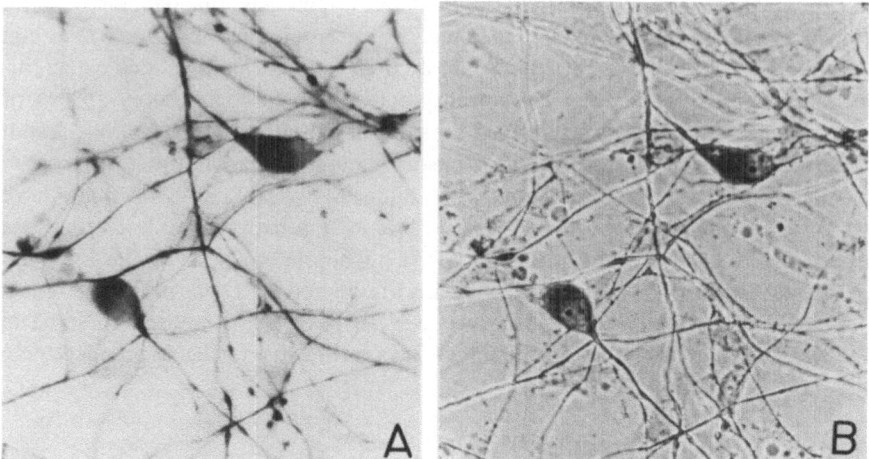

Figure 1. Monoclonal antibody binding to cultured DRG neurons. Monoclonal antibody E/C8 was produced against 7-day chick embryo DRG and screened for ability to bind to neurons but not to glial cells or fibroblasts in cultured 10-day avian DRG. (A) Bright-field micrograph of *p*-formaldehyde-fixed DRG culture stained with horseradish-peroxide-conjugated E/C8 antibody in the presence of 0.1% Triton X-100. Peroxidase activity was manifested using the Hanker–Yates reagent (see Ciment and Weston, 1982). (B) Phase-contrast micrograph of same field shown in (A). Note the staining of nerve-cell bodies and fibers and the presence of staining of nonneuronal cells.

of neurons, many of the cultured cells fail to survive. Many of those that do, however, appear to express the E/C8 antigen (Girdlestone, unpublished observation). This observation is consistent with the inference by Ziller *et al.* (1983) that there is selective expression (survival?) of precociously determined neurons in crest cell populations cultured in defined medium. Similar crest cell cultures that have been tested for both E/C8 immunoreactivity and glyoxylate-induced catecholamine fluorescence (GIF) reveal another operationally defined subpopulation (E/C8$^-$, GIF$^+$). It is not yet known whether such a neural crest-derived subpopulation exists *in vivo*.

3.2. Monoclonal Antibodies against Specific Gangliosides Reveal Other Subpopulations with Neuronal and Glial Traits in Crest Cell Cultures

Gangliosides, a class of cell-surface glycolipids, have characteristic distributions in neural tissue and act as receptors for various bacterial toxins (Mirsky *et al.*, 1978) (GD$_{1b}$, GT$_{1b}$, and GQ bind tetanus toxin; GM$_1$ binds cholera toxin). The oligosaccharide chains characteristic of specific gangliosides might also account for the reported selective binding of lectins to neurons and glia (Denis-

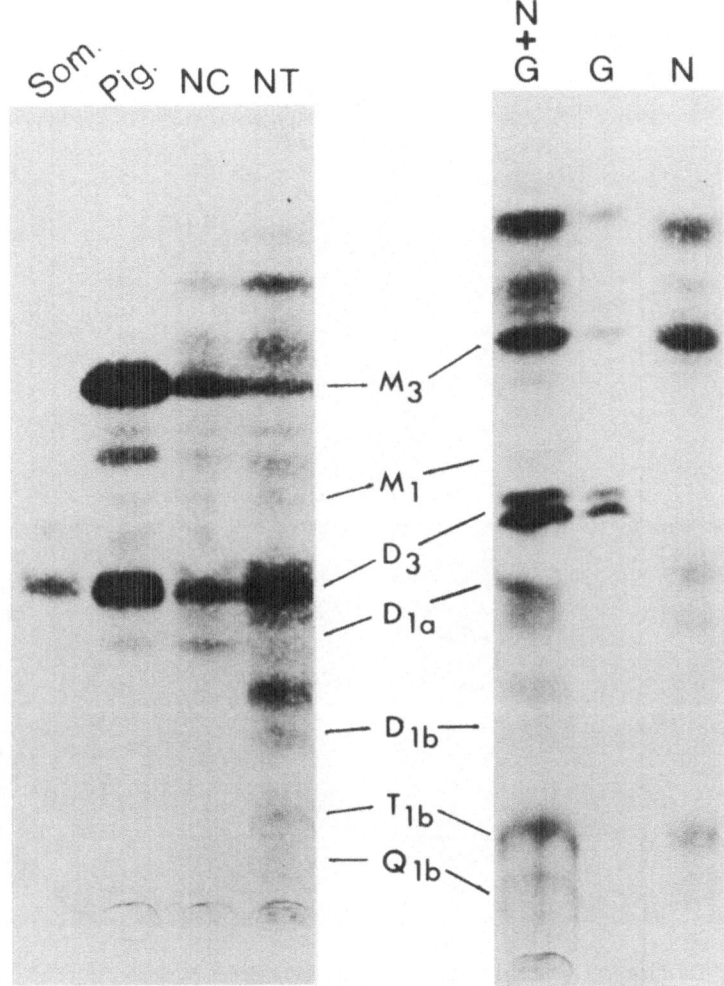

Figure 2. Ganglioside synthetic patterns in cultured embryonic cells. 3-Day cultures of 7-day DRG (N + G), or purified neuronal (N) and nonneuronal (G) cells from such ganglia, were extracted and the ganglioside fractions run on TLC plates. Radioactive gangliosides separated on such chromatographs, and detected by radioautography, were identified by their R_f values compared to ganglioside standards. Note that in this experiment GD$_3$ is present in lanes containing extracts of whole ganglia (N + G) and purified nonneuronal cells (G), but absent in lanes containing extracts of neurons (N). Conversely, gangliosides GD$_{1a}$, GD$_{1b}$, GT$_{1b}$, and GQ$_{1b}$ were produced by cultures of whole ganglia and neurons, but not by purified nonneuronal-cell cultures. Cultured neural tubes (NT), young (1-day) neural-crest cultures (NC), and older (7-day) pigmented crest-cell cultures (Pig.) produced gangliosides characteristic of both nonneuronal cells (e.g., GD$_3$) and neurons (e.g., GD$_{1b}$ and other higher gangliosides not readily visible in the photograph). Embryonic somite cultures (Som.) also appear to produce some GD$_3$.

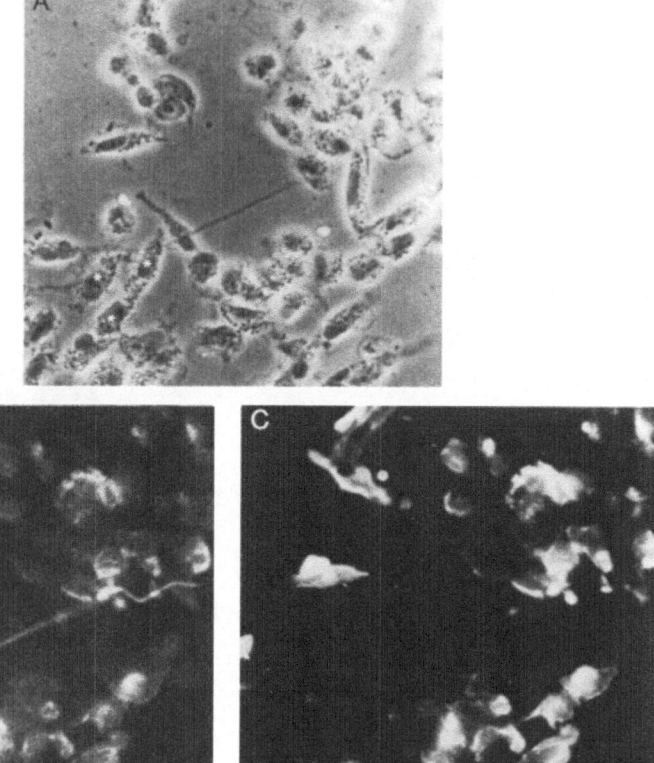

Figure 3. Dual labeling of cultured neural crest cells with monoclonal antibodies characteristic of neuronal and nonneuronal cells. Crest cultures were exposed to monoclonal antibody R24 [anti-GD_3, which is characteristic of nonneuronal cells (see Fig. 2)], washed and subsequently exposed to rhodamine-conjugated anti-mouse immunoglobulin. Cultures were then washed, fixed with 4% formaldehyde, and exposed to fluorescein-conjugated E/C8 [indicative of neuronal cells (see Fig. 1)] in the presence of 0.1% Triton X-100. (A) Phase-contrast photograph. (B) Same field viewed with epifluorescence optics for fluorescein (E/C8). (C) Same field viewed with rhodamine-epifluorescence optics. Note that there are cells in (A) that do not appear to bind either antibody (∗) and that there are E/C8$^+$ cells with neuronal morphology in (B) (◊) that are not stained with rhodamine-conjugated R24 (anti-GD_3). Conversely, many R24$^+$ cells in (C) do not bind fluorescein-conjugated E/C8 antibody. The cells that appear fluorescent in (A) and (B) exhibit pale yellow, rather than intense green fluorescence, which cannot be distinguished in these black-and-white photographs. Although it seems likely that this is nonspecific background fluorescence, the protocol used in this experiment does not rigorously exclude the possibility that the cells bind both antibodies.

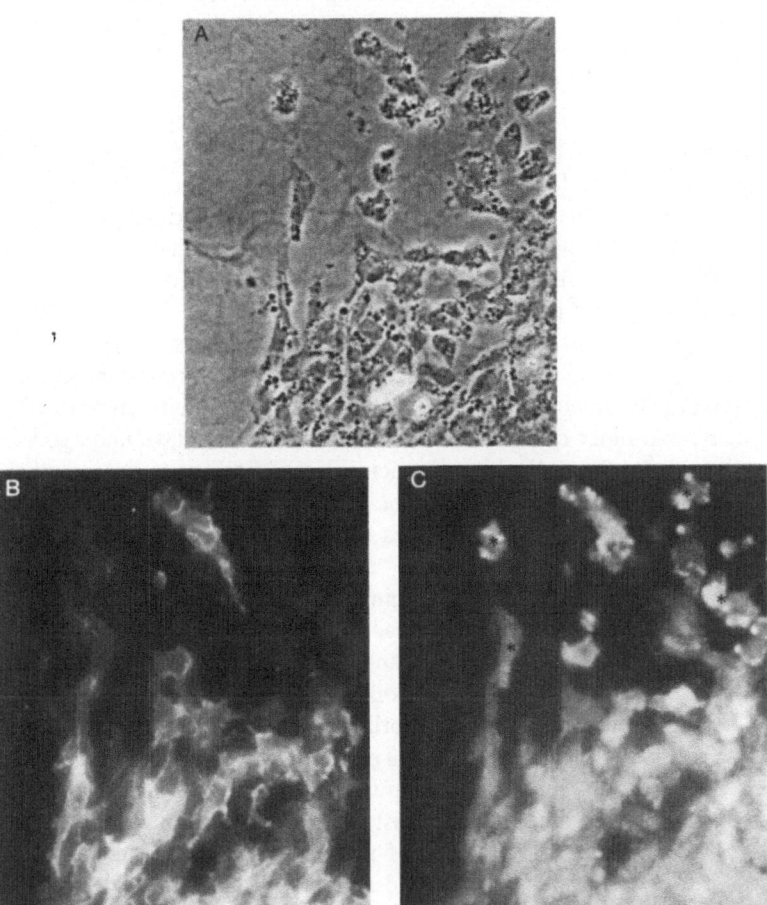

Figure 4. Cultured branchial arch cells binding neuron-specific monoclonal antibodies E/C8 and A2B5 [anti-GQ (see Fig. 2)]. The third branchial arches of 4-day chicken embryos were explanted, cultured for 2 days, and treated with monoclonal antibody A2B5. Cultures were washed, fixed in 4% *p*-formaldehyde, rinsed with phosphate-buffered saline containing 10 mM lysine, and then exposed to rhodamine-conjugated mouse immunoglobulin (Ig). After the cultures were rinsed and incubated with 10% mouse serum to saturate the remaining anti-mouse-Ig-binding sites, they were exposed to fluorescein-conjugated E/C8 antibody in the presence of 0.1% Triton X-100. (A) Phase-contrast photomicrograph of branchial arch cells with mesenchymal morphology. (B) Same field viewed with rhodamine-fluorescence optics reveals binding of A2B5 antibody to surfaces of some but not all of these cells. (C) Same field viewed under fluorescein-fluorescence optics shows E/C8$^+$ cells. Other cells in branchial arch cultures with neuronal morphology (not shown) also exhibited characteristic binding of both antibodies. Note that not all the E/C8$^+$ cells (*) were A2B5$^+$.

Donini *et al.*, 1978; Denis-Donini and Augusti-Tocco, 1980; Sieber-Blum and Cohen, 1978; see Weston, 1982). Although the neuronal gangliosides of peripheral neurons have been characterized (Cornbrooks *et al.*, 1980; Zurn, 1982), the contribution by nonneuronal components in sensory ganglia is unclear, and the pattern of gangliosides in crest cells has not been examined directly.

When cells from cultured dissociated sensory ganglia are labeled with [^{3}H]palmitic acid and the labeled gangliosides extracted from such cultures are characterized by thin-layer chromatographic (TLC) methods, the predicted differences between neurons and glia are revealed. Thus, purified glial cell cultures produce GM_3 and GD_3 but not neuron-specific gangliosides (GD_{1b}, GT_{1b}, and GQ_{1b}). In contrast, cultures enriched for DRG neurons produce abundant GD_{1b}, GT_{1b}, and GQ_{1b}, but produce little or no GD_3 (Fig. 2). These results suggest that the production of higher gangliosides may serve as unique and sensitive markers to distinguish neuronal cells, whereas GD_3 synthesis can very likely be used to mark the nonneuronal-cell populations. Short -term cultures of neural-crest cells, and older crest cultures in which most cells have undergone melanogenesis, produce gangliosides characteristic of both neurons and glial cells. Such metabolic labeling studies cannot, of course, reveal whether each cultured crest cell possesses all the gangliosides or whether subpopulations with different ganglioside complements exist.

Monoclonal antibodies that recognize specific gangliosides can be used to discriminate among these possibilities. Thus, monoclonal antibody A2B5 (Eisenbarth *et al.*, 1979) recognizes a ganglioside in the GQ fraction known to be diagnostic of neuronal cells. In addition, monoclonal antibody R24 (kindly provided by Dr. Kenneth Lloyd, Memorial Sloan–Kettering Cancer Center) recognizes GD_3, which is present in some embryonic nonneuronal cells but reduced or absent in neurons (see Fig. 2). Immunocytochemical staining, using E/C8 antibody and the two antibodies against gangliosides, has revealed that there are distinct subpopulations in cultures of neural-crest and crest-derived cells that express the neuron- and glial-specific antibody-binding patterns (Fig. 3). The A2B5$^+$ cells partially overlap the subpopulations identified *in vitro* with E/C8 (see Fig. 4).

4. CELL-TYPE-SPECIFIC MARKERS MAY PERMIT ANALYSES OF THE NORMAL TIME AND ORDER OF SEGREGATION OF CELLS IN THE NEURAL CREST LINEAGE WITH SPECIFIC DEVELOPMENTAL RESTRICTIONS

The E/C8 antibody binds not only to neuronal cells (see Section 3.1), but also to a limited number of apparently nonneuronal cells, including virtually all the mesenchymal cells of the posterior (3rd and 4th) branchial arches (Ciment and Weston, 1982; Ciment, 1983). Cultured branchial-arch cells bind both E/C8 and A2B5 antibodies (Fig. 4). These mesenchymal cells are known to be of neural-crest origin, arise from the same axial level of the neural tube that gives rise to the enteric neurons of the gut (Le Douarin and Teillet, 1974; Le Lièvre and Le Douarin, 1975), and are physically contiguous with the mesenchyme of the gut. It seems reasonable to suppose, therefore, that the branchial-arch cells, which show immunoreactivity characteristic of neuronal cells, are precursors of enteric neurons.

Recently, we have tested the hypothesis that these E/C8$^+$ cells in the posterior branchial arch are a partially restricted population of crest cells, able to give rise to enteric neurons, but not to various other known crest derivatives. Branchial arch cells exhibit neuronal traits in culture, as judged by their immunoreactivity (see above) and, occasionally, their distinctive neuronal morphology (not shown). Moreover, in coculture with aneural intestine (dissected from the postumbilical gut of 4-day chick embryos—before crest cells have entered this region), quail-derived branchial-arch cells migrate into the mesenchyme of the gut and form what appear to be normal E/C8$^+$ enteric ganglia and fibers (Fig. 5). The results obtained by Newgreen *et al.* (1980) in cocultures of aneural guts with neural-crest cells are consistent with these observations.

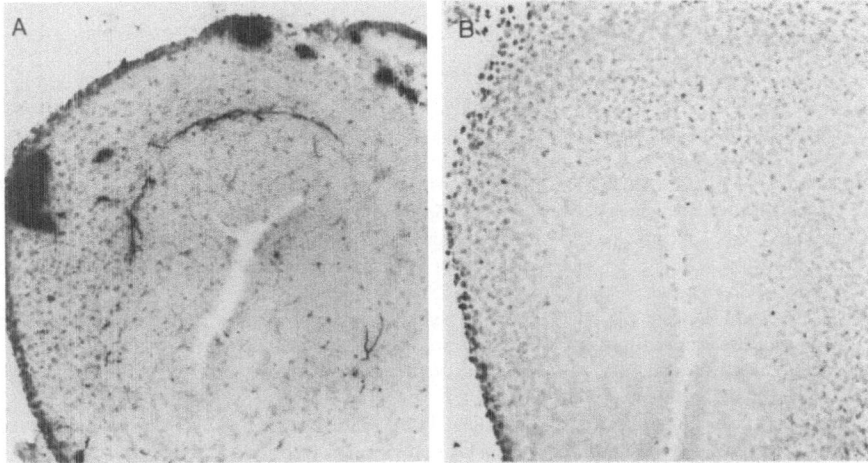

Figure 5. E/C8$^+$ enteric ganglia in aneural embryonic intestine cultured with branchial-arch mesenchyme. Aneural guts from 4-day chick embryos were cultured on chorioallantoic membranes (CAM) alone or with mesenchyme from the third branchial arch of 4-day chick embryos. Cultures were harvested after 7 days on the CAM, cryostat-sectioned, and exposed to horseradish-peroxidase-conjugated E/C8 antibody. Note the E/C8-immunoreactive material in the region of the myenteric plexus of guts cocultured with branchial arch (A) and its absence in control guts cultured alone (B).

Branchial arch and neural crest cells appear to be equivalent in their ability to form neuronal elements. The melanogenic ability of these cell types, however, is clearly different. It is known that all axial levels of the neural tube give rise to crest cells that become pigmented after several days in culture (see Weston, 1982; Le Douarin, 1982). In contrast, branchial-arch cells, which are derived from rhombencephalic neural crest (Le Lièvre and Le Douarin, 1975), do not form pigment when cultured under comparable environmental conditions *in vivo* and *in vitro* (Ciment and Weston, 1983). This suggests, at least, that these cells and the neural crest from which they originate differ in responsiveness to environmental cues.

We conclude, therefore, that the E/C8$^+$ (and A2B5$^+$) cells within the posterior branchial arches are a partially restricted population of crest-derived cells. They join a growing list of operationally defined, crest-derived, specifically localized, developmentally restricted subpopulations that are present in the embryo after the onset of crest-cell migration. The details of how subpopulations arise and respond to identified environmental cues, and hence the apparent functional roles of such cues during normal development, remain to be elucidated.

ACKNOWLEDGMENTS. Work from the authors' laboratory has been supported by grants from the NSF (PCM-7904577), the NIH (DE-04316), The Dysautonomia Foundation, Inc., and The National Neurofibromatosis Foundation, Inc.

REFERENCES

Ayer-Le Lièvre, C.S., and Le Douarin, N.M., 1982, The early development of cranial sensory ganglia and the potentialities of their component cells studied in quail–chick chimeras, *Dev. Biol.* **94**:291–310.

Bottenstein, J.E., and Sato, G.H., 1979, Growth of a rat neuroblastoma cell line in serum-free supplemented medium, *Proc. Natl. Acad. Sci. U.S.A.* **76**:514–517.

Ciment, G., 1983, Neurogenesis in the neural crest-derived branchial arch mesenchyme of avian embryos, in: *Developing and Regenerating Vertebrate Nervous Systems*, (D.W. Coates and R.R. Mashwald, eds.) Alan R. Liss, New York (in press).

Ciment, G., and Weston, J.A., 1982, Early appearance in neural crest and crest-derived cells of an antigenic determinant present in avian neurons, *Dev. Biol.* **93**:355–367.

Ciment, G., and Weston, J.A., 1983, Enteric neurogenesis by neural crest-derived branchial arch mesenchymal cells, *Nature* **305**:424–427.

Cornbrooks, C.J., Bunge, R.P., and Gottlieb, D.I., 1980, Neurites and somata of sensory and sympathetic neurons in culture contain multiple species of gangliosides, *J. Neurochem.* **34**:800–807.

Denis-Donini, S., and Augusti-Tocco, G., 1980, Molecular and lectin probe analyses of neuronal differentiation, *Curr. Top. Dev. Biol.* **16**:323–348.

Denis-Donini, S., Estenoz, M., and Augusti-Tocco, G., 1978, Cell surface modifications in neuronal maturation, *Cell Differ.* **7**:193–201.

Duband, J.L., and Thiery, J.P., 1982, Distribution of fibronectin in the early phase of avian cephalic neural crest cell migration, *Dev. Biol.* **94**:308–323.

Eisenbarth, G.S., Walsh, F.S., and Nirenberg, M., 1979, Monoclonal antibody to a plasma membrane antigen of neurons, *Proc. Natl. Acad. Sci. U.S.A.* **76**:4913–4917.

Erickson, C.A., Tosney, K.W., and Weston, J.A., 1980, Analysis of migratory behavior of neural crest and fibroblastic cells in embryonic tissues, *Dev. Biol.* **77**:142–156.

Furshpan, E.J., MacLeish, P.R., O'Lague, P.H., and Potter, D.D., 1976, Chemical transmission between rat sympathetic neurons and cardiac myocytes developing in microcultures: Evidence for cholinergic, adrenergic and dual-function neurons, *Proc. Natl. Acad. Sci. U.S.A.* **73**:425–429.

Le Douarin, N.M., 1982, *The Neural Crest*, Cambridge University Press, Cambridge.

Le Douarin, N.M., and Teillet, M.-A., 1974, Experimental analysis of the migration and differentiation of neuroblasts of the autonomic nervous system and of neuroectodermal mesenchymal derivatives, using a biological cell marking technique, *Dev. Biol.* **41**:162–184.

Le Douarin, N.M., Renaud, D., Teillet, M.-A., and Le Douarin, G., 1975, Cholinergic differentiation of presumptive adrenergic neuroblasts in interspecific chimeras after heterotypic transplantations, *Proc. Natl. Acad. Sci. U.S.A.* **72**:728–732.

Le Douarin, N.M., Teillet, M.-A., Ziller, C., and Smith, J., 1978, Adrenergic differentiation of cells of the cholinergic ciliary and Remak ganglia in avian embryos following *in vivo* transplantation, *Proc. Natl. Acad. Sci. U.S.A.* **75**:2030–2034.

Le Lièvre, C.S., and Le Douarin, N.M., 1975, Mesenchymal derivatives of the neural crest: Analysis of chimaeric quail and chick embryos, *J. Embryol. Exp. Morphol.* **34**:125–154.

Le Lièvre, C.S., Schweizer, G.G., Ziller, C.M., and Le Douarin, N.M., 1980, Restrictions of developmental capabilities in neural crest cell derivatives as tested by *in vivo* transplantation experiments, *Dev. Biol.* **77**:362–378.

Loring, J., Glimelius, B., and Weston, J., 1982, Extracellular matrix materials influence quail neural crest cell differentiation *in vitro*, *Dev. Biol.* **90**:165–174.

Mayer, B.W., Hay, E.D., and Hynes, R.O., 1981, Immunocytochemical localization of fibronectin in embryonic chick trunk and area vasculosa, *Dev. Biol.* **82**:267–286.

Mirsky, R., Wendon, L., Black, D., Stokin, C., and Bray, D., 1978, Tetanus toxin: A cell surface marker for neurons in culture, *Brain Res.* **148**:251–259.

Newgreen, D.F., Jahnke, I., Allan, I.J., and Gibbins, I.L., 1980, Differentiation of sympathetic and enteric neurons of the fowl embryo in grafts to the chorio-allantoic membrane, *Cell Tissue Res.* **208**:1–19.

Noden, D., 1975, An analysis of the migratory behavior of avian cephalic neural crest cells, *Dev. Biol.* **42**:106–130.

Noden, D., 1980, The migration and cytodifferentiation of cranial neural crest cells, in: *Current Research Trends in Prenatal Craniofacial Development* (R. Pratt, and R. Christiansen, eds.), Elsevier/North-Holland, Amsterdam, pp. 3–25.

Patterson, P.H., Reichardt, L.F., and Chun, L.L.Y., 1975, Biochemical studies on the development of primary sympathetic neurons in cell culture, *Cold Spring Harbor Symp. Quant. Biol.* **40**:389–397.

Sieber-Blum, M., and Cohen, A.M., 1978, Lectin binding to neural crest cells: Changes of the cells surface during differentiation *in vitro*, *J. Cell Biol.* **76**:228–238.

Sieber-Blum, M., and Cohen, A.M., 1980, Clonal analysis of quail neural crest cells: They are pluripotent and differentiate *in vitro* in the absence of non-crest cells, *Dev. Biol.* **80**:96–106.

Sieber-Blum, M., Sieber, F., and Yamada, K., 1981, Cellular fibronectin promotes adrenergic differentiation of quail neural crest cells *in vitro*, *Exp. Cell Res.* **133**:285–295.

Weston, J.A., 1982, Motile and social behavior of neural crest cells, in: *Cell Behaviour* (R. Bellairs, A. Curtis, and G. Dunn, eds.), pp. 429–470, Cambridge University Press.

Weston, J.A., 1983, Regulation of neural crest cell migration and differentiation, in: *Cell Interactions and Development: Molecular Mechanisms* (K.M. Yamada, ed.), Wiley, New York, pp. 153–184.

Weston, J.A., Derby, M., and Pintar, J., 1978, Changes in the extracellular environment of neural crest cells during their early migration, *Zoon* **6**:103–113.

Ziller, C., Dupin, E., Brazeau, P., Paaulin, D., and Le Douarin, N.M., 1983, Early segregation
      of a neuronal precursor cell line in the neural crest as revealed by culture in a chemically defined
      medium, *Cell* **32:**627–638.
Zurn, A.D., 1982, Identification of glycolipid binding sites for soybean agglutinin and differences
      in the surface glycolipids of cultured adrenergic and cholinergic sympathetic neurons, *Dev.
      Biol.* **94:**483–498.

Chapter 4

The First Growth Cones in the Central Nervous System of the Grasshopper Embryo

Michael J. Bastiani and Corey S. Goodman

1. INTRODUCTION

What factors guide the advancing tips of individual growth cones during embryonic development? One way to answer this question is to examine and manipulate a developing embryo in which the growth cones of different neurons are confronted with the same environment and yet make different and stereotyped choices of which way to grow. The best such divergent choice to study would be one in which the individual growth cones are identifiable and highly accessible and their environment is stereotyped and relatively simple. We have been studying several different examples in the grasshopper embryo that meet these criteria: identified neurons with growth cones that make divergent choices (e.g., Goodman *et al.*, 1982; Taghert *et al.*, 1982; Raper *et al.*, 1983a,b). In the example described in this chapter, the growth cones make divergent choices very early in embryogenesis when the terrain is relatively simple, the distances short, and the number of possible cells involved small. We will discuss the very first growth cones and the very first axonal pathways in the central nervous system (CNS) of the grasshopper embryo.

MICHAEL J. BASTIANI and COREY S. GOODMAN ● Department of Biological Sciences, Stanford University, Stanford, California 94305.

2. DIVERGENT CHOICES BY THE FIRST GROWTH CONES IN THE CENTRAL NERVOUS SYSTEM

What guides the very first neuronal growth cones within the CNS of the grasshopper embryo? On the basis of the pioneering work of Bate and Grunewald (1981), we know the identities of these cells: the first growth cones extend from the midline precursor 1 (MP1), dorsal MP2 (dMP2), and ventral MP2 (vMP2) neurons. The growth cones of these three individually identified neurons pioneer the very first longitudinal axonal pathways in each segment of the grasshopper CNS (Bate and Grunewald, 1981; Goodman *et al.*, 1982; Taghert *et al.*, 1982). They are confronted with the same environment and yet make divergent choices; the vMP2 growth cone turns anteriorly and pioneers one pathway and the dMP2 and MP1 growth cones turn posteriorly and pioneer a second pathway. These initial two axonal pathways are soon followed by many other axonal pathways as the neuropil of the CNS quickly unfolds, giving rise to the basic axonal scaffold of a 40% grasshopper embryo (Fig. 1).

This axonal scaffold develops just under the extracellular basement membrane that covers the dorsal surface of the neuroepithelium. This epithelium consists of a germinal zone on the ventral surface where the neuronal stem cells, called neuroblasts (NBs), divide repeatedly and generate chains of progeny that extend toward the dorsal surface. There is also a second class of neuronal precursor cells, called midline precursors (MPs), that divide only once and thus generate many fewer progeny. In each segment, the neuroepithelium contains 61 identified NBs and 7 identified MPs. The dorsal surface of the neuroepithelium, covered by the extracellular basement membrane, also contains the endfeet of numerous epithelial cells, the cell bodies of which lie near the ventral surface and the processes of which extend up to the dorsal surface. Spaces, or channels, appear under the dorsal basement membrane just as the very first growth cones (from the MP1, dMP2, and vMP2 neurons) extend anteriorly and posteriorly along the dorsal surface. The neuroepithelium thickens as more neurons are born, and the spaces under the dorsal basement membrane enlarge to become occupied by the developing axonal scaffold.

The first growth cones in the CNS arise from the progeny of MP1 and MP2; there are two MP2s (one on each side) and one MP1 (at the midline) in each segment (Fig. 2). Each MP2 divides once to give rise to a vMP2 and a dMP2 daughter cell. The single MP1 gives rise to a pair of bilaterally symmetric daughters, each of which comes to lie dorsal to the two MP2 progeny, thus forming a trio of cells on each side (Fig. 3A). All three cells (on each side) send growth cones up to the dorsal basement membrane. The growth cone of the vMP2 extends anteriorly, while the growth cones of the dMP2 (its sibling) and the MP1 extend posteriorly (Figs. 3B, C, 4, and 5). The cell bodies and often the axons and growth cones of the MP1 and MP2 progeny are visible with

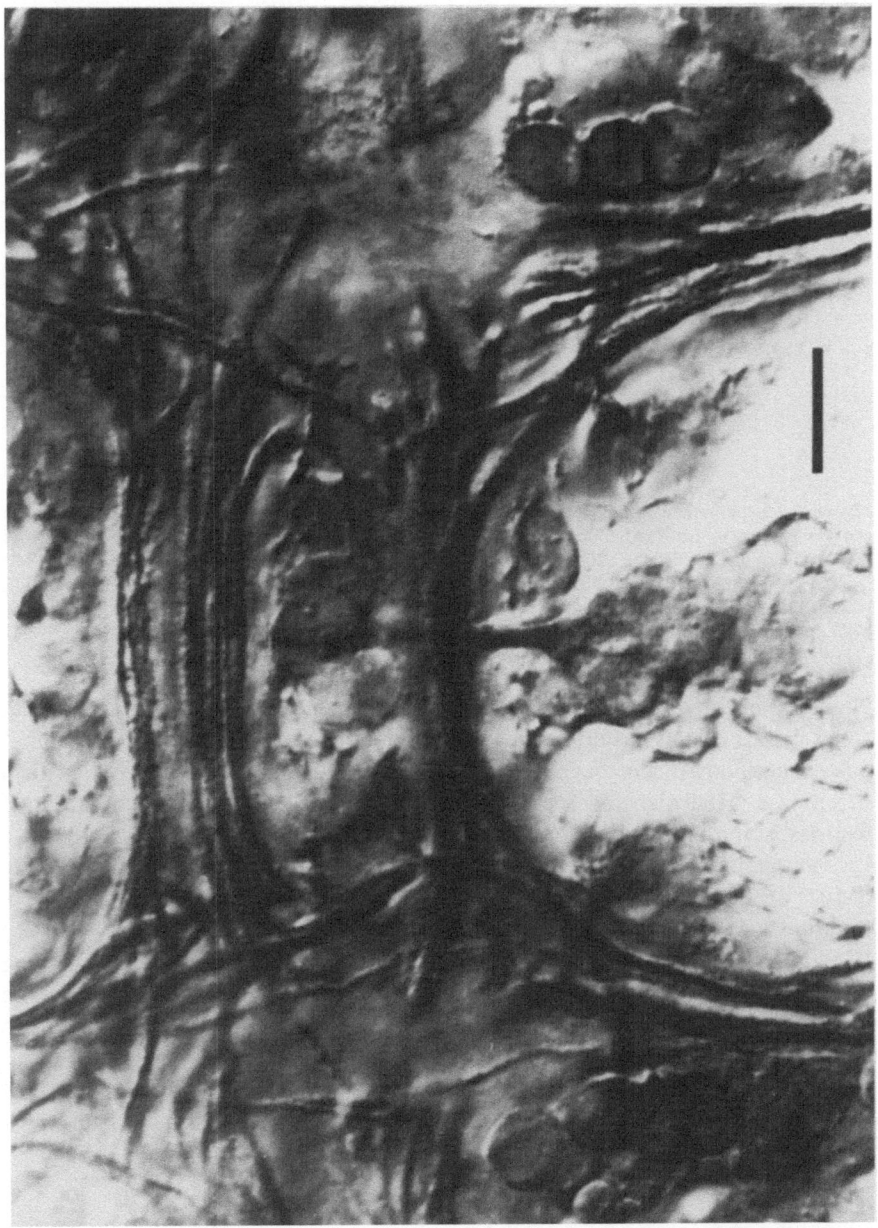

Figure 1. Photograph of whole-mount preparation of the T2 segment in a 40% grasshopper embryo stained with the I-5 monoclonal antibody and an HRP-labeled second antibody. Some of the commissural and longitudinal axon bundles in the 40% axon scaffold are seen in this single focal plane. Scale bar: 20 μm. Anterior is to the left, posterior to the right.

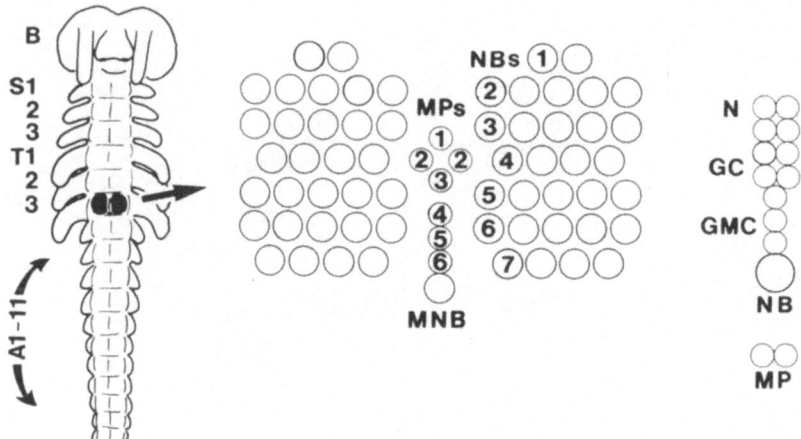

Figure 2.   Schematic diagram of the 30% grasshopper embryo showing the segmental organization and the pattern of neuronal precursor cells, including neuroblasts (NBs) and midline precursors (MPs). There are 30 NBs on each side of each segment, arranged in seven rows, and 1 median NB (MNB); there are 7 MPs. The NBs divide repeatedly, giving rise to ganglion mother cells (GMC), which divide once more to generate two ganglion cells (GC) that differentiate into neurons (N). The MPs divide once to give rise to two neuronal progeny.

Nomarski interference contrast optics in the living embryo. The axons and growth cones can be clearly visualized in fixed whole-mount preparations using a monoclonal antibody [I-5 mab, followed by immunocytochemistry with a horseradish peroxidase (HRP)-labeled second antibody] that selectively stains a subset of neurons in the grasshopper embryo (Fig. 4) (Chang *et al.*, 1983).

Preparations using the I-5 mab show that the growth cones of the MP1, dMP2, and vMP2 reach the basement membrane at about the same time and in a variety of orientations relative to each other and to the body axis (Fig. 5) (Goodman *et al.*, 1982; Taghert *et al.*, 1982). For example, of the two sibling cells (vMP2 and dMP2), in some cases the vMP2 growth cone is the more anterior and in other cases the dMP2 is the more anterior; similarly, either growth cone can be the more lateral of the two. Within several hours, and irrespective of their initial orientation along the membrane, their growth cones begin to make divergent choices: the vMP2 turns anteriorly and the dMP2 and MP1 turn posteriorly (Fig. 5).

3. CHANNELS AND SPACES

What guides the growth cones of the MP1, dMP2, and vMP2? Perhaps the growth cones passively follow mechanical guidance cues. For example, anatom-

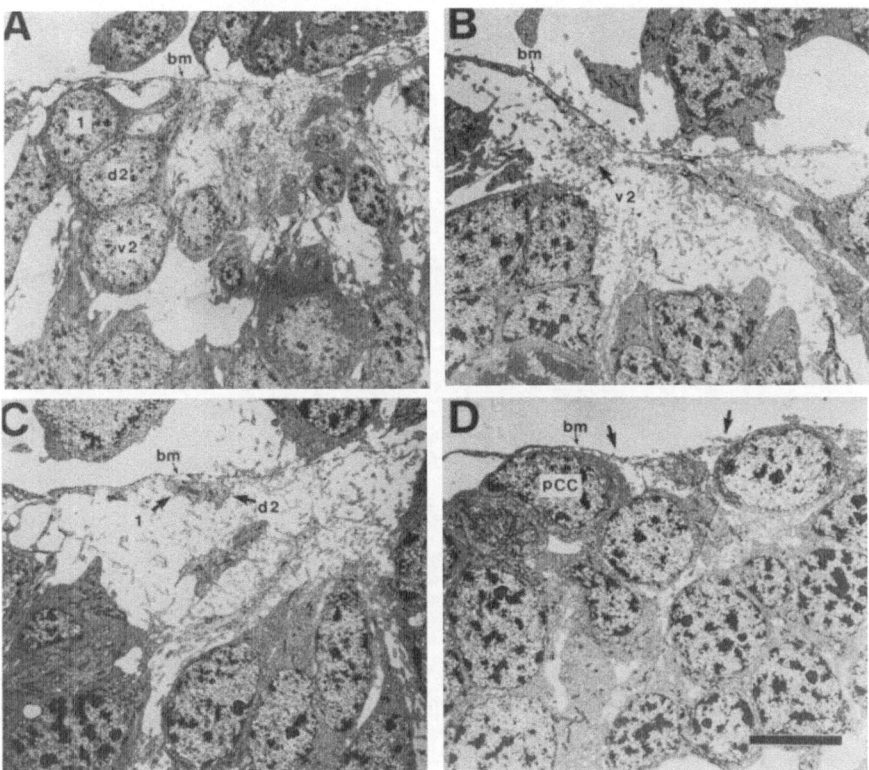

Figure 3. Environment of the first growth cones in the CNS of the grasshopper embryo. (A) Transmission-electron-microscopic (TEM) cross section through the middle of the segment. The trio of cells, MP1 (1), dMP2 (d2), and vMP2 (v2), extend their growth cones dorsally to the dorsal basement membrane (bm), as clearly seen for MP1 and dMP2 (the vMP2 extends its axon dorsally a few microns more anteriorly). Notice the relatively large space filled with a high density of filopodia lateral to the MP1 and dMP2 axons. (B) TEM section about 10 μm anterior to (A). The axon of vMP2 (v2) extends anteriorly near the dorsal basement membrane (bm) to pioneer one of the first axonal pathways in the longitudinal connective. Note the large space filled with filopodia and the growth cones from lateral cell bodies extending medially (toward the left) to pioneer the B commissure. (C) TEM section 5 μm posterior to (A). The MP1 (1) and dMP2 (d2) axons are surrounded by a large space filled with filopodia. Within 24 hr (5%), this space will become filled with several different axon bundles forming the longitudinal connective. (D) TEM section 40 μm posterior to (A). The growth cones of the MP1 and dMP2 ended 20 μm anterior to this point; the leading filopodia from the MP1 and dMP2 are present at this level and are in contact with either the posterior corner cell (pCC), epidermal cells, or dorsal basement membrane (bm). Note the relatively small space (between the arrows) that is occupied by processes of epidermal cells. This space will soon enlarge as in (C) to become the channel in which the longitudinal connective forms. The space opens up as the leading filopodia contact the cells surrounding the space. Scale bar: (A) 15 μm; (B–D) 10 μm.

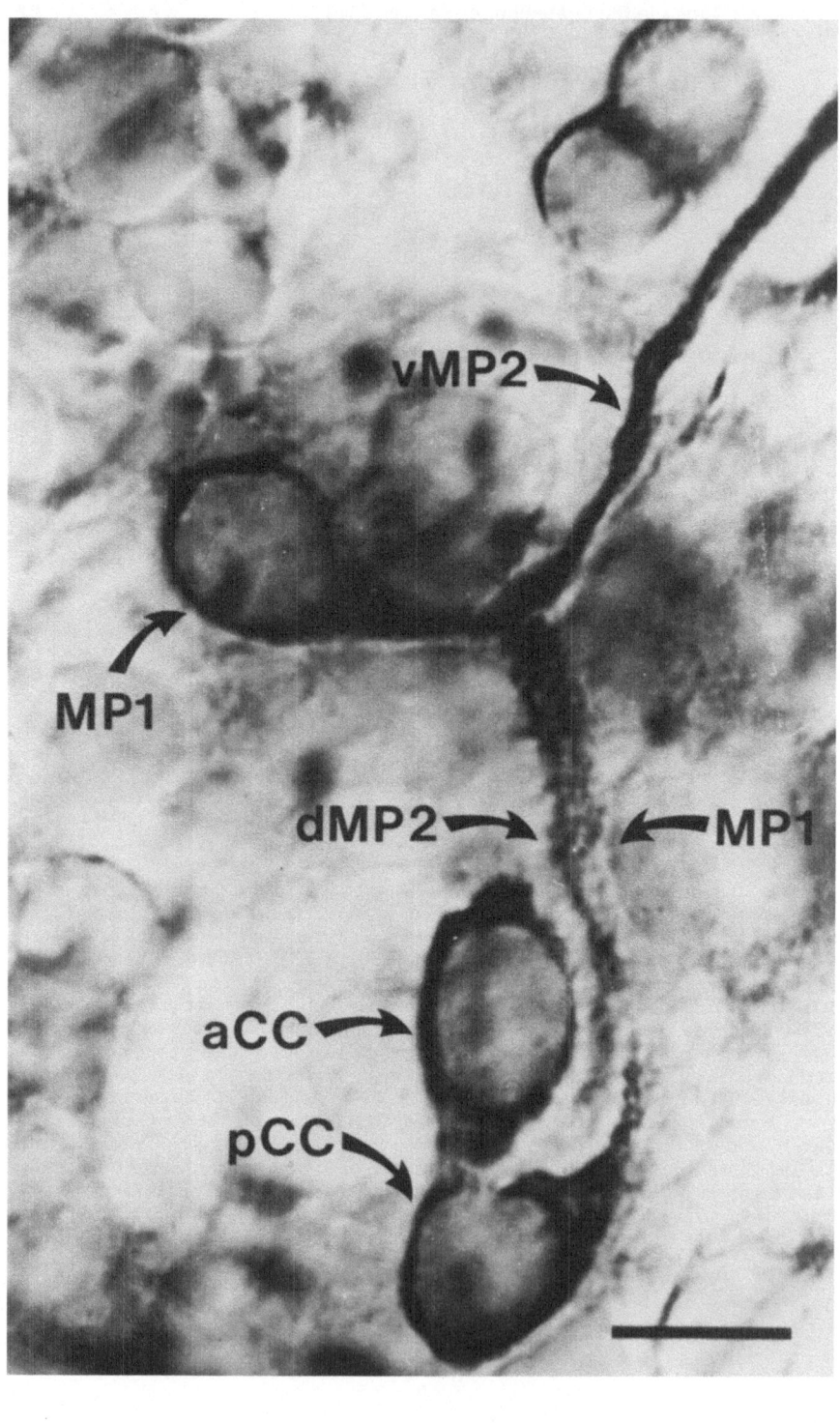

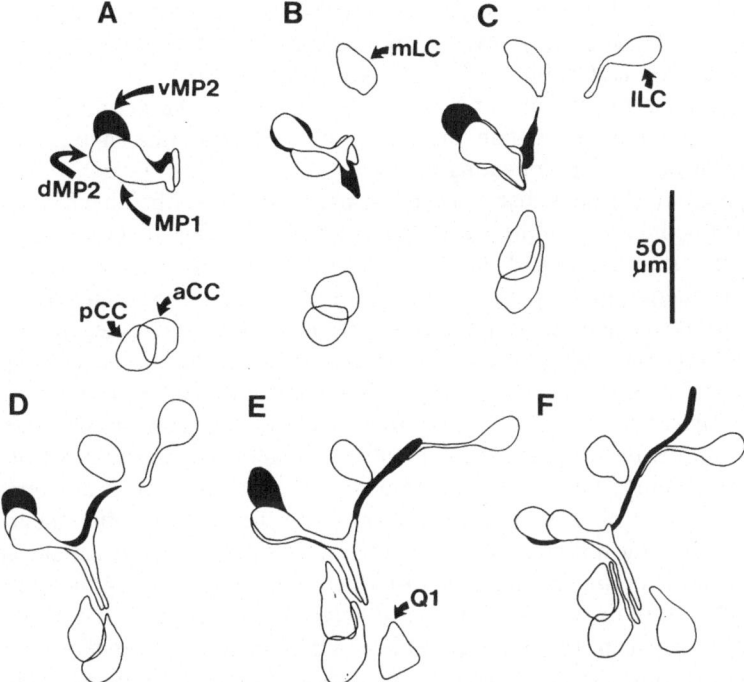

Figure 5. Camera-lucida drawings of the MP1, dMP2, vMP2, anterior corner cell (aCC), posterior corner cell (pCC), medial landmark cell (mLC), lateral landmark cell (lLC), and first cell division (Q1) cell bodies based on whole-mount preparations using the I-5 mab and HRP-labeled second antibody. The MP1 and dMP2 growth cones turn posteriorly and point toward the aCC and pCC bodies, which are migrating anteriorly (A–C); in particular, the tips of their growth cones point toward the pCC leading edge and subsequent growth cone (D). In time, the MP1 and dMP2 growth cones grow between these two cell bodies and the Q1 cell as they extend posteriorly (E, F). The vMP2 growth cone turns anteriorly and points toward the mLC body (C) and, shortly thereafter, toward the lLC growth cone (D). In time, the vMP2 growth cone grows between these two cell bodies as it extends anteriorly (E, F). Interestingly, the lLC growth cone extends posteriorly along the vMP2 axon (E, F). Each drawing represents about 1%, or approximately 5 hr of development. Note that there is no fixed orientation when the three growth cones initially reach the dorsal basement membrane (A, B). However, within several hours and irrespective of their initial orientation, they begin to make their cell-specific choices (C, D).

←——————————————————————————————————

Figure 4. Photograph of whole-mount preparation of grasshopper embryo showing some of the midline precursor progeny and other key cells during formation of the first axonal pathways in the CNS, as stained by the I-5 mab (Chang *et al.*, 1983). Photograph of a dorsal focal plane in which the MP1 cell body is in focus (the dMP2 and vMP2 cell bodies are directly below the MP1 cell body and thus not in focus). The darkly stained vMP2 axon extends anteriorly, whereas the more lightly stained dMP2 and MP1 axons extend posteriorly. Also stained are two conspicuous posterior cell bodies [the sibling anterior (aCC) and posterior (pCC) corner cells]. The sibling aCC and pCC are thought to be landmark cells for posteriorly extending growth cones of the MP1 and dMP2 (Goodman *et al.*, 1982; Taghert *et al.*, 1982). Scale bar: 20 μm.

ical studies in amphibian embryos have revealed channels between ependymal cells that are subsequently invaded by neuronal growth cones (e.g., Singer *et al.*, 1979; Silver and Robb, 1979; Silver and Sidman, 1980). These observations have given rise to the "blueprint" hypothesis whereby the germinal neuroepithelium is thought to contain the pattern for channels that in turn serve as mechanical guides for elongating axons.

Spaces or channels also appear just under the dorsal basement membrane in the neuroepithelium of the grasshopper embryo (see Fig. 3). Two observations are of interest in this regard. First, the divergent and cell-specific choices made by the pioneering growth cones in the CNS of the grasshopper embryo suggest the independent existence of active guidance mechanisms. In this regard, it should be noted that the blueprint hypothesis also requires the presence of specific biochemical cues to direct the elongating axons during pathway selection. Second, the spaces open up to about 50μm in front of the leading growth cone (Fig. 3D). It is difficult to determine, without manipulative experiments, whether the spaces are opening due to intrinsic properties of the epithelium or to interactions with filopodia of the leading growth cones. The filopodia are often longer than 50 μm and in transmission-electron-microscopic (TEM) reconstructions are observed to contact the cell surfaces just where the spaces are beginning to open (Fig. 3D). It is to be hoped that cell-ablation experiments in progress will distinguish between the intrinsic and interactive models of channel formation.

Studies of these pioneering growth cones in the grasshopper embryo support the notion that there is "information" in their environment and that they are differentially determined to respond to that information in cell-specific ways. Their environment could provide polarity information, positional information, or both. This information could be in the form of a diffusible gradient, a substrate gradient (on either the dorsal basement membrane or the epidermal or neuronal cell surfaces), or specific positional information on particular cell surfaces. In trying to define sources of guidance information in the environment of the growth cone, it is important to understand just how large that environment is.

4. GROWTH CONES AND FILOPODIA

The key to understanding the environment of the growth cone resides in the environment within the grasp of its filopodia. Growth-cone motility involves three phases: extension, adhesion, and contraction (Bray, 1982; Letourneau, 1982). Most of our present knowledge about the structure and behavior of growth cones comes from studies in tissue culture. Growth cones extend numerous long fingerlike filopodia, approximately 0.1 μm in diameter and up to 50 μm or more in length. These filopodia radiate in many directions from the growth cone, transiently exploring their environment. Most of these filopodial extensions are transient and dynamic structures that can increase in length, move about, and retract in a matter of minutes. Although most are short-lived and regress into the growth cone, those that adhere to other surfaces persist and play a significant role in neurite extension.

Some of the filopodia contact other cell surfaces or extracellular basement membranes; they adhere strongly to some of these surfaces and much more weakly to others. These noncovalent adhesive bonds may involve selective molecular recognition, such as between an antibody and an antigen or a receptor and a ligand. Filopodia are retracted in a contractile cycle. If adhesion is weak, the filopodium is retracted; if, however, its adhesion is strong, then tension in that direction is increased during the contractile cycle and the leading tip of the growth cone advances toward the point of attachment (Bray, 1982).

For example, several reports have observed that if a filopodium becomes attached to an axon across its path, it can pull on this axon repeatedly at intervals of about 2 min (Nakai, 1960; Wessells *et al.*, 1980). Furthermore, Letourneau (1975) grew neurons on patterned substrates of different adhesivity in cell culture. He observed that contact with an area of greater adhesiveness by a single filopodium was sufficient to guide the growth cone toward that surface. The direction of advance of the growth cone depends on the vector of the tension; if the direction of tension is changed, the direction of growth-cone extension changes accordingly. For example, if individual filopodia are lifted from the surface of the culture, they affect the extension of the growth cone in a manner consistent with the contraction–tension mechanism. If those of the left are detached, the growth cone turns right; if those in the center are detached, the growth cone branches (Wessells and Nuttall, 1978). Thus, filopodia test the adhesiveness of the available substrate (the terrain within filopodial grasp). Those filopodia with strong adhesion to other cells or surfaces will more effectively guide the growth cone during the contractile phase.

5. MP1 FILOPODIA AND LANDMARK CELLS

Our previous light-level analysis suggested that one source of guidance for the MP1, dMP2, and vMP2 growth cones is likely to be positional information in the form of "landmark cells" (Goodman *et al.*, 1982; Taghert *et al.*, 1982). We define a landmark cell as an identified cell the differentially labeled surface of which distinguishes it from the cells around it. In the anterior direction, likely candidates for landmark cells for the vMP2 growth cone are the medial landmark cell (mLC) and lateral landmark cell (lLC) neurons (Fig. 5). In the posterior direction, likely candidates for landmark cells are the anterior corner cell (aCC) and posterior corner cell (pCC) neurons (Fig. 5). The aCC and pCC are sibling progeny from the first division of NB 1-1 from the next posterior segment. As the corner cells migrate anteriorly, they and the growth cones of the dMP2 and MP1 appear to point and grow directly toward one another (Fig. 5). In particular, the MP1 and dMP2 growth cones point at the growth-cone-like leading edge of the pCC. In time, the corner cells finish migrating and the growth cones of the MP1 and dMP2 then pass between them and the (Q1) cell body (Q1 is from the first cell division of NB 7-4 see Fig. 2). Interestingly, the dMP2 and MP1 growth cones first become selectively dye-coupled to the corner cells and later to Q1 (Goodman *et al.*, 1982; Raper and Goodman, 1982). [Dye coupling refers to the spread of the fluorescent dye Lucifer Yellow (LY) (molecular weight 450) from the interior of one cell to another.]

To examine the filopodia of the MP1 growth cone and their possible adhesive interactions with the pCC, we made rabbit serum antibodies to LY (Fig. 6) (Taghert *et al.*, 1982). LY is injected into a cell, and the embryo is fixed, incubated with the anti-LY antibody, and then incubated with an HRP-labeled second antibody. The filopodia are visualized in the fixed whole-mount preparations after HRP immunocytochemistry.

Our light-level analysis using the anti-LY antibody suggested that the filopodia of the MP1 and dMP2 growth cones selectively adhere to a small number of identified cells along their pathway as they make their posterior turn, including in particular the pCC (Goodman *et al.*, 1982; Taghert *et al.*, 1982). A disproportionate number of the MP1 filopodia adhere to the surface of the pCC body and its growth cone (which has just been initiated) (Figs. 6A and 7). Furthermore, those filopodia that do contact the pCC tend to continue running along its surface

---→

Figure 6. Filopodia and dye coupling from MP1 at three stages of development, as visualized with an anti-LY antibody, in the CNS of the grasshopper embryo. Only a very thin plane is in focus with Nomarski optics; thus, only some of the filopodia are visible in the photographs. (A) The filopodia from the MP1 growth cone extensively cover the pCC; note dye coupling to pCC mediated via the filopodia. The anteriorly extending axon is the dye-coupled vMP2, and the cell bodies out of focus around the MP1 are those of the dye-coupled dMP2 and vMP2. (B) The MP1 growth cone has just grown between the pCC and Q1 cells; note the MP1's long leading filopodia that extend posteriorly beyond the pCC and Q1 cells and the relative amounts of dye coupling to the pCC and Q1 cell bodies. (C) Dye coupling is observed from the MP1 to the pCC, Q1, and Q2 (Q1's sibling, the cell body just lateral to Q1), and to two other identified cell bodies further posterior and out of the plane of focus (◁). The dye-coupled vMP2 axon has extended anteriorly and is dye-coupled to a conspicuous identified cell, the ILC, via that cell's axon. The MP1 growth cone has entered the next posterior segment [▶ segmental boundaries]; the dMP2 growth cone has extended about the same distance. The dye-coupled vMP2 axon has extended anteriorly into the next anterior segment. The growth cones of the dye-filled, posteriorly directed MP1 and dMP2, and the unfilled, anteriorly directed vMP2 of the next posterior segment, pass each other near the segmental boundary and continue extending within a few microns of each other. Yet, while their filopodia overlap extensively, they do not become dye-coupled to one another. Similarly, the dye-filled, anteriorly directed vMP2 and the unfilled, posteriorly directed MP1 and dMP2 of the next anterior segment do not become dye-coupled to one another, although their filopodia overlap extensively. Scale bar: (A) 15 μm; (B) 20 μm; (C) 30 μm. From Taghert *et al.* (1982).

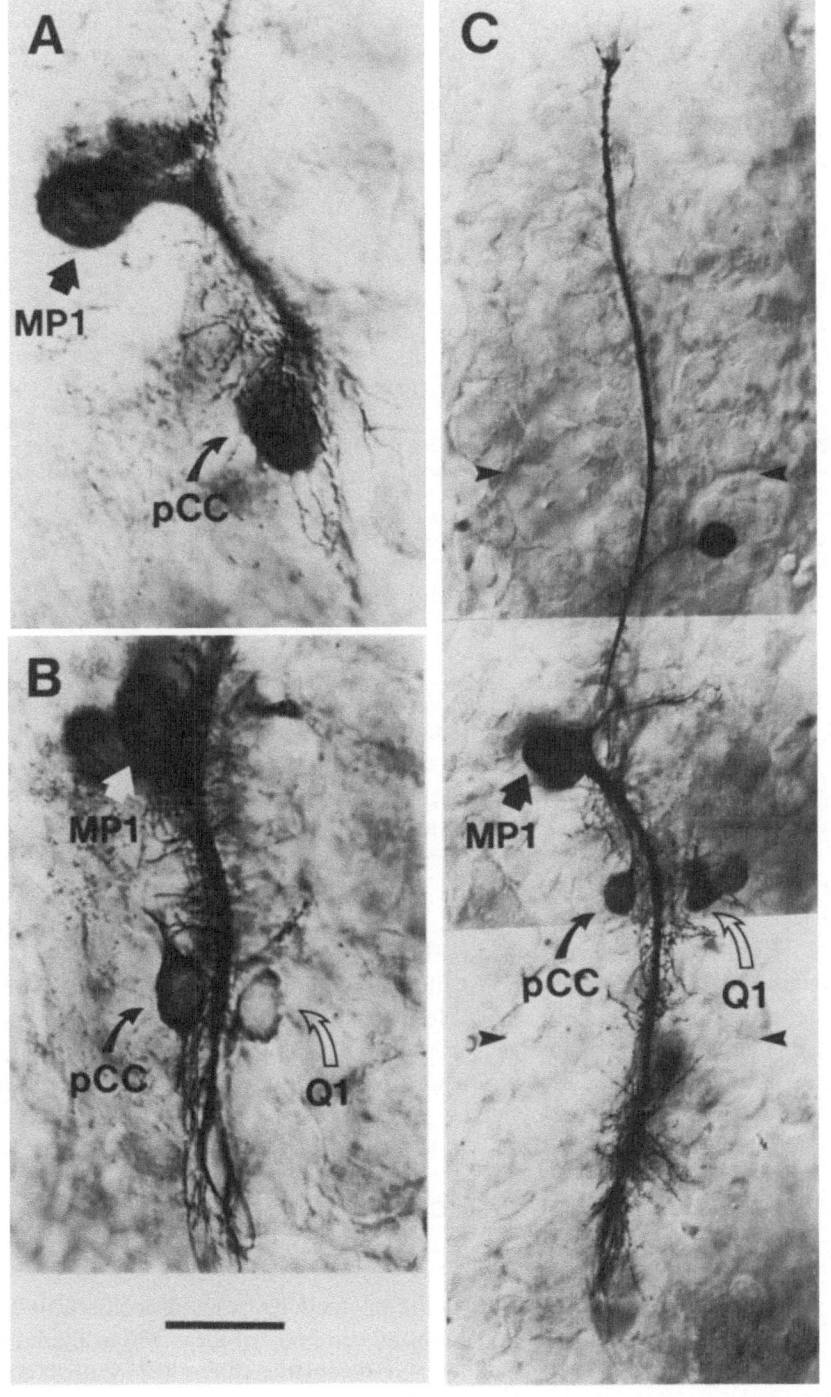

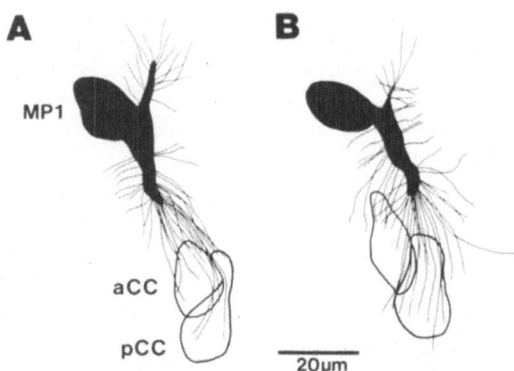

Figure 7. Apparent selective filopodial adhesion from the MP1 growth cone to the aCC and pCC in the developing CNS of the grasshopper embryo. Two examples of the MP1 growth cone and its filopodia at the same stage of development: LY was injected into the cell body and the embryo was processed with the anti-LY antibody. Note the extensive filopodial contact, particularly with the pCC body. Selective dye coupling was observed to both these cells, although it was stronger to the pCC; the only sites of cellular contact that could mediate this coupling are the filopodia. Note several long filopodia in (B) that are not contacting the aCC or pCC bodies.

(Fig. 7). We interpret this to imply selective adhesion, although we have not directly measured adhesive forces. It may be that one or a few filopodia contacting the pCC are sufficient to mediate the cell-specific turn. When we examined 11 preparations at the stage of development just after the MP1 growth cone turned posteriorly, we consistently found extensive filopodial contacts from the MP1 growth cone to the pCC, strongly suggesting selective filopodial adhesion to the surface of this cell. On average, 45% of the filopodia contacted the surface of the pCC and aCC (mostly the pCC), although these cells constitute less than 10% of the available cell surfaces. Thus, the MP1 growth cone at this stage demonstrates a strong tendency for filopodial contact with the pCC, less of a tendency with the aCC, and very little tendency for selective contact with any of the other cells in its immediate environment. Within a short period, however, some of the MP1 filopodia begin to selectively adhere to the Q1 cell body, and to other cell bodies along its pathway (Fig. 6).

6. TRANSMISSION-ELECTRON-MICROSCOPIC RECONSTRUCTIONS OF MP1 FILOPODIA

We have now examined the specific interactions of the filopodia from the MP1 growth cone with the growth cone and cell body of the pCC landmark cell by TEM serial-section reconstructions. We reconstructed the MP1 growth cone

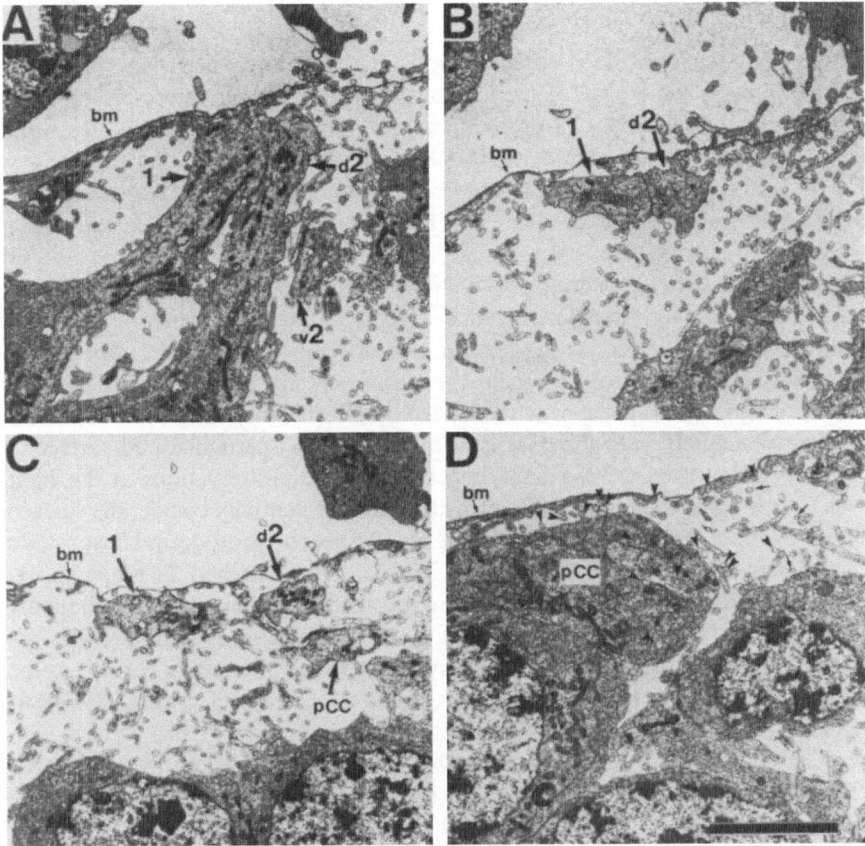

Figure 8. Examples of electron micrographs taken from serial sections of the first growth cones in the CNS of the grasshopper embryo. The growth cones of MP1 and dMP2 extends up to the dorsal basement membrane, turn, and grow posteriorly toward the pCC, suspended by their lamellipodia and filopodia from the basement membrane. (A) The axons of the MP1 (1), dMP2 (d2), and vMP2 (v2) extend from the cell bodies just out of the picture to the lower left. Note that the axons of MP1 and dMP2 are in close apposition to each other near the basement membrane (bm). The vMP2 axon extends dorsally at a slightly more anterior position, so that only part of its axon is cut in cross section. (B) This section is taken approximately 10 μm posterior to (A). The axons of MP1 (1) and dMP2 (d2) are closely adherent to each other. Notice the lamellipodium extending dorsally to the basement membrane (bm) from dMP2 (d2). (C) This section, approximately 20 μm posterior to (A), shows the growth cones of MP1 (1), dMP2 (d2), and pCC. The growth cones of MP1 (1) and dMP2 (d2) are not extensive contact with each other, in contrast to (A) and (B), or with any other structure in their environment. The growth cones are suspended in this space by their filopodial contacts with other cells and the basement membrane. (D) Approximately 40 μm posterior to (A), the growth cone of the pCC is merging into the cell body. The growth cones of the MP1 and dMP2 have broken up into an array of filopodia. The filopodia from MP1 (▼) are contacting the basement membrane (bm) (two), endfeet of epidermal cells (two), and the pCC (seven). Notice especially the four filopodia that are embedded within the pCC. These are filopodia that first contacted the pCC at its growth cone and have inserted deep into the cell. (←) Filopodia from dMP2. The dMP2 growth cone ends several microns before the MP1 growth cone and extends fewer filopodia this far posteriorly. Scale bar: 2 μm.

and its filopodia from TEM serial sections at a developmental stage just after the MP1 growth cone turns posteriorly and before it comes into direct contact with the pCC growth cone. Approximately 60 μm of serial thin sections were taken from the anterior edge of the MP1 cell body to the posterior edge of the pCC body in the mesothoracic (T2) segment. The cell bodies and axons of MP1, dMP2, vMP2, aCC, and pCC were identified in thin section by their characteristic positions and shapes (Fig. 8).

There are two interesting observations about the MP1 and dMP2 growth cones. First, although their axons stick together (Fig. 8B), their growth cones need not (Fig. 8C). Second, the leading tips of their growth cones need not be in contact with the dorsal basement membrane or any cell surface (Fig. 8C). Rather, the growth cones appear suspended by their leading filopodia, which are in extensive contact with the basement membrane and particular cell surfaces.

Our TEM serial-section analysis confirmed our observations at the light level. The filopodia from the MP1 growth cone preferentially contact the surface of the pCC body and growth cone. However, the second most desirable substrate is the dorsal basement membrane. The MP1 growth cone had 28 filopodia extending from its leading 4 μm; all 28 were reconstructed. Some of these filopodia contacted the dorsal basement membrane, the endfeet of epidermal cells, the surfaces of unidentified cells, and many unidentified filopodia. However, the majority of filopodia from MP1 (19 of 28) contacted the pCC [Fig. 8D (▼)]. Of these 19, 6 came into intermittent and short (less than 1 μm) contact with the pCC body, growth cone, or filopodia. Another 6 contacted the pCC at its cell body and stayed in close contact for several microns. No obvious membrane specializations were abundantly present between these filopodia and the pCC. Although we have occasionally found TEM sections in which filopodia form tight, gap, and desmosomelike juctions, they were seen rarely (tight and desmosomelike) or never (gap) between the filopodia of MP1 and the pCC. The remaining 7 filopodia from MP1 were the most interesting. They all first contacted the pCC at its growth cone and were found inserted anywhere from 0.1 to 7 μm into the cell, as described below.

7. FILOPODIAL INSERTION AND INDUCTION OF COATED VESICLES

We have discovered a novel and highly specific interaction between these cells (Bastiani and Goodman, 1983). In addition to further supporting our previous results on the extensive adhesion by the MP1 filopodia to the pCC surface, our TEM analysis revealed that the filopodia from the MP1 growth cone insert deep into the pCC growth cone (Figs. 8D and 9) and induce the formation of coated vesicles (Figs. 10 and 11). This interaction is highly specific, since filopodia from other nearby growth cones that contact the surface of the MP1 and pCC growth cones do not penetrate them or induce coated vesicles.

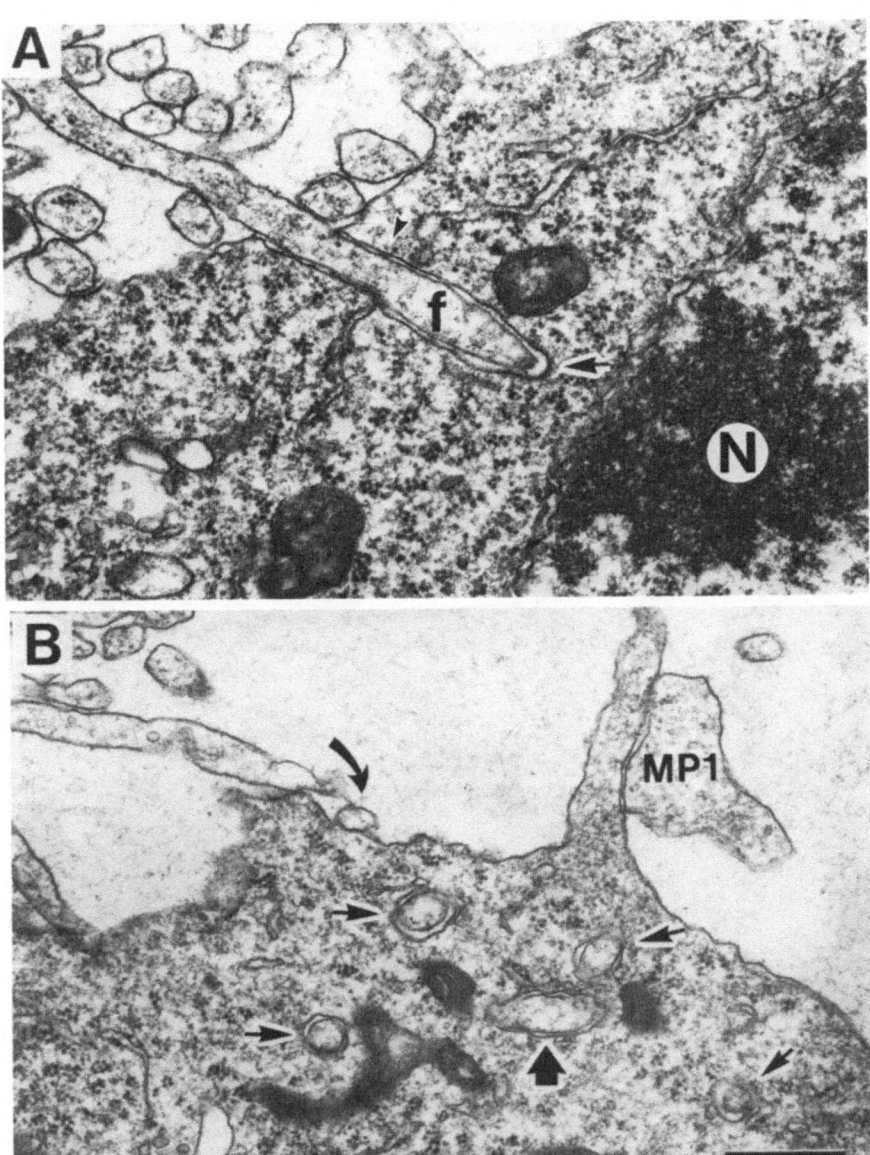

Figure 9. Filopodia from neuronal growth cones inserting deep into specific cell bodies and growth cones and inducing the formation of coated pits and vesicles. (A) A filopodium (f) is cut in longitudinal section to show its insertion into a neuronal cell body almost to the nucleus (N) of the cell. A coated pit is seen at the tip of the insertion (←), and punctate junctions are seen along the sides of the filopodium (▼). (B) Five filopodial insertions are seen in this cross section of a portion of the pCC growth cone. Four of the insertions are from the MP1 (→) and one from the MP2 (↘). The filopodium on the surface of the growth cone (⬆) is from the MP1. This filopodium is in very close apposition to the growth cone membrane and a few microns distally inserts into the pCC growth cone and induces coated pits and vesicles. The larger profile (MP1) is the terminal filum of the MP1 growth cone. Scale bar: 0.5 μm.

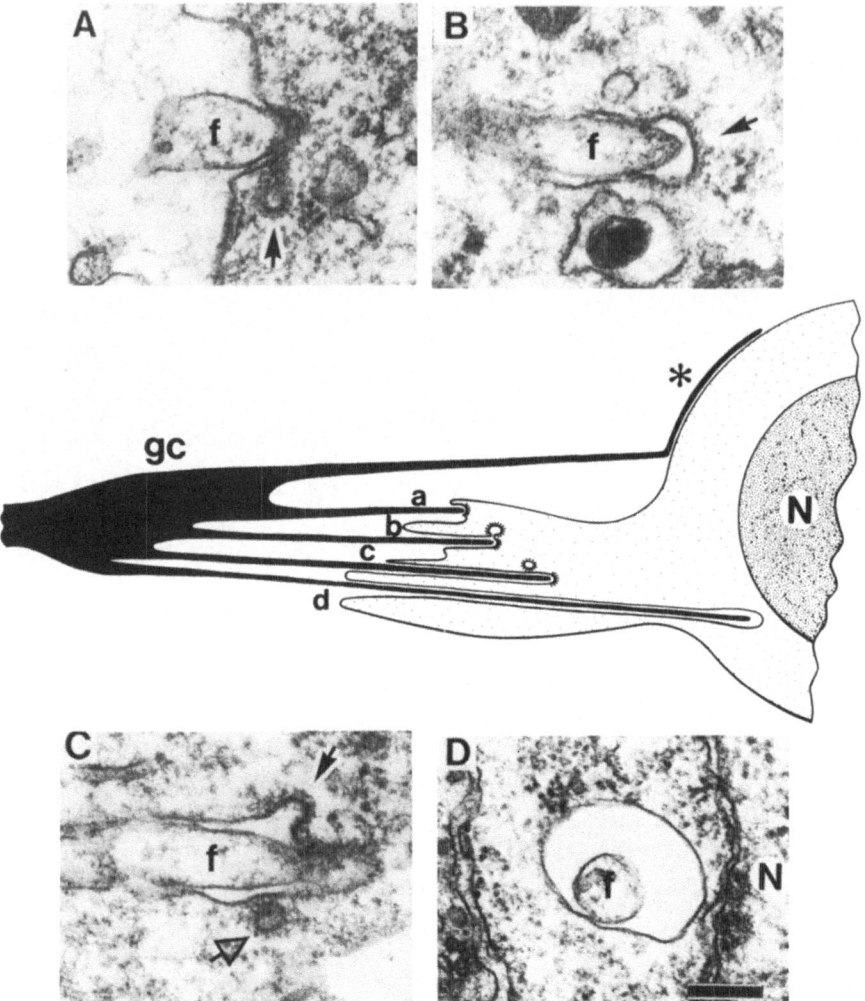

Figure 10. Proposed time–course of filopodial insertion and induction of coated pits and vesicles (see the text for discussion of the assumptions made in constructing this model). A filopodium from a specific identified growth cone (gc) contacts the growth cone of another specific cell (a). (A) The filopodium (f) initially makes a desmosomelike junction with the growth cone. Soon a coated pit is seen opposite the tip of the filopodium and coated vesicles begin forming (↑). (B) The filopodium inserts deeper into the growth cone (b), punctate junctions form along the sides of the filopodium, and a coated-pit (←) region is very prominent at the tip of the insertion. Coated vesicles are seen in all stages of formation. (C) The filopodium penetrates further into the growth cone (c), and multiple sites of coated-vesicle formation (←) are often seen. A coated vesicle has almost completely pinched off from the side of the insertion (▷). Coated-vesicle formation occurs only within 0.2 μm of the tip of the insertion. (D) The filopodium (d) has inserted to within a few tenths of a micron of the nucleus (N). Most of the punctate junctions along the sides of the filopodium are lost beginning near the tip. Concomitantly, the sleeve surrounding the filopodium enlarges and a region of coated membrane is no longer seen at the tip. At this time, the filopodium retracts out of the growth cone and the sleeve collapses. Few filopodia penetrate all the way to the nucleus; most complete the whole cycle after penetrating only a short distance (0.5–2.0 μm) into a growth cone. Notice that

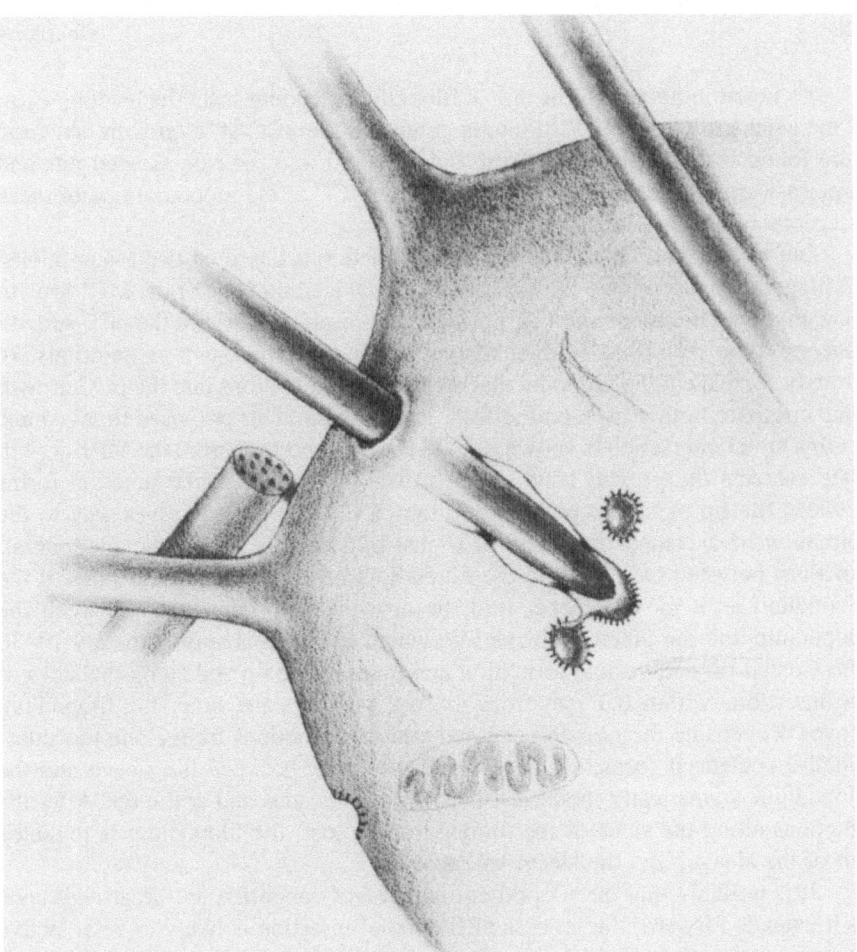

Figure 11. Drawing illustrating many of the important features of filopodial insertion and the induction of coated vesicles. The induction of coated-vesicle formation is localized to the membrane just opposite the tip of the filopodium. The tip of the filopodium is morphologically distinct from its sides, with a more densely staining matrix on both the extracellular and cytoplasmic sides of its membrane. Punctate junctions are seen between the filopodium and the growth cone along the sides of the insertion. The insertion process is probably active; the filopodium grows forward at the same time that junctions are forming along the sides and the growth cone is advancing. Filopodia from other cells contact the growth cone, but do not insert, again suggesting that insertion is not simply a passive engulfment of filopodia by the advancing growth cone. Coated-vesicle formation occurs on the surface of the growth cone, but not in any particular relationship to membrane folds, surface filopodia, or any other observable feature of the environment of the growth cone.

---

filopodia (*) from the same growth cone (gc) that contacts the cell body do not insert or induce coated pits even though they are in close apposition to the cell body for several microns. The interaction is specific; filopodia from other nearby cells contact the growth cone, but do not insert. (A–C) Longitudinal sections; (D) cross section within 0.1 μm of the tip of the filopodium. Scale bar: 0.2 μm.

As noted in Section 6, of the 28 filopodia extending from the leading 4 μm of the MP1 growth cone, 7 filopodia contacted the pCC at its growth cone and were found inserted anywhere from 0.1 to 7 μm into the cell. Coated pits and vesicles were present in the membrane of the pCC at the filopodial tips of these insertions in all but one case.

Our proposed time–course for this interaction is based on two assumptions: (1) filopodia that originate farther back from the leading edge of the MP1 growth cone are generally older and (2) filopodia that insert deeper into the pCC growth cone are older than those with shallower insertions. These two assumptions are strongly correlated; the filopodia that make deeper insertions into the pCC growth cone originate farther back on the MP1 growth cone. Our proposed time–course for this novel interaction is shown in Fig. 10. A filopodium from the MP1 growth cone contacts the growth cone of the pCC. A desmosomelike junction forms between the tip of the filopodium and the pCC. This quickly gives way to the formation of a coated pit in the pCC just opposite the tip of the filopodium; junctions between the pCC and the filopodium now form along the sides of the filopodium as it moves deeper into the growth cone. The space between the filopodium and the tubelike sleeve into which it inserts is approximately 10–30 nm. Coated-pit and vesicle formation continues at the tip and along the sides of the insertion, within 0.1 μm from the tip. At the same time, the filopodium moves deeper into the growth cone and punctate junctions form along the sides. Finally, coated-pit formation ceases and the space between the sleeve and the filopodium dramatically increases, both along the sides and at the tip. After the junctions along the sides of the filopodium are lost, the filopodium is retracted out of the sleeve, and the sleeve collapses.

It is unlikely that the filopodium is passively engulfed by the growth cone as it extends forward; the process of filopodial insertion is likely to be an active process on the part of both cells. First, filopodia have been observed that insert at the growth cone and penetrate all the way to the nucleus of the cell. Second, filopodia have been observed to insert into the cell bodies that have not yet extended growth cones and to penetrate all the way to their nucleus (see Fig. 9A).

The filopodial insertion and induction of coated pits is a remarkably specific cell interaction. Coated pits and vesicles are induced only at the tip of the penetrating filopodium (Figs. 10 and 11). Other filopodia from the MP1 growth cone that contact the surface of the pCC body do not induce coated pits; rather, they must contact the surface of the growth cone to induce coated pits. Although coated pits form in other places on the pCC growth cone, there is no obvious relationship between their position and membrane invaginations, surface filopodia, or contact with other cells. It is quite striking that the pCC growth cone is riddled with ten filopodial insertions; seven of these are from MP1, two from dMP2, and one from an unidentified source. There are three growth cones closer to the pCC growth cone than either the MP1 or the dMP2. Filopodia from these other growth cones contact the pCC growth cone, yet do not insert and induce coated pits. The aCC (the sibling of the pCC) lies just anterior to the pCC. Filopodia from the MP1 contact the aCC growth cone, but do not insert or induce coated pits.

Although we have focused on the interaction of the MP1 filopodia with the pCC growth cone, the same phenomenon occurs between the pCC filopodia and the MP1 growth cone and reciprocally between the MP1 and dMP2 growth cones. Thus far, the occurrence of this phenomenon has correlated with the observation of dye coupling between unrelated embryonic neurons. It is tempting to suggest that the dye coupling observed between these unrelated cells is mediated via gap junctions along the sides of the filopodial insertions. Filopodia are certainly competent to form gap junctions along their sides. Unfortunately, the small size of the gap junctions so far observed (less than 40 nm) relative to the section thickness makes them observable only rarely even in serial sections.

The specificity of the filopodial interactions we describe here has several interesting implications and several possible roles during development. Two observations are of particular interest: First, only filopodia from particular identified growth cones (in this case, MP1 and dMP2) are capable of inserting into another identified growth cone (in this case, pCC), even though the filopodia of many other growth cones contact their surfaces. This suggests that the filopodia from different identified neurons have different cell surfaces that allow other growth cones to distinguish among them. This specificity of filopodia insertion appears to be identical to the specificity of filopodial adhesions that is likely to mediate growth-cone guidance. The numerous adhesive junctions around the filopodial shank and the proposed delay in their retraction may allow the inserted filopodia to mediate increased tension relative to surface filopodia contacting the same cell.

Second, coated pits are induced only at the tips of these inserting filopodia. This suggests that the tips of filopodia are biochemically different from their shanks, a notion further supported by the different profiles of filopodial tips vs. shanks seen in our serial TEM analysis (Fig. 12). Quite possibly, specific informational molecules used to trigger this insertion and induce coated pits (and perhaps also involved in the initial surface adhesion) may be concentrated at the tips of the filopodia relative to their shanks. Thus, when the filopodium comes into contact with the correct cell that has the appropriate receptors, it may interact with the receptors to induce the formation of coated pits and the subsequent receptor-mediated endocytosis as an inductive interaction.

What role do these specific filopodial interactions play during neuronal development? Growth cones typically make a series of cell-specific pathway choices on the way to their appropriate target. Growth cones appear to be guided through each of these choice points by the selective adhesion of their filopodia

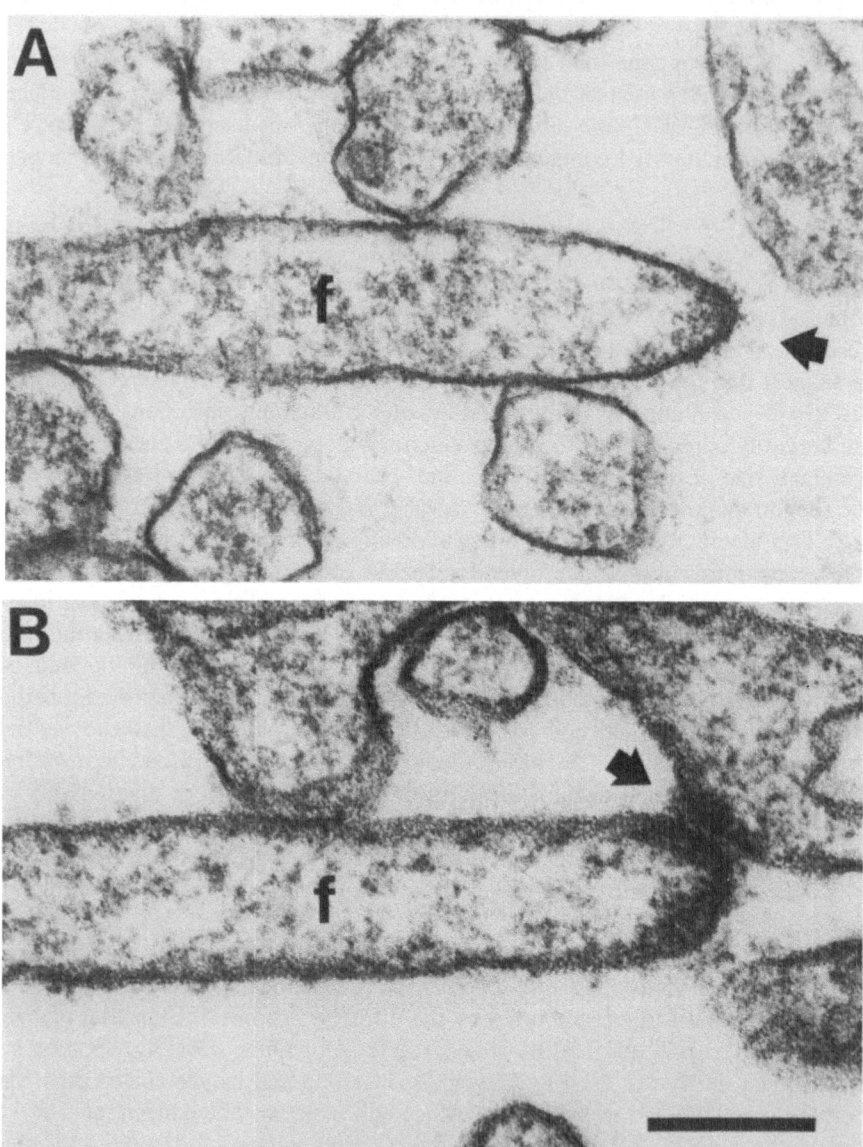

Figure 12. Morphologically distinct tips and shanks of filopodia. (A) The tip of the filopodium (f) has a more densely staining intracellular matrix than regions along the shank. In addition, the extracellular membrane at the tip of the filopodium often has a more extensive coating (◀) than other areas along the shank. (B) This filopodium (f) has made a desmosomelike junction with another cell. Notice the extensive dense-staining material at the site of the junction (➡) and also intracellularly at the tip of the filopodium. Desmosomelike junctions have occasionally been observed along the shank of a filopodium; gap and tight junctions have been observed only along the shanks of filopodia, never at the tips. Scale bar: 0.2 μm.

to specific surfaces. It is quite possible that the adhesive properties of growth cones and axons change during the course of these growth-cone navigations. The interactions involved in navigating through one choice point might induce the cell to change its expression of cell-surface molecules involved in filopodial adhesion at a subsequent choice point, either by the disappearance of old receptors or by the appearance of new receptors. The specific interactions described herein might mediate such inductive changes during neuronal development.

The observation of filopodial insertions is not limited to the very first growth cones in the grasshopper CNS; we have observed this phenomenon between many different identified growth cones at different stages of development. Whereas filopodial adhesion takes place onto the surfaces of axons and growth cones, filopodial insertion takes place only into growth cones (or cell bodies that have not yet initiated growth cones). Interestingly, the presence of filopodial insertions is not limited to grasshopper growth cones. Nordlander and Singer (1978) have reported similar inserted structures in the growth cones of amphibian embryos, but did not identify the source of these structures. They also noticed an association between growth cones and coated pits in glial cells. Slavkin and Bringas (1976) describe long filopodia-like processes from mesenchymal cells protruding into epithelial cells of the developing tooth germ during the suggested inductive interaction between these two tissues. It is likely that a closer examination of these and other developing systems will show the presence of filopodial insertions and induced coated pits between specific cells or tissues at important times during development. The specific interaction of filopodial insertion and induction of coated vesicles (see Fig. 11) may be a general mechanism underlying inductive events not only in the developing nervous system, but also in many other developing tissues.

ACKNOWLEDGMENTS. We thank Francis Thomas for valuable assistance with the electron microscopy. This research is supported by an NIH postdoctoral fellowship (M.J.B.) and by an NSF grant and McKnight Foundation Scholars Award (C.S.G.).

REFERENCES

Bastiani, M.J., and Goodman, C.S., 1983, Neuronal growth cones: Specific interactions mediated by filopodial insertion and induction of coated vesicles, *P.N.A.S.* (in press).

Bate, C.M., and Grunewald, E.B., 1981, Embryogenesis of an insect nervous system. II. A second class of precursor cells and the origin of the intersegmental connectives, *J. Embryol. Exp. Morphol.* **61**:317–330.

Bray, D., 1982, Filopodial contraction and growth cone guidance, in: *Cell Behavior* (R. Bellairs, A. Curtis, and G. Dunn, eds.), pp. 299–317 Cambridge University Press, London and New York.

Chang, S., Ho, R.K., and Goodman, C.S., 1983, Monoclonal antibodies that reveal specific patterns of neurons and mesodermal cells early in grasshopper embryogenesis, *Dev. Brain Res.* **9**:297–304.

Goodman, C.S., Raper, J.A., Ho, R.K., and Chang, S., 1982, Pathfinding by neuronal growth cones during grasshopper embryogenesis, *Symp. Soc. Dev. Biol.* **40**:275–316.

Letourneau, P., 1975, Cell-to-cell substratum adhesion and guidance of axonal elongation, *Dev. Biol.* **66**:183–196.

Letourneau, P., 1982, Nerve fiber growth and its regulation by extrinsic factors, in: *Neuronal Development* (N.C. Spitzer, ed.), Plenum Press, New York, pp. 213–254.

Nakai, J., 1960, Studies on the mechanism determining the course of nerve fibers in tissue culture. II. The mechanism of fasciculation, *Z. Zellforsch.* **51**:427–449.

Nordlander, R.H., and Singer, M., 1978, The role of ependyma in regeneration of the spinal cord in the urodele amphibian tail, *J. Comp. Neurol.* **180**:349–374.

Raper, J.A., and Goodman, C.S., 1982, Transient dye coupling between developing neurons reveals patterns of intracellular communication during embryogenesis, in: *Cellular Communication during Ocular Development* (J.B. Sheffield and S.R. Hilfer, eds.), Springer-Verlag, New York, Heidleberg, and Berlin, pp. 85–96.

Raper, J.A., Bastiani, M.J., and Goodman, C.S., 1983a, Pathfinding by neuronal growth cones in grasshopper embryos. I. Divergent choices made by the growth cones of sibling neurons, *J. Neurosci.* **3**:20–30.

Raper, J.A., Bastiani, M.J., and Goodman, C.S., 1983b, Pathfinding by neuronal growth cones in grasshopper embryos. II. Selective fasciculation onto specific axonal pathways, *J. Neurosci.* **3**:31–41.

Silver, J., and Robb, R.M., 1979, Studies on the development of the eye cup and optic nerve in normal mice and in mutants with congenital optic nerve aplasia, *Dev. Biol.* **68**:175–190.

Silver, J., and Sidman, R.L., 1980, A mechanism for the guidance and topographic patterning of retinal ganglion cell axons, *J. Comp. Neurol.* **189**:101–111.

Singer, M., Nordlander, R.H., and Egar, M., 1979, Axonal guidance during embryogenesis and regeneration in the spinal cord of the newt: The bluprint hypothesis of neuronal pathway patterning, *J. Comp. Neurol.* **185**:1–22.

Slavkin, H.C., and Bringas, P., 1976, Epithelial–mesenchyme interactions during odontogenesis. IV. Morphological evidence for direct heterotypic cell–cell contacts, *Dev.Biol.* **50**:428–442.

Taghert, P.H., Bastiani, M.J., Ho, R.K., and Goodman, C.S., 1982, Guidance of pioneer growth cones: Filopodial contacts and coupling revealed with an antibody to Lucifer Yellow, *Dev. Biol.* **94**:391–399.

Wessells, N.K., and Nuttall, R.P., 1978, Normal branching, induced branching and steering of cultured parasympathetic motor neurons, *Exp. Cell Res.* **114**:111–122.

Wessells, N.K., Letourneau, P.C., Nuttall, R.P., Luduena-Anderson, M., and Geiduschek, J.M., 1980, Responses to cell contacts between growth cones, neurites and ganglionic non-neuronal cells, *J. Neurocytol.* **9**:647–664.

II

Developmental Expression of Neuronal Phenotypic Characters

Session Chairman: Ira B. Black

Chapter 5

Surface-Bound and Released Neuronal Glycoconjugates

Paul H. Patterson

1. INTRODUCTION

The establishment of synaptic arrays during vertebrate neural development is the result of a series of cellular interactions between neurons and their targets. These interactions occur throughout the stages of cell migration, axonal outgrowth, recognition of target cells, differentiation of pre- and postsynaptic elements, competition between and reorganization of connections, and stabilization of the final connections in maturity (cf. Patterson and Purves, 1982). A number of these stages, particularly the later ones, are controlled primarily by synaptic transmission. For example, distribution of acetylcholine receptors (AChR) in adult skeletal muscle is determined largely by the activity induced in the muscle by synaptic transmission (Lømo and Westgaard, 1976). On the other hand, a number of interactions, particularly during the early phases of synapse formation, do not appear to be mediated by the transmitter. The initial organization of muscle AChR under the nerve, for example, can occur in the absence of ACh binding to its receptor (Anderson et al., 1977). Cellular interactions in this case must involve other secreted or cell-surface molecules. The identification, purification, and characterization of these molecules are essential for at least two reasons: (1) purification of such molecules and production of antibodies against them will demonstrate the physical reality of the often formal concept of "trophic"

PAUL H. PATTERSON ● Biology Division, California Institute of Technology, Pasadena, California 91125.

interactions, and (2) manipulation of these molecules is essential for studies of the mechanisms that underlie such interactions.

2. SURFACE GLYCOCONJUGATES

We have been using monoclonal antibodies as well as standard biochemical methodology to identify molecules that neurons secrete or have on their surfaces that could be used in interactions with other cells. Glycoproteins or lipids that are highly enriched in specific neuron types are candidates for this role. For example, cultured adrenergic neurons bind the lectin soybean agglutinin (SBA) at a 5-fold higher density than cholinergic sympathetic neurons (Schwab and Landis, 1981). Zurn (1982) identified the neuronal receptors for SBA as two neutral glycolipids, one comigrating with globoside in thin-layer chromatography (TLC). SBA binds specifically to these lipids on polyacrylamide gels, TLC plates, and the surfaces of living neurons. The latter deduction was made from a protection experiment in which the surface membranes of neurons were labeled using the galactose oxidase–sodium [^{3}H]borohydride reduction technique, with and without prior incubation with the lectin. Prior incubation with 5×10^{-7} M SBA inhibited the labeling of the two glycolipids. Neuronal proteins extracted and separated on gels labeled poorly with SBA. Ricin agglutinin (RCA) has the same hapten sugar specificity as SBA, but binds equally well to adrenergic and cholinergic neurons (Schwab and Landis, 1981). RCA bound to proteins on gels quite well, but did not protect the glycolipids on living cells from labeling with the galactose oxidase technique (Zurn, 1982). In addition, when the surfaces of living neurons were labeled with galactose oxidase, Zurn found that the adrenergic neurons had more of the two neutral glycolipids accessible for labeling than did the cholinergic neurons, which is consistent with the lectin-binding data. Of direct relevance to these findings is the work of Raff et al. (1979), which showed that only half of dissociated rat dorsal-root-ganglion neurons in culture bound an antigloboside serum. Thus, there is a difference in glycolipid composition or accessibility among sensory neurons as well.

Cultured adrenergic and cholinergic sympathetic neurons also exhibit several differences in their profiles of major surface glycoproteins. Braun et al. (1981) labeled neuronal surface proteins by metabolic and surface-specific methods and then analyzed the proteins by two-dimensional gel electrophoresis and autoradiography. A total of approximately 35 glycoproteins could be resolved, of which at least 14 were exposed on the cell surface. The expression of two of the surface proteins was correlated with the transmitter phenotype, one being greater in adrenergic neurons (A155) and one elevated in cholinergic neurons (C55). These two proteins could be labeled with periodate, lactoperoxidase, or galactose oxidase treatments, and the pIs of the proteins were modified by neuraminidase treatment of intact cells.

In addition to these biochemical methods, monoclonal antibodies have been developed that bind preferentially to either adrenergic or cholinergic sympathetic neurons grown in culture (Chun *et al.*, 1980; in preparation). The advantage of these markers is that they can potentially be used as probes for determining the function of the surface molecules to which they bind. To this end, Chun *et al.* (1980, in preparation) grew sympathetic neurons in the presence of a number of monoclonal antibodies that had been selected for their ability to bind to the surfaces of these cultured neurons. Thus far, 14 of the antibodies have been used in chronic incubations and their effects on development assessed by biochemical analysis of a variety of neuronal functions. Of these, 3 had no detectable effects on the neurons, 5 had marginal effects, and 6 caused reproducible and specific functional changes in the neurons. Alterations, both positive and negative, were seen in catecholamine (CA) uptake as well as synthesis and accumulation. Only one of these antibodies had detectable deleterious effects on neuronal survival or growth. For the most part, therefore, nonspecific, gross effects caused by antibody binding were not a problem. We do not know, however, whether the developmental alterations being studied were caused by the antibody blocking the activity of a given antigen or causing the internalization of the antigen.

The one of these antibodies that has been studied in the most detail, ASCS4, binds to a variety of neurons in both the peripheral and central nervous systems. Immunohistochemical analysis of staining of frozen sections also revealed that the antibody does not bind to a variety of glands, including the adrenal medulla (Chun *et al.*, in preparation). When sympathetic neurons are grown in the presence of this antibody, their ability to synthesize and accumulate both CA and ACh is increased. How could an antibody binding on the cell surface have a positive effect on transmitter synthesis and storage? One possibility is that its binding alters the process of vesicle turnover during the exocytosis–endocytosis cycle. This notion gained a hint of plausibility when Chun found that the surface antigen to which the antibody binds is a protein that had been independently studied by Sweadner, whose findings (discussed below) suggested the hypothesis that this protein may be involved in vesicle turnover (Sweadner and Patterson, 1981; Sweadner, 1983a).

The protein in question is one of a pair of acidic, major surface-membrane glycoproteins of high molecular weight (230 and 200 daltons, called B1 and B3, respectively). B1 is recognized by Chun's monoclonal antibody ASCS4, and its size, charge, and precipitability by monoclonal and conventional antibodies identify it as the sympathetic-neuron equivalent of the PC12 cell line protein termed NGF-inducible, large external protein (NILE) by Greene and colleagues (Salton *et al.*, 1983; McGuire *et al.*, 1978). (This name may now be inappropriate, since antibodies against the protein bind to a variety of neurons.) B1 and B3 are first detected in pulse–chase experiments in more basic, lower-molecular-weight precursor forms. The B1 precursor is precipitated by antibody ASCS4, and is termed P1 in Fig. 1. The precursors can initially be labeled by incubation of the sympathetic neurons with [^{3}H]mannose, but this label is lost with a half-life of about 1 hr. As the [^{3}H]mannose is lost, the precursors are inserted into the surface membrane, and labeling with [^{3}H]fucose can be detected (Fig. 1) (Sweadner, 1983a). The conversion of P1 to B1 involves a shift to a more acidic pI and the addition of an apparent 15 k daltons. The actual change in molecular weight could be less than 15 k daltons, because the changes in carbohydrate composition indicated by the labeling experiments could alter the behavior of P1 and B1 in the sodium dodecyl sulfate gels. A parallel sequence of labeling steps occurs in the synthesis of B3, but no antibody is yet available to confirm a relationship between P3 and B3.

The appearance of B1 and B3 in the surface membrane is indicated by susceptibility to a number of external agents added to cultures of living neurons (trypsin, neuraminidase, lactoperoxidase, and galactose oxidase) (Sweadner and Patterson, 1981; Sweadner, 1983a). In Fig. 1, B1 is depicted as having a tail inserted into the membrane. This is hypothetical, since we have no information as yet on the actual disposition of this protein in or on the membrane. The labeling studies suggest, however, that B1 is an intrinsic membrane glycoprotein, rather than an extracellular matrix proteoglycan (cf. Lennarz, 1980). The pulse–chase studies further show that after their appearance on the cell surface, B1 and B3 each undergo apparently spontaneous conversion ($t_{1/2} \sim 5$ hr) to two lower-molecular-weight forms, termed B2 and B4, respectively. Again, the apparent change in molecular weight is about 15 k daltons. In Fig. 1, this conversion is depicted as a loss of the tail in the membrane, but this again is hypothetical; we have no information as yet on the chemical nature of the conversion.

The final stage in the life cycle of these surface proteins occurs when the neurons are depolarized. When transmitter release is evoked with any of a variety of agents (50 mM K^+ + Ba^{2+}, A23187, black widow spider venom, veratridine, p-chloromercuribenzene sulfonic acid (pCMBS), alomethecin, or monensin), B2 and B4 are released from the cell at the expense not only of B2 and B4, but also of B1 and B3. The loss of these proteins is selective; many other membrane and cytoplasmic proteins are not released under the same conditions. The interpretation that the released proteins (S2 and S4) are in fact derived from B2 and B4 is supported by the finding that antibody ASCS4 can precipitate S2 (as well as B1 and B2). It is interesting that although the apparent sequence of P3, B3, B4, S4 completely parallels the P1, B1, B2, S2 sequence, with very similar changes in molecular weights and charge, antibody ASCS4 fails to recognize any of the P3, B3, B4, and S4 molecules.

The release of S2 and S4 requires Ca^{2+} [Co^{2+} blocks the effect of high

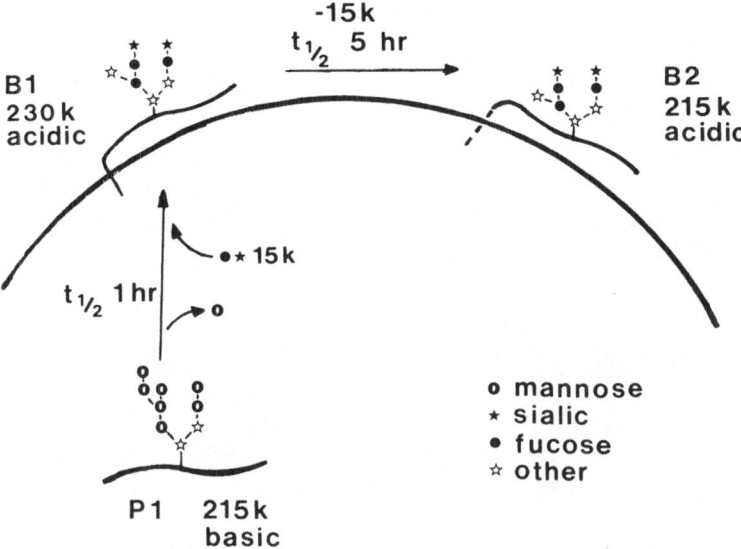

Figure 1. Diagram summarizing the processing of one of the two sets of proteins described in the text. Monoclonal antibody ASCS4 specifically precipitates proteins P1, B1, and B2. Pulse–chase experiments suggest that P1 is a precursor of B1. P1 can be labeled with [³H]mannose at early times, but loses the label during a chase. At the same time it loses its mannose label, P1 is apparently inserted into the surface membrane, acquiring fucose, sialic acid, an apparent 15 kdaltons of mass, and a more acidic charge. The tail inserted into the membrane is hypothetical. With a slower half-time, B1 loses about 15 kdaltons of mass, yielding B2. B2 is then released from the neuronal surface after depolarization. The loss of the tail (- - -) is again hypothetical. The carbohydrate structures are shown solely for illustrative purposes; nothing is yet known about the number or structure of these chains.

K^+, A23187, and black widow spider venom (BWSV); tetrodotoxin blocks the veratridine effect]. This would be expected if the protein release were related to exocytosis. However, Sweadner also found that protein release requires Ca^{2+} even under circumstances in which transmitter release does not. That is, while norepinephrine release evoked from the neurons by BWSV can occur in the absence of Ca^{2+} or in the presence of La^{3+}, protein release is blocked by La^{3+} or low Ca^{2+}. Thus, while protein release is always accompanied by transmitter release, the reverse is not true. Protein release is therefore not required for exocytosis. That La^{3+} blocks protein release is interesting in light of the observation that La^{3+} also blocks the recovery of motor-nerve terminals after heavy stimulation and vesicle depletion (Clark *et al.*, 1972). Our working hypothesis is that protein release follows exocytosis and may be involved in vesicle-membrane recovery during endocytosis. The idea that release of these proteins is required for endocytosis is fancifully represented in Fig. 2. Since the surface and released proteins have hydrodynamic properties compatible with either an

resting

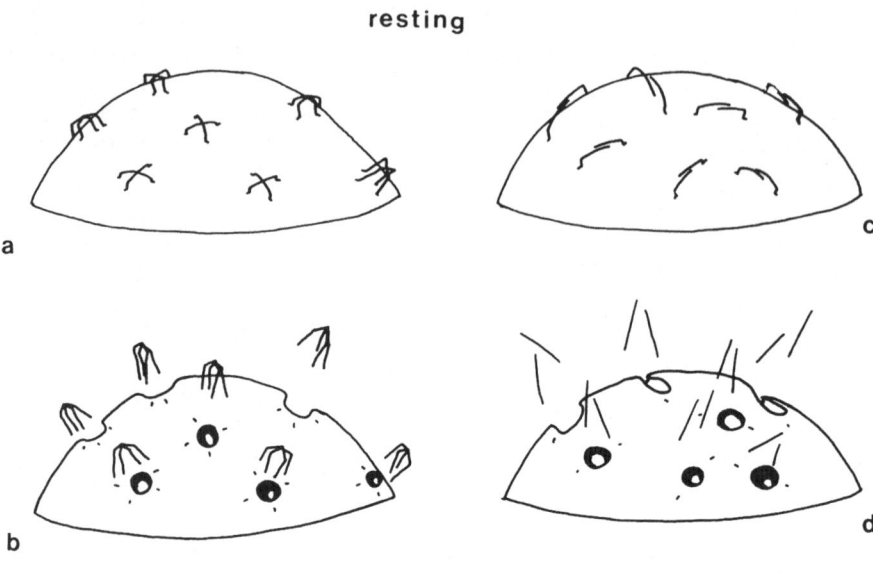

depolarized

Figure 2.   Model depicting the release of surface proteins (described in the text) following depolarization. It is hypothesized that the protein release allows vesicles to invaginate (endocytosis). The proteins are depicted as "baskets" (a, b) or rods (c, d).

extended, rodlike shape or baskets trapping water (Sweadner, 1983b), two versions of the model are depicted in Fig. 2 a,b and c,d. Since B2 and S2 behave in parallel with B4 and S4, we imagine that the two sets of glycoproteins are complementary in function.

To test the endocytosis hypothesis, W. Matthew (unpublished findings) has mutagenized PC12 cells and selected for cells that lack B1 and B2, using antibody ASCS4 and complement. These cells are being cloned, and their exo- and endocytotic capabilities will be assayed. It will be important to determine the nature and content of the surface changes in these mutant lines, as well as to attempt the potentially difficult experiment of adding back B2 or S2 to the mutant cells in an effort to correct whatever deficits they display.

3. SPONTANEOUSLY RELEASED PROTEINS

In addition to the release of selected surface proteins associated with depolarization, cultured neurons also spontaneously release or secrete well-defined and restricted sets of intracellular glycoproteins. Sweadner (1981) observed striking differences in the families of glycoproteins spontaneously secreted into the medium by cultured adrenergic vs. cholinergic sympathetic neurons. After being labeled with [^{3}H]leucine or [^{3}H]fucose, the neurons secrete 16–18 major soluble proteins that differ from the surface-membrane glycoproteins in two-dimensional gel analysis. Their secretion is spontaneous and is not affected by increased or decreased transmitter release. Five of these proteins are correlated with the neuronal transmitter phenotype and are thus candidates for extracellular functions specific to one or the other type of neuron.

Recently, R. Pittman (1983) has found that several types of sensory neurons also spontaneously secrete characteristic families of glycoproteins into the culture medium. Neuronal cultures from rat dorsal root, nodose, and trigeminal ganglia each secrete at least one protein that does not overlap on two-dimensional gels with proteins from the other sensory ganglia or with the proteins secreted by sympathetic ganglia. These findings may be of interest in the context of developmental trophic interactions. It is well known that a variety of sensory receptors are dependent on their sensory innervation for initial differentiation as well as for maintenance of differentiated phenotype in the adult (cf. Patterson and Purves, 1982). In addition, there is evidence that various sensory neurons differ in their ability to induce particular receptors. Gustatory neurons in the geniculate ganglion can induce taste buds in the tongue epithelium, whereas the skin sensory fibers of the auricular branch of the vagus (cell bodies in the superior ganglion of the vagus nerve) do not appear to be capable of inducing taste-bud formation (Zalewski, 1969). It is possible that some of the spontaneously secreted proteins subserve trophic roles such as these.

To examine this possibility further and to develop probes for localizing these proteins *in vivo*, Matthew and Pittman are attempting to raise monoclonal antibodies against the secreted proteins. Although these proteins are major cellular products, comprising 1–3% of total ongoing protein synthesis, the actual amounts of protein released from primary neuron cultures is quite small. Thus, alternative approaches are being used for this purpose. *In vitro* immunization (Luben and Mohler, 1980) has the advantage that very small amounts of antigen are necessary. Another strategy is to use neuronal cell lines as the source of released protein. Matthew (unpublished findings) has successfully used the *in vitro* immunization method to obtain monoclonal antibodies that bind to molecules secreted by PC12 cells. This approach therefore offers promise in the search for developmental signals produced by neurons.

ACKNOWLEDGMENTS. The recent research reported in this chapter was supported by the NINCDS, the Rita Allen and McKnight Foundations, the Charles A. King Trust, the Jane Coffin Childs Memorial Fund, and the Swiss National Foundation for Scientific Research.

REFERENCES

Anderson, M.J., Cohen, M.W., and Zorchyta, E., 1977, Effects of innervation on the distribution of acetylcholine receptors on cultured muscle cells, *J. Physiol.* **268:**731–756.

Braun, S., Sweadner, K.J., and Patterson, P.H., 1981, Neuronal surfaces: Distinctive glycoproteins of cultured adrenergic and cholinergic sympathetic neurons, *J. Neurosci.* **1:**1397–1406.

Chun, L.L.Y., Patterson, P.H., and Cantor, H., 1980, Preliminary studies on the use of monoclonal antibodies as probes for sympathetic development, *J. Exp. Biol.* **89:**73–83.

Clark, A.W., Mauro, A., Longevecker, H.E., and Hurbut, W.P., 1972, Changes in the fine structure of the neuromuscular junction of the frog caused by black widow spider venom, *J. Cell Biol.* **52:**1–14.

Lennarz, W.J. (ed.), 1980, *The Biochemistry of Glycoproteins and Proteoglycans,* Plenum Press, New York.

Lømo, T., and Westgaard, R.H., 1976, Control of ACh sensitivity in rat muscle fibers, *Cold Spring Harbor Symp. Quant. Biol.* **40:**263–274.

Luben, R.A., and Mohler, M.A., 1980, *In vitro* immunization as an adjunct to production of hybridoma producing antibodies against the lymphokine osteoclast activating factor, *Mol. Immunol.* **17:**635–639.

McGuire, J.L., Greene, L.A., and Furano, A.V., 1978, NGF stimulates incorporation of fucose or glucosamine into an external glycoprotein in cultured rat PC12 pheochromocytoma cells, *Cell* **15:**357–365.

Patterson, P.H., and Purves, D., 1982, *Readings in Developmental Neurobiology,* Cold Spring Harbor Press, Cold Spring Harbor, New York, p. 700.

Pittman, R.N., 1983, Spontaneously released proteins from cultures of sensory ganglia include plasminogen activator and a calcium dependent protease, *Soc. Neurosci. Abstr.* **9:**5.4.

Raff, M.C., Fields, K.L., Hakomori, S., Mirsky, R., Pruss, R., and Winter, J., 1979, Cell-type specific markers for distinguishing and studying the major classes of neurons and the major classes of glial cells in culture, *Brain Res.* **174:**283–308.

Schwab, M., and Landis, S.C., 1981, Membrane properties of cultured rat sympathetic neurons: Morphological studies of adrenergic and cholinergic differentiation, *Dev. Biol.* **84:**67–78.

Salton, S.R.J., Richter-Landsberg, C., Greene, L.A., and Shelanski, M.L., 1983, *J. Neurosci.* **3:**441–454.

Sweadner, K.J., 1981, Environmentally regulated expression of soluble extracellular proteins of sympathetic neurons, *J. Biol. Chem.* **256:**4063–4070.

Sweadner, K.J., 1983a, Post-transitional modification and evoked release of two large surface proteins of sympathetic neurons, *J. Neurosci.* (in press).

Sweadner, K.J., 1983b, Size, shape, and solubility of a class of releasable cell surface proteins of sympathetic neurons, *J. Neurosci.* (in press).

Sweadner, K.J., and Patterson, P.H., 1981, Neuronal surface protein release accompanies transmitter release, *Soc. Neurosci. Abstr.* **7:**227.15.

Zalewski, A.A., 1969, Combined effects of testosterone and motor, sensory, or gustatory nerve reinnervation in the regeneration of taste buds, *Exp. Neurol.* **24:**285–297.

Zurn, A., 1982, Identification of glycolipid binding sites for soybean agglutinin and differences in the surface glycolipids of cultured adrenergic and cholinergic sympathetic neurons, *Dev. Biol.* **94:**483–498.

Chapter 6

The Differentiation of Membrane Properties of Spinal Neurons

Nicholas C. Spitzer

1. INTRODUCTION

Studies of neural induction involve assessments of neuronal differentiation following experimental manipulations. These studies have in the past relied principally on neuroanatomical descriptions of neurite outgrowth, which assess an important neuronal phenotype. In recent years, it has become possible to analyze the differentiation of key cytoplasmic specializations of neurons, such as their neurotransmitter synthetic capacity, as well as the differentiation of their characteristic membrane properties. Some of the recent progress in understanding the development of neuronal-membrane properties will be reviewed. These assays are likely to be useful in studies of neural induction for several reasons: (1) The increasing ease of application of the techniques involved invites their general use. (2) Acquisition of neuronal membrane properties occurs very early in normal embryonic development, recommending them for rapid evaluation of neuronal induction. (3) As Fig. 1 illustrates, different neurons exhibit different constellations of properties that appear in particular sequences (Spitzer and Lamborghini, 1981). A broader characterization of neuronal development affords the opportunity to distinguish the induction of different neuronal types. (4) The expression of neurotransmitter synthetic enzymes or membrane proteins could be the results of activities of single genes, in contrast to the more complex phenomenon of

Nicholas C. Spitzer • Department of Biology, University of California, San Diego, La Jolla, California 92093.

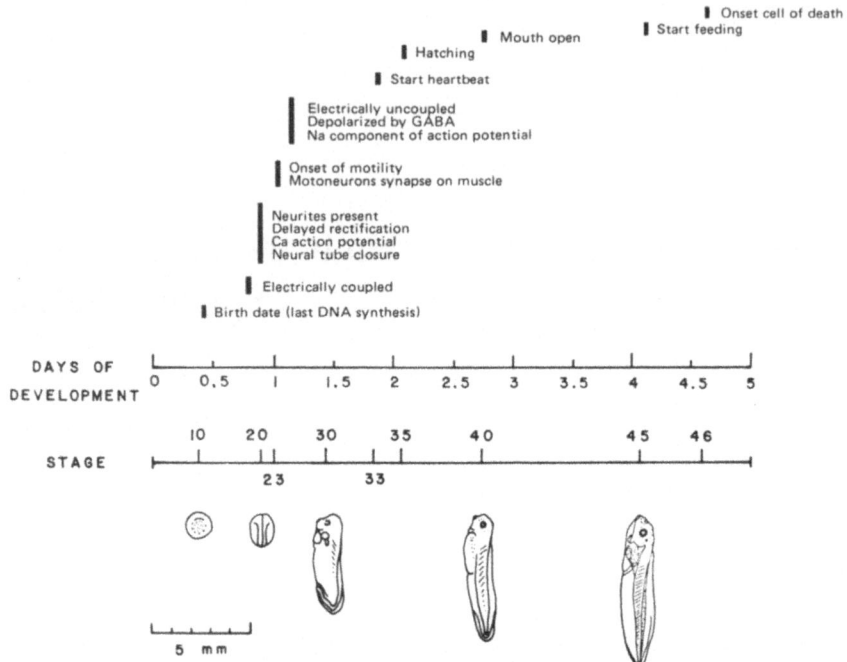

Figure 1. A developmental timetable for Rohon–Beard neurons in the *Xenopus* embryo *in vivo*. A vertical bar denotes the time of onset, or first appearance, of a phenotype in these neurons. The drawings at bottom illustrate the morphology of the embryo at different stages. From Spitzer and Lamborghini (1981).

neurite extension, which involves multiple gene products. The analysis of the molecular events involved in neuronal induction may be correspondingly facilitated.

Neurons possess a number of types of ion channels, which account for some of the functions of these cells. Voltage-dependent channels open in response to changes in membrane potential and allow cells to make action potentials. Chemically sensitive channels open in response to neurotransmitters and mediate chemical synaptic transmission. Other channels involved in the production of electrical synapses permit currents to flow directly between cell interiors. Precursor cells often lack the capacity to make action potentials or to respond to neurotransmitters, although there are exceptions, most notably among egg cells (Hagiwara and Jaffe, 1979; see alo Baud *et al.*, 1982). Precursor cells are often electrically coupled, however.

A general conclusion from the studies of the differentiation of nerve and muscle membranes is that their characteristic properties first appear in an immature form and change during the course of further development. When cells first become electrically excitable, they often make Ca^{2+}-dependent action potentials that are of long duration; these are subsequently converted to brief Na^+-dependent impulses. Cells that are coupled become uncoupled. When cells begin to respond to neurotransmitters, the localization of receptors is often diffuse; the receptors aggregate densely during differentiation. However, the ionic basis of neurotransmitter responses is constant during development.

2. DEVELOPMENT OF THE ACTION POTENTIAL

Studies of the development of neuronal-membrane properties have been expedited by choice of favorable preparations. The Rohon–Beard neurons of the amphibian spinal cord have been very useful for this work. There are about two hundred of these cells, which are primary sensory neurons (Roberts and Clarke, 1982). Further, they are located on the dorsal aspect of the spinal cord, where they are accessible for electrophysiological examination. They have their birth date during the gastrula stage of development, along with five other populations of neurons, which are the first in the embryo to exhibit this step in neural differentiation (Lamborghini, 1980). Birth dates are remarkably synchronous in the population, with 90% of Rohon–Beard cells completing their final round of DNA synthesis in a period of 6 hr. This event occurs close to the time of primary induction (Tarin, 1971). The precursor cells from which Rohon–Beard neurons arise have been identified at the 16-cell stage of the embryo (Jacobson, 1981).

Cells of the neural plate do not appear to be electrically excitable (Palmer and Slack, 1970; Slack and Warner, 1975). When Rohon–Beard cells can first be impaled with microelectrodes at the neural-tube stage, about 6 hr after their birth date, they are unable to generate action potentials. Within a few hours, they are capable of making impulses when depolarized. These impulses are hundreds of milliseconds in duration and depend chiefly on an influx of Ca^{2+}, as shown by experiments involving ion substitutions and the application of pharmacological blocking agents (Baccaglini and Spitzer, 1977). At the tailbud stage, roughly 1 day of development, the action potential acquires a Na^+ component and gradually shortens to an average duration of tens of milliseconds. The same tests show that the initial peak is Na^+-dependent, while the subsequent plateau is Ca^{2+}-dependent. As early as 3 days of development, on attainment of the larval stage, the impulse, which is now about 1 msec in duration, is principally Na^+-dependent.

The electrical excitability of amphibian dorsal-root-ganglion (DRG) neurons has also been investigated. These cells arise later than the Rohon–Beard neurons, over a prolonged period of time, and take over the role of primary sensory neurons as the Rohon–Beard cells die. Ca^{2+}-dependent impulses are observed in the smallest and presumably youngest neurons, and the frequency with which they are observed declines with increasing embryonic age, while that of Na^+-dependent impulses rises. The results suggest that these neurons follow the same developmental sequence as that seen in Rohon–Beard neurons (Baccaglini, 1978). The maturation of amphibian olfactory neurons also follows this pattern of development (Strichartz *et al.*, 1980). The outgrowing axons produce action potentials that are initially Ca^{2+}-dependent and later Na^+-dependent.

There are an increasing number of instances in which the development of electrical excitability of vertebrate neurons has been observed *in vitro*. Neurons from the presumptive spinal cord of the amphibian neural plate can be grown in dissociated cell culture. The action potentials elicited from the cell bodies of these neurons exhibit the same change in the ionic dependence of the inward current, from Ca^{2+} to Na^+, along the same time–course as the Rohon–Beard neurons *in vivo* (Spitzer and Lamborghini, 1976). Since these cultures contain neurons that innervate skeletal muscle cells and are presumably motor neurons, this aspect of differentiation is not unique to sensory cells. Furthermore, the excitability of the neurites of these cells has been studied in culture, where they can be clearly visualized at early stages, and the ionic dependence of their action potentials also shifts from Ca^{2+} to Na^+ (Willard, 1980). DRG cells of the mouse generate action potentials that depend first on Ca^{2+} and Na^+ and later on Na^+ alone (Matsuda *et al.*, 1978). Murine neuroblastoma cells can make Ca^{2+}-dependent action potentials, and acquire Na^+-dependent impulses during subsequent differentiation (Miyake, 1978). Cerebral cortical neurons from the chick embryo make action potentials that are Ca^{2+}- and Na^+-dependent; the Ca^{2+} component of the inward current apparently disappears during subsequent development (Mori-Okamoto *et al.*, 1983). However, there are a number of cases in which the ionic basis of the action potential has not been observed to change with increasing age in culture. In some cases, recordings could have been made sufficiently late that changes would have already occurred, or the culture medium could have been inadequate to support normal development. In others, the possiblity remains that development involves no change in ionic dependence of the action potential.

The binding of ion-channel-specific ligands such as toxins is useful in identifying the time of appearance of ion channels and quantitating their changing levels during development. Although probes for the voltage-sensitive Na^+-channels have been used successfully (Berwald-Netter *et al.*, 1981), the application of comparable ligands to the study of Ca^{2+} channels has not been described.

Among invertebrate nerve cells, ganglionic neurons in the grasshopper embryo are visually identifiable and accessible to microelectrode impalements at very early stages (Goodman and Spitzer, 1979; Heathcote, 1981). The action potentials elicited from the cell bodies of dorsal unpaired median (DUM) interneurons, as well as those derived from a different precursor neuroblast, depend on an influx of both Ca^{2+} and Na^+ when the cells first become excitable, and either ion is sufficient to support an impulse. At later stages, both ions are needed to produce an action potential. This change in the ionic mechanism of the action potential in the cell body stands in contrast to the absence of change in the Na^+-dependent action potentials arising in the axons (Goodman and Spitzer, 1981).

During the course of regeneration, neurons synthesize new membrane that becomes electrically excitable. This process seems to recapitulate the pattern of differentiation seen during development. The regenerating neurites of embryonic guinea pig DRG cells *in vitro* initially contain voltage-dependent Ca^{2+} channels that later seem to disappear (Fukuda and Kameyama, 1979). The giant axons of the adult cockroach initially produce Ca^{2+}-dependent action potentials in the proximal stump close to the site of transection; later, the membrane generates normal Na^+-dependent impulses (Meiri *et al.*, 1981).

The shift in ionic dependence of the impulse from Ca^{2+} to Na^+ during embryogenesis has also been seen in some striated and cardiac muscle cells (for a review, see Spitzer, 1982b). In contrast, tunicate skeletal-muscle cells lose the Na^+ component of their impulse while retaining the Ca^{2+} component (Takahashi *et al.*, 1971; Takahashi and Yoshii, 1981). Thus, a change in ionic dependence, rather than in its direction, may be the more general phenomenon in the development of action potentials.

3. DEVELOPMENT OF ELECTRICAL UNCOUPLING

Cells with different developmental fates are frequently coupled by low-resistance junctions at an early stage and become uncoupled at a particular time during their development. This phenomenon was initially observed in a variety of electrically inexcitable cells (e.g., Potter *et al.*, 1966; Sheridan, 1968; Spitzer, 1970; Lo and Gilula, 1979). More recently, it has been described in the differentiation of nerve and muscle. Rohon–Beard neurons are electrically coupled at the early neural-tube stage, before they are able to make action potentials. Current injected into one cell spreads to others, presumably via gap junctions. This coupling is voltage-dependent, in that the degree of coupling depends on the voltage difference between the cells (Spitzer, 1982a). A shift in the membrane potential of one cell, away from that of its neighbors, markedly reduces the strength of coupling for as long as the potential difference is maintained. Other cells in the spinal cords of the same embryos exhibit electrical coupling that is not voltage-dependent. The voltage-dependent coupling in Rohon–Beard cells appears to be the same process first described for isolated pairs of amphibian blastomeres (Spray *et al.*, 1979).

Rohon–Beard neurons are still electrically coupled when Ca^{2+} action potentials can first be elicited, and an impulse in one cell can be sufficient to trigger an impulse in a nearby coupled cell. More commonly, the Ca^{2+} action potential in one cell causes a subthreshold depolarization of the second that decreases during its course, perhaps reflecting voltage-dependent uncoupling. These cells become completely uncoupled from one another around the time of appearance of the Na^+ component of the action potential, although other cells are still coupled. The Rohon–Beard cells remain uncoupled during their further development. The coupling and uncoupling of amphibian spinal neurons in cell culture has not yet been investigated.

The uncoupling of DUM neurons from their progenitor neuroblast, and the uncoupling of other tissues from one another, have been described for the grasshopper embryo (Goodman and Spitzer, 1979, 1981). The disappearance of gap junctions between ganglion cells of the amphibian retina has been reported (Dixon and Cronly-Dillon, 1972). Mesoderm cells of amphibian embryos and striated muscle cells in rat embryos become uncoupled at specific times (Blackshaw and Warner, 1976b; Dennis *et al.*, 1981). Thus, stage-specific uncoupling appears to be a general process.

4. DEVELOPMENT OF NEUROTRANSMITTER SENSITIVITY

In contrast to the developmental changes in features of ion channels that mediate action potentials and electrical coupling, the channels that mediate responses to neurotransmitters have some properties that appear to be constant from the time of their first appearance. The application of a variety of neurotransmitters to Rohon–Beard neurons at neural-tube stages is without effect on their membrane potential or conductance. However, these cells begin to respond at the early tailbud stage and are depolarized by γ-aminobutyric acid (GABA); roughly half the cells are also depolarized by glycine (Bixby and Spitzer, 1982a). The responses to both these neurotransmitters seem to appear at the same time, and the cells remain insensitive to a host of other compounds. Examination of the ionic basis of the response to iontophoretic application of GABA in mature cells reveals that it is the result of a conductance increase to Na^+ and K^+, with a reversal potential of -30 mV. The same reversal potential is observed in newly sensitive cells, strongly suggesting that the ionic dependence of the response is constant. At all stages of development examined, this response is blocked by picrotoxin or curare and exhibits desensitization like that seen for many other neurotransmitter receptors.

Some features of this chemosensitivity do change, however. The sensitivity of cells to GABA increases about 10-fold during their maturation, probably as a result of an increase in the number of receptors or in some properties of single channels, or both. Furthermore, the glycine response is transient and can no longer be elicited by $3\frac{1}{2}$ days of development. Since the number of Rohon–Beard cells is stable during this time, the response seems to be lost from the existing population.

The development of the neurotransmitter sensitivity of amphibian spinal neurons in culture parallels development *in vivo* in most respects. The time of first appearance, initial sensitivity, reversal potential, pharmacology, and desensitization of the response to GABA are the same for one class of cultured neurons as they are for the Rohon–Beard neurons in the spinal cord. Other neurons, which are hyperpolarized by GABA and glycine and depolarized by glutamate, and thus likely to include motor neurons, first begin to respond at the same age in culture (Spitzer and Bixby, 1982). The reversal potential of the response to GABA in these cells is -60 mV, which is close to the value reported for mature motor neurons *in vivo* (Curtis *et al.*, 1968).

Ganglionic DUM neurons of the grasshopper become sensitive to both acetylcholine (ACh) and GABA at an early stage of embryogenesis. The appearance of responses to both neurotransmitters seems to occur at the same time. The reversal potentials are constant during development, implying that the ionic dependences of the responses are unchanging (Goodman and Spitzer, 1979, 1980).The pharmacology of the responses, as well as the lack of desensitization, also appear invariant with development.

Vertebrate skeletal muscle cells are sensitive to ACh. The reversal potential of this response is constant from the time of its first appearance early in the differentiation of these cells (Fambrough and Rash, 1971; Fischbach, 1972; Cohen and Kullberg, 1974; Steinbach, 1975; Blackshaw and Warner, 1976a; Ohmori and Sasaki, 1977), although there are dramatic changes in the distribution of ACh receptors (AChR). Again, constancy of the ionic selectivity of the channels seems indicated. There may be exceptions to this widespread phenomenon, however. The response of the chick atrium to Ach has been reported to change its ionic dependence during the course of development (Pappano, 1972).

5. DEVELOPMENTAL SIGNIFICANCE OF CHANGING MEMBRANE PROPERTIES

The functions of Ca^{2+}-dependent action potentials at early stages of development remain to be identified. These impulses of long duration allow the influx of large amounts of Ca^{2+} that could have profound effects on cellular metabolism. Alternatively, the importance of these impulses could be electrical rather than ionic. The reversal of the sign of the membrane potential for substantial periods of time could be essential for the insertion of membrane components. Agents that selectively block Ca^{2+} channels may allow resolution of this issue (Dunlap and Fischbach, 1978; Bixby and Spitzer, 1982b). It has recently been shown that conditions that promote Ca^{2+} entry into the growth cones of cultured neuroblastoma cells lead to morphological changes that are probably associated with neurite elongation (Anglister et al., 1982). Na^+-dependent impulses cannot play a role in early development in those cases in which they appear after considerable differentiation has occurred (e.g., Goodman and Spitzer, 1979) or when their blockage by tetrodotoxin (TTX) has no effect on some aspects of later development (Obata, 1977; Harris, 1980, 1981). The chronic application of TTX is toxic to some cells with Na^+-dependent action potentials, however, perhaps as a consequence of interference with trophic interactions (Bergey et al., 1980), and such blockage may alter normal synapse elimination (Van Essen, 1982).

The precise role of electrical coupling and subsequent uncoupling between embryonic cells is unknown. Metabolic cooperativity has been demonstated between coupled cells in culture (Gilula, et al., 1972). Induction could involve cellular interactions via these specialized junctions in vivo, which would not be needed at later stages. Although Rohon–Beard cells develop their action-potential mechanism and become sensitive to neurotransmitters while they are becoming uncoupled, the relationship is not obligatory (Bixby and Spitzer, 1982a). The development of amphibian spinal neurons in dissociated cell culture indicates that if coupling is necessary for early development, it must be required prior to the neural-plate stage at which the cultures are prepared. The significance of the voltage-dependent feature of electrical coupling is still obscure. The Ca^{2+}-dependent action potentials could cause transient uncoupling, which might be important for independent development of the cells or as a prelude to permanent uncoupling. Neurons can also become funtionally uncoupled although the electrical synapse persists (Rayport and Kandel, 1980).

The early appearance of neurotransmitter sensitivity in embryonic cells could reflect the specalization of the membrane required for the formation of chemical synapses. However, there is no evidence for synapses on the cell bodies of neurons, the development of which has been studied, as is also the case for a number of mature neurons, the somal sensitivity of which has been examined. The increase in GABA sensitivity of Rohon–Beard cell bodies is unlike the development of embryonic muscle fibers, which have been seen to accumulate AChR at the neuromuscular junction and lose them elsewhere (e.g., Blackshaw and Warner, 1976a; Ohmori and Sasaki, 1977). The existence of the early, developmentally transient glycine sensitivity of Rohon–Beard neurons, like the transient neurotransmitter sensitivity of rat Purkinje cells (Crepel, et al., 1982), raises the possibility that these ion channels may have some role in events other than synaptogenesis.

6. ROLES OF RNA AND PROTEIN SYNTHESIS IN DIFFERENTIATION OF NEURONAL MEMBRANE PROPERTIES

There is at present little information about the molecular basis for the appearance of different membrane properties during embryogenesis and the subsequent changes in these phenotypes during maturation. These processes may involve the synthesis and insertion of new channel proteins or modification of existing membrane components. This issue has been addressed by examining the effects of specific metabolic inhibitors applied to cells during a restricted period of their development, to determine the timing of RNA or protein synthesis required for the expression of different properties. There have been several studies of the temporal dependence of the effects of RNA-synthesis inhibitors on the cellular differentiation of amphibian nerve and muscle (Duprat *et al.*, 1966; Stocker and Bride, 1980). Experiments with specific metabolic inhibitors have also demonstrated that the RNA and protein synthesis needed for the expression of neurotransmitter synthesis follows that required for neurite outgrowth in the development of the fruit fly and the mouse (Seecof, 1977; Bloom and Black, 1979).Dissociated cell cultures containing amphibian spinal neurons are attractive for such studies, since neuronal differentiation *in vitro* parallels that *in vivo*, in several respects discussed above. The timing of the macromolecular syntheses necessary for the development of the Na^+-dependent impulse in these cultured neurons has been investigated. The early application of actinomycin D, to block RNA synthesis, or cycloheximide or puromycin, to suppress protein synthesis, prevents the appearance of the mature impulse (O'Dowd, 1981; Blair, 1981). The neurons continue to make Ca^{2+}-dependent action potentials of long duration, and the basal state of the membrane (resting potential, input resistance) is unaffected, as is the development of other cellular properties (neurite extension, voltage-dependent K^+ channels). This effect is further limited, in that application of these inhibitors at later times does not prevent the appearance and maturation of the Na^+-dependent action potentials. There is similar evidence for transcriptional control of the development of this phenotype in cultured chick muscle cells (Kano and Suzuki, 1982). If biochemical excision of single phenotypes can be achieved by the appropriately timed application of reversible inhibitors, the removal of the block should permit determination of the effects of inhibiting the expression of a single phenotype on subsequent development.

More specific probes are needed to distinguish between the roles of new synthesis and modification of existing molecules in the differentiation of membrane properties. Antibodies to membrane proteins and techniques for identifying their messenger RNAs should reveal the times of onset of transcription and translation. One may hope to obtain developmental timetables for the appearance of RNAs and proteins that give rise to various membrane properties. Advances in molecular biology and immunology suggest that such a detailed understanding of the early program of neuronal development will be achieved in the relatively near future.

ACKNOWLEDGMENTS.I thank Drs. John Bixby, Leslie Henderson, and Janet Lamborghini for their comments on the manuscript. Grant support was provided by the National Institutes of Health (NS 15918).

REFERENCES

Anglister, L., Farber, I.C., Shahar, A., and Grinvald, A., 1982, Localization of voltage sensitive calcium channels along developing neurites: Their possible role in regulating neurite elongation, *Dev. biol.* **94**:351–365.

Baccaglini, P.I., 1978, Action potentials of embryonic dorsal root ganglion neurones in *Xenopus* tadpoles, *J. Physiol.* **283**:585–604.

Baccaglini, P.I., and Spitzer, N.C., 1977, Developmental changes in the inward current of the action potential of Rohon–Beard neurones, *J. Physiol.* **271**:93–117.

Baud, C., Kado, R.T., and Marcher, K., 1982, Sodium channels induced by depolarization of the *Xenopus laevis* oocyte, *Proc. Nat'l. Acad. Sci.* **79**:3188–3192.

Bergey, G.K., Fitzgerald, S.C., Schrier, B.K., and Nelson, P.G., 1980, Neuronal maturation in mammalian cell culture is dependent on spontaneous electrical activity, *Brain Res.* **207**:49–58.

Berwald-Netter, Y., Martin-Moutot, N., Koulakoff, A., and Couraud, F., 1981, Na$^+$-channel-associated scorpion toxin receptor sites as probes for neuronal evolution *in vivo* and *in vitro*, *Proc. Natl. Acad. Sci. U.S.A.* **78**:1245–1249.

Bixby, J.L., and Spitzer, N.C., 1982a, The appearance and development of chemosensitivity in Rohon–Beard neurones of the *Xenopus* spinal cord, *J. Physiol.* **330**:513–536.

Bixby, J.L., and Spitzer, N.C., 1982b, Enkephalin shortens Ca^{++}-spikes at early stages of embryonic spinal neurons *in vivo*, *Soc. Neurosci. Abstr.* **8**:229.

Blackshaw, S., and Warner, A., 1976a, Onset of acetylcholine sensitivity and endplate activity in developing myotome muscles of *Xenopus, Nature (London)* **262**:217–218.

Blackshaw, S.E., and Warner, A.E., 1976b, Low resistance junctions between mesoderm cells during development of trunk muscles, *J. Physiol.* **255**:209–230.

Blair, L., 1981, The timing of protein synthesis necessary for the acquisition of the Na$^+$ action potential during development, *Soc. Neurosci. Abstr.* **7**:245.

Bloom, E.M., and Black, E.B., 1979, Metabolic requirements for differentiation of embryonic sympathetic ganglia cultured in the absence of exogenous nerve growth factor, *Dev. Biol.* **68**:568–578.

Cohen, M.W., and Kullberg, R.W., 1974, Temporal relationship between innervation and appearance of acetylcholine receptors in embryonic amphibian muscle, *Proc. Can. Fed. Biol. Soc.* **17**:176.

Crepel, F., Dupont, J.L., and Gardette, R., 1982, Connectivity and chemosensitivity of Purkinje cells in the immature rat cerebellum: An *in vitro* study, *J. Physiol.* **332**:62.

Curtis, D.R., Hösli, L., Johnston, G.A.R., and Johnston, I.H., 1968, The hyperpolarization of spinal motoneurones by glycine and related amino acids, *Exp. Brain Res.* **5**:235–258.

Dennis, M.J., Ziskind-Conhaim, L., and Harris, A.J., 1981, Development of neuromuscular junctions in rat embryo, *Dev. Biol.* **81**:266–279.

Dixon, J.S., and Cronly-Dillon, J.R., 1972, The fine structure of the developing retina in *Xenopus laevis*, *J. Embryol. Exp. Morphol.* **28**:659–666.

Dunlap, K., and Fischbach, D.G., 1978, Neurotransmitters decrease the calcium component of sensory neurone action potentials, *Nature (London)* **276**:837–839.

Duprat, A.-M., Zalta, J.-P., and Beetschen, J.-C., 1966, Action de l'actinomycine D sur la différenciation de divers types de cellules embryonnaires de l'amphibien *Pleurodeles waltlii* en culture *in vitro*, *Exp. Cell Res.* **43**:358–366.

Fambrough, D., and Rash, J.E., 1971, Development of acetylcholine sensitivity during myogenesis, *Dev. Biol.* **26**:55–68.

Fischbach, G.D., 1972, Synapse formation between dissociated nerve and muscle cells in low density cell cultures, *Dev. Biol.* **28**:407–429.

Fukuda, J., and Kameyama, M., 1979, Enchancement of Ca spikes in nerve cells of adult mammals during neurite growth in tissue culture, *Nature (London)* **279**:546–548.

Gilula, N.B., Reeves, O.R., and Steinbach, A., 1972, Metabolic coupling, ionic coupling and cell contacts, *Nature (London)* **235**:262–265.

Goodman, C.S., and Spitzer, N.C., 1979, Embryonic development of identified neurones: Differentiation from neuroblast to neurone, *Nature (London)* **280**:208–214.

Goodman, C.S., and Spitzer, N.C., 1980, Embryonic development of neurotransmitter receptors in grasshoppers, in: *Receptors for neurotransmitters, Hormones and Pheromones in Insects* (D.B. Satelle Hildebrand, J.G., and Hall, L.M., eds.), Elsevier/North-Holland, Amsterdam, pp. 195–207.

Goodman, C.S., and Spitzer, N.C., 1981, The development of electrical properties of identified neurones in grasshopper embryos, *J. Physiol.* **313**:385–403.

Hagiwara, S., and Jaffe, L.A., 1979, Electrical properties of egg cell membranes, *Annu. Rev. Biophys. Bioeng.* **8**:385–416.

Harris, W.A., 1980, The effect of eliminating impulse activity on the development of the retinotectal projection in salamanders, *J. Comp. Neurol.* **194**:303–317.

Harris, W.A., 1981, Neural activity and development, *Annu. Rev. Physiol.* **43**:689–710.

Heathcote, R.D., 1981, Differentiation of an identified sensory neuron (SR) and associated structures (CTO) in grasshopper embryos, *J. Comp. Neurol.* **202**:1–18.

Jacobson, M., 1981, Rohon–Beard neuron origin from blastomeres of the 16-cell frog embryo, *J. Neurosci.* **1**:918–922.

Kano, M., and Suzuki, N., 1982, Inhibition by α-amanitin of development of tetrodotoxin-sensitive spike induced by brain extract in cultured chick skeletal muscle cells, *Dev. Brain Res.* **3**:674–678.

Lamborghini, J.E., 1980, Rohon–Beard cells and other large neurons in *Xenopus* embryos originate during gastrulation, *J. Comp. Neurol.* **189**:323–333.

Lo, C.W., and Gilula, N.B., 1979, Gap junctional communication in the postimplantation mouse embryo, *Cell* **18**:411–422.

Matsuda, Y., Yoshida, S., and Yonezawa, T., 1978, Tetrodotoxin sensitivity and Ca component of action potentials of mouse dorsal root ganglion cells cultured *in vitro*, *Brain Res.* **154**:69–82.

Meiri, H., Spira, M.E., and Parnas, I., 1981, Membrane conductance and action potential of a regenerating axonal tip, *Science* **211**:709–712.

Miyake, M., 1978, The development of action potential mechanism in a mouse neuronal cell line *in vitro*, *Brain Res.* **143**:349–354.

Mori-Okamoto, J., Ashida, H., Maru, E., and Tatsuno, J., 1983, The development of action potentials in cultures of explanted cortical neurons from chick embryos, *Dev. Biol.*, **97**:408–416.

Obata, K., 1977, Development of neuromuscular transmission in culture with a variety of neurons and in the presence of cholinergic substances and TTX, *Brain Res.* **119**:141–153.

O'Dowd, D.K., 1981, The timing of RNA synthesis necessary for the development of the Na^+-dependent action potential in cultured neurons, *Soc. Neurosci. Abstr.* **7**:245.

Ohmori, H., and Sasaki, S., 1977, Development of neuromuscular transmission in a larval tunicate, *J. Physiol.* **269**:221–254.

Palmer, J.F., and Slack, C., 1970, Some bioelectric parameters of embryos of *Xenopus laevis*, *J. Embryol. Exp. Morphol.* **24**:535–553.

Pappano, A.J., 1972, Sodium-dependent depolarization of noninnervated embryonic chick heart by acetylcholine, *J. Pharmacol. Exp. Ther.* **180**:340–350.

Potter, D.D., Furshpan, E.J., and Lennox E., 1966, Connections between cells of the developing squid as revealed by electrophysiological methods, *Proc. Natl. Acad. Sci. U.S.A.* **55**:328–336.

Rayport, S.C., and Kandel, E.R., 1980, Developmental modulation of an identified electrical synapse: Functional uncoupling, *J. Neurophysiol.* **44**:555–567.

Roberts, A., and Clarke, J.D.W., 1982, The neuroanatomy of an amphibian embryo spinal cord, *Philos. Trans. R. Soc. London Ser. B* **296**:195–212.

Seecof, R.L., 1977, A genetic approach to the study of neurogenesis and myogenesis, *Am. Zool* **17**:577–584.

Sheridan, J.D., 1968, Electrophysiological evidence for low resistance electrical connections between cells of the chick embryo, *J. Cell Biol.* **37**:650–659.

Slack, C., and Warner, A.E., Properties of surface and junctional membranes of embryonic cells isolated from blastula stages of *Xenopus laevis, J. Physiol.* **248**:97–120.

Spitzer, N.C., 1970, Low resistance connections between cells in the developing anther of the lily, *J. Cell Biol.* **45**:565–575.

Spitzer, N.C., 1982a, Voltage- and stage-dependent uncoupling of Rohon–Beard neurones during embryonic development of *Xenopus* tadpoles, *J. Physiol.* **330**:145–162.

Spitzer, N.C., 1982b,The development of electrical excitability, in: *Neuronal–Glial Cell Interrelationships* (T.A. Sears, ed), pp. 77–91, Springer-Verlag, Berlin.

Spitzer, N.C., and Bixby, J.L., 1982, Appearance and development of chemosensitivity of embryonic amphibian spinal neurons in vitro, *Soc. Neurosci, Abstr.* **8**:130.

Spitzer, N.C., and Lamborghini, J.E., 1976, The development of the action potential mechanism of amphibian neurons isolated in culture, *Proc. Natl. Acad. Sci. U.S.A.* **73**:1641–1645.

Spitzer, N.C., and Lamborghini, J.E., 1981, Programs of early neuronal development, in: *Studies in Developmental Neurobiology* (W.M. Cowan, ed.), Oxford University Press, New York, pp. 261–287.

Spray, D.C., Harris, A.L., and Bennett, M.V.L., 1979, Voltage-dependence of junctional conductance in early amphibian embryos, *Science* **204**:432–434.

Steinbach, J.H., 1975, Acetylcholine responses in clonal myogenic cells *in vitro, J. Physiol.* **247**:393–405.

Stocker, S., and Bride, M., 1980, Effects of α-amanitin and actinomycin D on *Xenopus laevis* (Daud.) heart in culture during cardiac differentiation, *Cell. Mol. Biol.* **26**:303–317.

Strichartz, G., Small, R., Nicholson, C., Pfenninger, K.H., and Llinas, R., 1980, Ionic mechanisms for impulse propagation in growing nonmyelinated axons: Saxitoxin binding and electrophysiology, *Soc. Neurosci, Abstr.* **6**:660.

Takahashi, K., and Yoshii, M., 1981, Development of sodium, calcium and potassium channels in the cleavage-arrested embryo of an ascidian, *J. Physiol.* **315**:515–529.

Takahashi, K., Miyazaki, S., and Kidokoro, Y., 1971, Development of excitability in embryonic muscle cell membranes in certain tunicates, *Science* **171**:415–418.

Tarin, D., 1971, Histological features of neural induction in *Xenopus laevis, J. Embryol. Exp. Morphol.* **26**:543–570.

Van Essen, D.C., 1982, Neuromuscular synapse elimination, in: *Neuronal Development* (N.,C. Spitzer, ed.), Plenum Press, New York, pp. 333–376.

Willard, A.L., 1980, Electrical excitability of outgrowing neurites of embryonic neurones in cultures of dissociated neural plate of *Xenopus laevis, J. Physiol* **301**:115–128.

Chapter 7

The Accumulation of Acetylcholine Receptors at Nerve–Muscle Synapses in Culture

Gerald D. Fischbach, Lorna W. Role, and Richard I. Hume

We are studying the formation of synapses between embryonic chick ciliary-ganglion neurons and muscle fibers maintained in cell culture. Our assumption is that certain rules that govern synapse formation can be best understood by observing the behavior of individual cells in a controlled environment over periods of time that range from seconds to days. This observation is difficult if not impossible to achieve in intact embryos. The ability to visualize and manipulate growing neurons *in vitro* is a major advantage in this regard. In addition, cell cultures provide convenient and relevant bioassays that might lead to an understanding of the phenomenology of synapse formation on a molecular level. In this summary report, we describe some recent experiements in our laboratory concerning the early release of acetylcholine (ACh) from presynaptic terminals and the initial accumulation of ACh receptors (AChR) in the postsynaptic muscle membrane.

Synaptic transmission begins soon after a cholinergic neuron contacts a receptive myotube (Fig. 1) (Frank and Fischbach, 1979; Cohen, 1980; Kidokoro and Yeh, 1982; Role *et al.*, 1983). Indeed, we have recently found that cholinergic growth cones can release ACh even before they contact a myotube (Hume *et al.*, 1983). This conclusion is based on a novel bioassay in which 2- to 3-μm-tip electrodes sealed with outside-out patches of muscle membrane (see Hammil *et al.*, 1981) were used to detect ACh in the vicinity of growth cones.

GERALD D. FISCHBACH, LORNA W. ROLE, and RICHARD I. HUME ● Department of Anatomy and Neurobiology, Washington University School of Medicine, St. Louis, Missouri 63110.

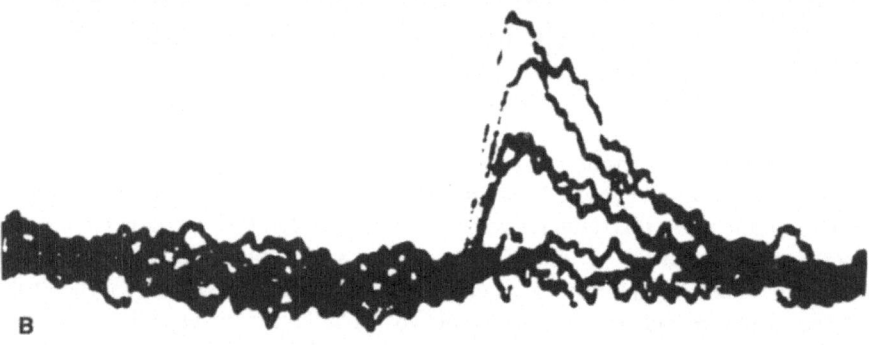

Figure 1.   A newly formed nerve–muscle synapse. (A) A ciliary ganglion neuron lying along the lower edge of a 5-day-old myotube. The neuron was added to the culture 24 hr previously. (←) Position of one of this neuron's growth cones. (B) Superimposed oscilloscope traces show synaptic potentials recorded with a microelectrode placed in the myotube following stimulation of the neuron cell body.

The patches were calibrated by iontophoresis of ACh, and judging from the size of single ACh channel currents (~3 pA) and the peak current produced by large doses of ACh, they contained as many as 300 AChR. When ciliary–ganglion neurons were stimulated with a short train of pulses, channel openings were detected by patch electrodes positioned next to a growth cone in about one third of the trials (Fig. 2). These experiments were performed in the presence of eserine to inhibit intracellular cholinesterases. Additional work is needed to determine whether ACh is released from growth cones under more physiological circumstances and whether it is released spontaneously—in the absence of imposed patterns of activity.

AChR accumulate at nerve–muscle synapses within a few hours after the onset of synaptic transmission (Frank and Fischbach, 1979; Anderson *et al.,*

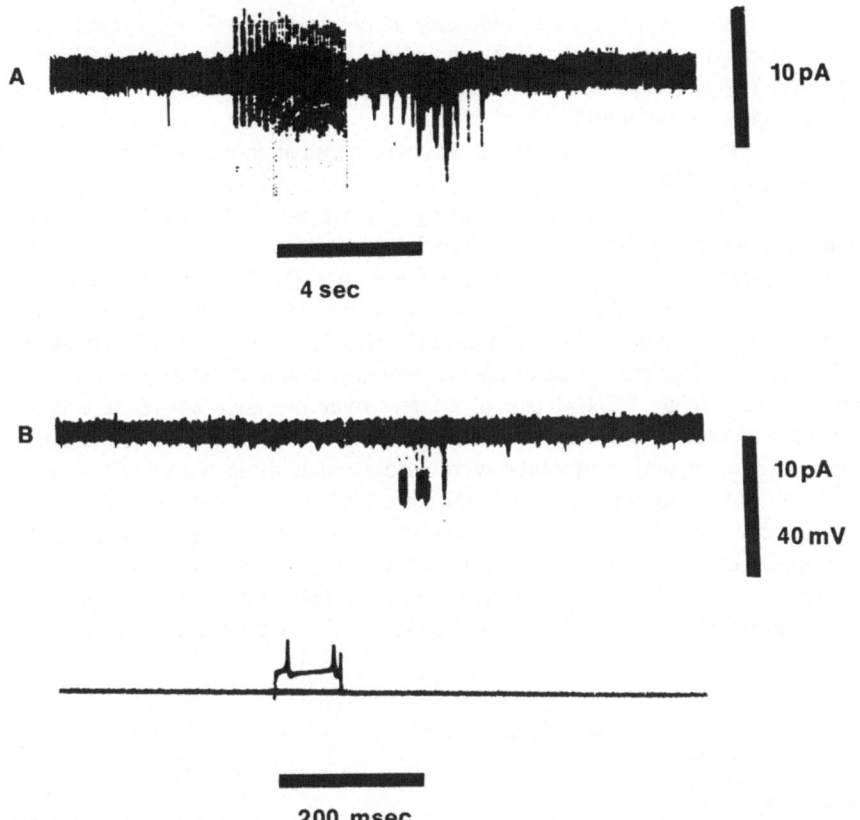

Figure 2. ACh release from a ciliary ganglion growth cone detected with an outside-out patch of myotube membrane. (A) The patch electrode was positioned close to a growth cone while the cell body was stimulated repetitively at about 10/sec for 3 sec. After the stimulus train, discrete channel openings (downward deflections following the stimulus artifacts) were detected. (B) In another experiment with a different cell, channel openings were observed following a single stimulus. The size and duration of the channel openings were identical to those evoked by iontophoretically applied ACh, and they were never observed in the presence of 10^{-4} M curare. We conclude, therefore, that they are due to ACh released from growth cones. These neurons were grown in cultures that did not contain myotubes.

1977). AchR clusters are present on uninnervated myotubes, but it is now clear that the clusters found at synapses are induced by the cholinergic neurites.

Ciliary-ganglion neurons extend several processes that grow rapidly (40–60 μm/hr) during the first 36 hr in culture. This initial growth spurt is followed by a period of slower extension and "pruning" during which all but one of the major processes are eliminated. We examined the relationship between individual neurites and AChR clusters by injecting the neurons with Lucifer Yellow CH (LY) and labeling the receptors with rhodamine–α-bungarotoxin (R-BTX) conjugates. LY-filled nerve processes could be visualized even when they passed beneath refractile myotubes or formed fascicles with other neurites. The number of receptor clusters on uninnervated chick myotubes is low, so if a cluster was observed within 5 μm of a neurite, it was considered an induced nerve-associated receptor patch (NARP).

Several observations are noteworthy: (1) All the major processes of young neurons induced NARPs (Fig. 3). Thus, by this criterion, no unique process that might represent an early axon was evident. (2) The neurons induced many NARPs: as many as 30/cell were observed within 24 hr after adding neurons to established myotubes (4–5 days in culture). The "density" of NARPs peaked at 2.75/100 μm of neurite–muscle contact between 8 and 15 hr after plating. The density declined to 1.2/100 μm of contact over the next 24–48 hr and then remained stable for the next 3 weeks. (3) Between 60 and 80% of the growth cones in contact with a myotube were associated with an NARP. Considering the mean NARP frequency of 1.2/100 μm, this result cannot have occurred by chance, so we conclude that NARPs are induced at the advancing tips of neurites. The appearance of NARPs at growth cones, together with the fact that growth cones can release ACh, indicates that the motile tips contain the tools necessary for synapse formation. (4) Between 3 days and 3 weeks of coculture, when the

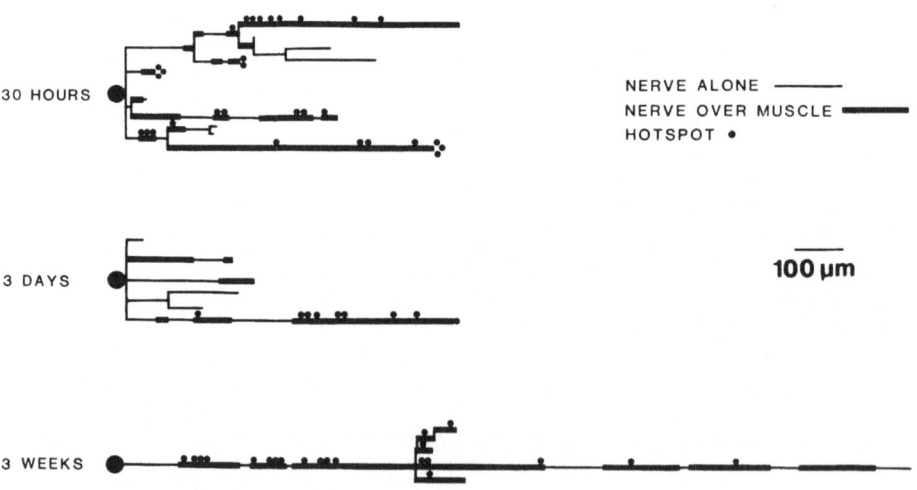

Figure 3. Representative diagrams of LY-injected neurons and NARPs at the indicated times of ciliary ganglion–muscle coculture. (━) Lengths of nerve–muscle contact; (—) neurite–collagen substrate contact; (●) NARPs. Note that more than one primary process can induce NARPs at early times, that neurons become unipolar with time in culture, and that the surviving neurite is associated with many NARPs.

ciliary neurons change from a multipolar to a unipolar shape, the longest process was always associated with many NARPs, whereas short, perhaps retracting, neurites from the same cell exhibited few NARPs or none at all (Fig. 3). Other experiments have shown that ciliary neurons become unipolar in culture even when they are grown without target myotubes. Thus, the cellular machinery for ciliary-neuron-process elimination can operate in the absence of an appropriate target. However, the possiblity that formation of NARPs or synapses or both influences neurite survival has not been ruled out. We do not mean to imply that receptor clusters *per se* impart a survival value. They may simply serve as indicators of retrograde as well as orthograde inductive interaction.

AChR are mobile in membranes of mammalian and amphibian muscle cells, and evidence has been presented that receptors in *Xenopus* myotubes migrate to new nerve–muscle contacts (Anderson and Cohen, 1977). Our data indicate that migration of receptors plays a relatively minor role in NARP formation on chick myotubes. We sought evidence for AChR migration by labeling receptors with R-BTX *before* adding dissociated ciliary neurons to muscle cultures (cf. Anderson and Cohen, 1977) and then examining nerve-associated patches of fluorescence 19–24 hr later (Mathews-Bellinger and Fischbach, unpublished observations). The number and intensity of prelabeled NARPs were compared with those of NARPs in sibling cultures labeled at the end of the incubation period. In five different experiments (platings), we found fewer neurons with NARPs in prelabeled cultures (mean = 33%) than in end-labeled cultures (67%). Moreover, NARPs containing prelabeled receptors were far less intense than end-labeled NARPs. The latter were still clearly visible when a 90% neutral-density filter was inserted in the optical patch, and in this situation, they appeared at least as intense as the prelabeled NARPs observed without the filter. The low intensity of NARPs formed from prelabeled receptors cannot be accounted for by receptor degradation. Chick receptors are degraded with a half-life of 20–30 hr at 37°C (Deverotes and Fambrough, 1975; Schuetze *et al.*, 1977; Jessell *et al.*, 1979; Buc-Caron *et al.*, 1983). Thus, on the average, receptor migration in chick membranes probably accounts for less than 20% of the total number of receptors present at 24-hr-old NARPs.

The same conclusion was reached following experiments in which receptors were blocked with uncoupled α-BTX (for 1 hr) and then relabeled with R-BTX 24 hr after. NARPs composed of the "new" receptors that were inserted during the 24-hr interval were far brighter than NARPs formed from receptors "pulse-labeled" with R-BTX prior to plating the neurons. In two experiments, bright NARPs were observed when α-BTX was present in the bath for 12 hr after the neurons were plated and the receptors were labeled with R-BTX 5 hr after washout.

We have no direct evidence for local insertion of newly synthesized receptors during synapse formation, but we have identified an intracellular vehicle that may ferry AChR from their site of synthesis to the surface membrane (Bursztajn and Fischbach, 1983). Coated vesicles are present in the myoplasm of embryonic myotubes *in vitro,* and they are five times more common immediately beneath physiologically identified sites of transmitter release than beneath other segments of the surface membrane. They are also more numerous beneath receptor clusters (hot spots) on uninnervated myotubes. We added horseradish peroxidase–α–bungarotoxin (HRP-BTX) conjugates to saponin-permeabilized myotubes and showed by electron microscopy that about 50% of the sub-hot-spot coated vesicles contain specific HRP-BTX binding sites. It is difficult to relocate and section identified synapses, so all experiments to date were performed on uninnervated myotubes treated for 4 days with saline extract of embryonic chick brain. Our working hypothesis is that brain extract contains the same factors present in cholinergic nerves that induce receptor synthesis or aggregation or both at newly formed synapses.

Three lines of evidence indicate that labeled coated vesicles are involved in exocytosis rather than endocytosis: (1) Their number was increased nearly 5-fold following treatment with brain extract. Extract increases the rate of myotube-receptor synthesis and incorporation 3- to 4-fold without significantly changing receptor degradation. (2) Only 10% of the vesicles contained HRP reaction product when HRP-BTX was added to intact (unpermeabilized) cells. (3) Exposure to puromycin (20 μg/ml) for 6 hr resulted in a 20-fold decrease in the number of labeled vesicles beneath hot spots. Inhibition of protein synthesis does not alter receptor degradation in chick myotubes (Devreotes and Fambrough, 1975).

We and others have attempted to purify the receptor-inducing activity and brain extract. Initial experiments (Jessell *et al.,* 1979) showed that factors in a 100,000g supernatant of a saline extract prepared from embryonic chick spinal cord and brain produced a dose-dependent increase in the number of AChR in cultured chick myotubes. Saline extract was added to multinucleated (fused) myotubes for 4 days beginning on the 4th day after plating mononucleated cells, and the number of receptors was assayed with $[^{125}I]$-α-BTX. The total number of receptors increased gradually, and after 4 days (8th day after plating), it amounted to 3–5 times control level. Autoradiography and ACh iontophoresis showed that the increase in AChR number reflected an increase in receptor density rather than an increase in myotube membrane area or number of cells. The same techniques revealed a large increase (≈40-fold) in the number of densely packed receptor aggregates. The increase in overall receptor density was not due to a decrease in the rate of receptor degradation and was independent of the level of muscle electrical or contractile activity or both.

Some degree of tissue specificity was evident in that extracts prepared from liver, heart, or cultured fibroblasts were without effect. Culture medium conditioned by spinal cord or ciliary-ganglion cells induced myotube AChR synthesis, but medium conditioned by sensory neurons did not. Thus, inducing factors are contained in cholinergic neurons and released from intact cells.

The effect of brain extract on cultured myotubes was not limited to AChR synthesis and aggregation in that we also measured a 3- to 4-fold increase in total acetylcholinesterase specific activity and a small increase in [^{3}H]thymidine incorporation into fibroblasts present in the muscle cultures. On the other hand, brain extract did not simply serve a general health-promoting role. We measured no change in total protein (or in the rate of incorporation of [^{3}H] amino acids into trichloroacetic-acid–precipitable material) or in the activity of creatine kinase, a myoplasmic enzyme.

Two peaks of activity were observed when the brain extract 100,000g supernatant was filtered through Biogel P150. One peak migrated with markers in the 50,000- to 70,000-dalton range, and the other appeared to be 2000–10,000 daltons in size. Activity in the first peak was retained following dialysis, whereas activity in the second peak was lost.

We decided to pursue the small material. Nearly all (90%) of the low-molecular-weight peak remained soluble in 2 N acetic acid, and about 85% of the acid-soluble activity was destroyed by 0.1 mg/ml L-(tosylamide-2-phenyl) ethyl chloromethyl ketone (TPCK)–trypsin. These data, together with the fact that inducing activity is extremely hydrophobic at low pH, lead us to the tentative conclusion that activity resides in one or more small peptides.

In subsequent experiments (Buc-Caron *et al.*, 1983), we found that acid and saline extracts produced a large increase in AChR number after 24 hr (instead of 3–4 days) if we used older myotubes (7 days instead of 4 days after plating). We also found that the initial rate of receptor incorporation into the surface membrane, measured with [^{125}I]-α-BTX, 1, 4, and 7 hr after blocking all exposed sites with cold BTX (Devreotes and Fambrough, 1975), is a more sensitive assay of receptor synthesis than is total AChR number. We used this assay and a new extraction protocol to further purify AChR-inducing activity by reverse-phase high-pressure liquid chromatography (RP-HPLC).

The extraction cocktail—which contains 2% trifluoroacetic acid (TFA), 5% formic acid, 1 N HCl, 0.1 N NaCl, 0.01% thiodiglycol, and the protease inhibitors pepstatin and leupeptin, each at 1 μg/ml—was introduced by Bennett *et al.* (1977, 1978) and is now widely used in peptide purification. The acid extract was delipidated with diethyl ether and passed through a large, preparative octadecasilyl silica (C_{18}) column to remove amino acids and salts. All receptor-inducing activity was retained on the C_{18} column when loaded in 0.1% TFA (pH 3.5) and was recovered when the column was eluted with an organic solvent (2-propanol). We tested several RP-HPLC analytical columns. To date, our most consistent results have been obtained with an Altex 5-μm Ultrasphere C_{18} column developed with a 0–60% acetronitrile gradient in 0.1% TFA. TFA is volatile and relatively transparent at OD_{210}. Material that elutes between 35 and 40% acetonitrile represents a 1000-fold purification compared to saline brain extract.

The most active C_{18} fractions produce a detectable increase in receptor incorporation rate in 3–5 hr, and the effect is maximal after 10 hr. This early action is remarkable considering that it takes 2–3 hr for newly synthesized receptors to traverse the intracellular pool and appear on the cell surface (Devreotes and Fambrough, 1975; Devreotes et al., 1977). It is an important finding because new AChR clusters can appear at synapses within a few hours after neurite–muscle contact.

In addition to promoting AChR synthesis, the most active C_{18} fractions also promote the aggregation of receptors. Aggregation occurs over the same time–course as increased incorporation, so the two phenomena cannot be distinguished on this basis. Aggregation can occur in the absence of receptor synthesis (Christian et al., 1978), but further experiments are needed to determine whether synthesis is invariably linked to aggregation.

In sum, nerve–muscle synapses form rapidly in vitro and AChR accumulate in the postsynaptic membrane within a few hours after nerve–muscle contact. The accumulation of postsynaptic receptors is probably mediated by one or more soluble factors released from competent growth cones. Predictably, several important questions remain to be answered. What is the relationship between the low-molecular-weight and the high-molecular-weight activity? Do the same molecules induce receptor synthesis and receptor aggregation? Is the activity (small or large) present in brain extract the same as the activity present in and released from embryonic motoneurons? These and related questions will be answered definitively only after the activity is completely purified and specific antibodies are obtained and used to explore its distribution in the nervous system and its biological role.

REFERENCES

Anderson, M.J., and Cohen, M.W., 1977 Nerve-induced and spontaneous redistribution of acetylcholine receptors on cultured muscle cells, J. Physiol. **268**:757–773.

Anderson, M.J., Cohen, M.W., and Zorychta, E., 1977 Effects of innervation on the distribution of acetylcholine receptors on cultured muscle cells, J. Physiol. **268**:731–756.

Bennett, H.P.J., Hudson, A.M., McMartin, C., and Purdon, G.E., 1977, Use of ODS-S for the extraction and purification of peptides in biological samples. Application to the identification of circulating metabolites of corticotropin-(1-24)-tetracospeptide and hematostatin in vivo, Biochem. J. **168**:9–13.

Bennett, H.P.J., Hudson, A.M., Kelly, L., McMartin, C., and Purdon, G.E., 1978, A rapid method, using octadecasilyl-silica, for the extraction of certain peptides from tissues, Biochem. J. **175**:1139–1141.

Buc-Caron, M.-H., Nystrom, P., and Fischbach, G.D., 1983, Induction of acetylcholine receptor synthesis and aggregation: Partial purification of low-molecular-weight activity, *Dev. Biol.* **95:**378–386.

Bursztajn, S., and Fischbach, G.D., 1983 Evidence that coated vesicles transport acetylcholine receptors to the surface membrane of chick myotubes, *J. Cell Biol.* (in press).

Christian, C.N., Daniels, M.P., Sugiyama, H., Vogel, Z., Jacques, L., and Nelson, P.G., 1978 A factor from neurons that increases the number of acetylcholine receptor aggregates on cultured muscle cells, *Proc. Natl. Acad. Sci. U.S.A.* **75:**4011–4015.

Cohen, S.A., 1980 Early nerve–muscle synapses *in vitro* release transmitter over postsynaptic membrane having low ACh sensitivity, *Proc. Natl. Acad. Sci. U.S.A.* **77:**644–648.

Devreotes, P.N., and Fambrough, D.M., 1975, Acetylcholine receptor turnover in membranes of developing muscle fibers, *J. Cell Biol.* **65:**335–358.

Devreotes, P.N., Gardner, J.M., and Fambrough, D.M., 1977 Kinetics of biosynthesis of acetylcholine receptor and subsequent incorporation into plasma membrane of cultured chick skeletal muscle, *Cell* **10:**365–373.

Frank, E., and Fischbach, G.D., 1979 Early events in neuromuscular junction formation *in vitro:* Induction of acetylcholine receptor clusters in the postsynaptic membrane and the morphology of newly formed synapses, *J. Cell Biol.* **83:**143–158.

Hammil, O.P., Marty, A., Neher, E., Sackmann, B., and Sigworth, F.J., 1981 Improved patch-clamp techniques for high resolution current recording from cells and cell-free membrane patches, *Pfluegers Arch.* **391:**85–100.

Hume, R.I., Role, L.W., and Fischbach, G.D., 1983 ACh release from growth cones detected with patches of ACh receptor-rich membranes, *(London) Nature* (in press).

Jessell, T.M., Siegel, R.E., and Fischbach, G.D., 1979 Induction of acetylcholine receptors on cultured skeletal muscle by a factor extracted from brain and spinal cord, *Proc. Natl. Acad. Sci. U.S.A.* **76:**5397–5401.

Kidokoro, Y., and Yeh, E., 1982 Initial synaptic transmission at the growth cone in *Xenopus* nerve–muscle cultures, *Proc. Natl. Acad. Sci. U.S.A.* **79:**6727–6731.

Role, L.W., Hume, R.E., and Fischbach, G.D., 1983 Transmitter release and receptor aggregation at ciliary neuron–muscle synapses, *Soc. Neurosci. Abstr.* **9:**129.

Schuetze, S.M., Frank, E.F., and Fischbach, G.D., 1978 Channel open time and metabolic stability of synaptic and extrasynaptic acetylcholine receptors on cultured chick myotubes, *Proc. Natl. Acad. Sci. U.S.A.* **75:**520–523.

Transmitter Phenotypic Plasticity In Developing and Mature Neurons *in Vivo*

Ira B. Black, Joshua E. Adler, Martha C. Bohn,
G. Miller Jonakait, John A. Kessler, and Keith A. Markey

1. INTRODUCTION

Experimental manipulation has revealed the *potential* for remarkable plasticity in the expression and development of neurotransmitter phenotypic characters (Black and Patterson, 1980; Le Douarin, 1980; Black, 1982). While autonomic neurons, which have been examined in particular detail, may alter a number of transmitter characters, and even acquire new transmitters in culture (for reviews, see Patterson, 1978; Varon and Bunge, 1978), relevance to *in vivo* events has not been fully determined. Two issues of central interest may be identified: Do neurons or their progenitors *normally* exhibit phenotypic plasticity during ontogeny *in vivo*? If they do, does phenotypic plasticity persist into adulthood in the postmitotic neuron?

We have been examining neuronal phenotypic plasticity during development and maturity, employing rat autonomic and sensory neurons. Using a number of specific catecholamine (CA) and peptide transmitter characters, we are seeking to determine whether transmitter mutability normally occurs during development *in vivo* and whether mutability persists into adulthood. By monitoring the CA biosynthetic enzymes, tyrosine hydroxylase (TH), dopamine-β-hydroxylase (DBH),

IRA B. BLACK, JOSHUA E. ADLER, MARTHA C. BOHN, G. MILLER JONAKAIT, JOHN A. KESSLER, and KEITH A. MARKEY ● Division of Developmental Neurology, Cornell Medical College, New York, New York 10021.

and phenylethanolamine-*N*-methyltransferase (PNMT), as well as the specific, high-affinity uptake (U_1) system for norepinephrine (NE), and the CA transmitters themselves, we have been able to characterize the expression and development of individual CA phenotypic characters. Simultaneously, we have studied the expression and regulation of substance P (SP) and somatostatin (SS), putative peptide transmitters. This approach has allowed us to define transmitter plasticity of developing and mature individual neurons *in vivo*.

Our studies suggest that neurons do indeed exhibit transmitter phenotypic lability during development in the embryo and fetus. Studies of neonates have provided evidence that plasticity occurs during the postnatal period as well (Landis, 1983). Moreover, our work suggests that transmitter phenotypic plasticity *in vivo* is not restricted to the developmental period. Rather, recent experiments indicate that striking plasticity persists into adulthood and perhaps for the life of the neuron. Expression of transmitter characters by mature neurons *in vivo* appears to be a dynamic changing process governed by the physiological state of the neuron. In turn, the status of the neuron is markedly influenced by a variety of environmental stimuli during adulthood as well as during development. Consequently, extracellular factors, which contribute to the regulation of transmitter expression and ontogeny, continue to regulate transmitter metabolism in the mature, postmitotic neuron.

Studies performed to date have not determined definitively whether plasticity of transmitter characters *in vivo* consists of qualitative as opposed to marked quantitative shifts in molecular species elaborated. The immunochemical, radiochemical, and chromatographic methods employed are all subject to very real limits of sensitivity. Further, the transmitter characters described herein are the protein (and peptide) transmitter gene products. Apparent qualitative changes in these traits may simply reflect marked alteration in translational or posttranslational processing, without indicating qualitative alterations in transcription. In view of these uncertainties, terms such as "phenotypic plasticity" are used in this review without prejudice regarding underlying molecular mechanisms. Rather, "plasticity" is simply used in the generic sense to denote change, or the potential thereof, in the metabolism or expression of transmitter phenotypic characters.

2. TRANSMITTER PLASTICITY IN THE EMBRYO

During embryonic development, the neural crest gives rise to a variety of derivatives, including autonomic and sensory neurons, support cells of the peripheral nervous system, thyroid calcitonin-producing cells, melanocytes, and mesenchymal derivatives of the cephalic region (Horstadius, 1950; Weston, 1963, 1970; Weston and Butler, 1966; Coulombre *et al.*, 1974; Le Douarin,

1980). The crest, a transient aggregate of cells, lies dorsal to the tube; cells detach and migrate in lateral and medial (ventral) sheets in a rostrocaudal wave (Weston, 1963). The latter contains precursors of autonomic and sensory neurons.

Studies performed to date suggest that the crest itself does not express transmitter traits prior to migration. Recent work, however, is consistent with the contention that at least some cells express the cholinergic enzyme choline acetyltransferase and are capable of acetylcholine (ACh) synthesis at the outset of migration (for a review, see Le Douarin, 1980). Subsequently, however, the vast majority of sympathetic neurons (>95%) and virtually all crest-derived sensory neurons do not exhibit cholinergic characters, raising the possibility that migrating cells express these transmitter traits only transiently. In addition, of course, a variety of cholinergic parasympathetic neurons are also derived from the crest. CAs and individual CA phenotypic characters are undetectable in the crest itself or in cells migrating from the crest to autonomic ganglion primordia (Cochard *et al.*, 1978, 1979). TH and DBH immunoreactivity and CA histofluorescence are initially detectable as cells aggregate dorsolateral to the dorsal aorta to form the sympathetic ganglion anlage on embryonic day 11.5 (E11.5; 17–30 somites). Subsequent to initial expression of these characters in sympathetic precursors, each trait increases quantitatively in a continuous fashion through the neonatal period to reach adult levels by 21 days of postnatal life (Black, 1982). Other populations, however, dramatically alter transmitter expression in the embryo.

Adrenomedullary precursors and presumptive neuroblasts in the embryonic intestine appear to change transmitters during normal ontogeny. The former convert from the *noradrenergic* to *adrenergic* phenotypes, whereas the latter transiently express CA characters, but mature transmitter status is unknown. Adrenomedullary progenitors express TH and DBH immunoreactivity on E13.5, as the cells lie lateral to the dorsal aorta, prior to secondary ventrad migration into the adrenal anlage (Bohn *et al.*, 1981). However, PNMT, the epinephrine-forming enzyme used to monitor *adrenergic* expression, is not detectable until E17.5, days after formation of the adrenal primordium (Bohn *et al.*, 1981). Although the pituitary–adrenocortical axis, through the mediation of glucocorticoid hormones, regulates the quantitative, ontogenetic increase in PNMT *after* it has been initially expressed, the mechanisms that evoke initial expression remain to be defined (Bohn *et al.*, 1981; Black, 1982). The developmental *increase* in PNMT molecule number, after initial expression, is associated with an increase in messenger RNA (mRNA) coding for the enzyme, suggesting that adrenergic development is regulated by glucocorticoids at the transcriptional level (Sabban *et al.*, 1982). Regardless of underlying mechanisms, it is apparent that most embryonic adrenomedullary precursors change from the noradrenergic to adrenergic phenotypes, by expressing new CA characters such as PNMT. In other developing populations, phenotypic plasticity is more striking, involving apparent selective loss of phenotypic characters.

Neural crest derivatives also populate the embryonic gut, giving rise to at least some of the enteric ganglia (Yntema and Hammond, 1945; Andrew, 1969, 1970; Le Douarin and Teillet, 1973, 1974). A population of cells in the intestine transiently expresses noradrenergic phenotypic characters. At E11.5, TH and DBH immunoreactivity, as well as CA histofluorescence, are detectable (Fig. 1) (Cochard *et al.*, 1978; Jonakait *et al.*, 1979). However, by E14.5, these endogenous CA traits are lost (Jonakait *et al.*, 1979). Nevertheless, the U_1 system for NE persists, suggesting that (1) this population does not die, but remains in the gut, and (2) selectively loses certain CA phenotypic characters while retaining others (Jonakait *et al.*, 1979). Such selectivity of expression strongly suggests that individual CA phenotypic traits are independently regulated. Finally, the intestinal cells exhibit responses to environmental factors that are typical of peripheral CA neurons: administration of nerve growth factor (NGF) increases levels and prolongs expression of CAs, while elevation of maternal glucocorticoids also prolongs catecholaminergic expression (Kessler *et al.*, 1979; Jonakait

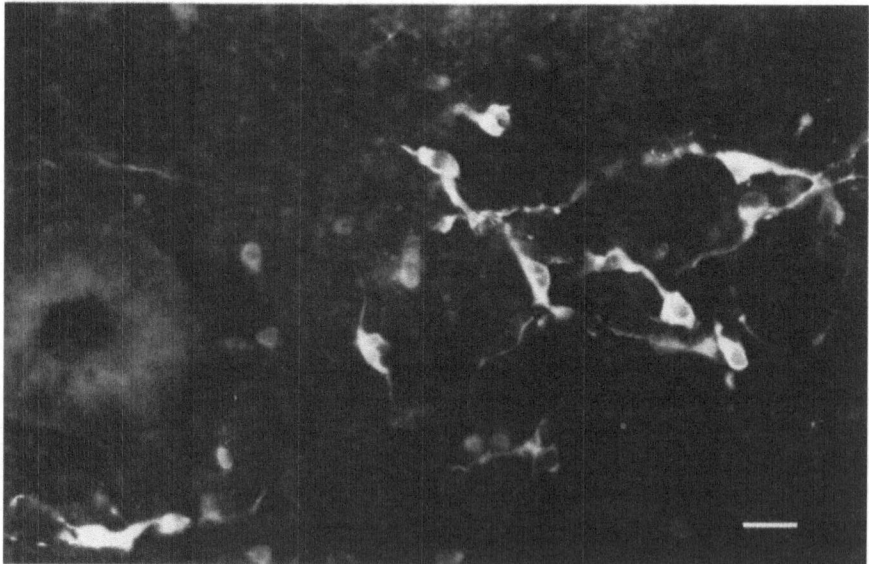

Figure 1. H-containing cells in the gut of a rat embryo at E12.5 (35 somites). Note the large cells with fluorescent processes and large, negatively staining nuclei. The cells are located in the gut mesenchyme, and the mucosa is faintly visible to the left. Scale bar: 25 μm. Photomicrograph by G.M. Jonakait.

et al., 1980, 1981). Consequently, this population of presumptive gut neuroblasts exhibits diverse characters shared by other peripheral CA neurons. We are at present attempting to identify the mature transmitter status of these cells.

Current studies in our laboratory suggest that transient phenotypic expression in the embryo is rather widespread, and not restricted to populations of neural crest origin. Cells in cranial sensory ganglion primordia, as well as rostral dorsal-root ganglion (DRG) anlagen, transiently express TH immunoreactivity (Jonakait *et al.*, 1982). Bright TH-immunofluorescent, bipolar cells, which extend processes into the primitive brainstem, are detectable in the trigeminal anlage at E11.0 (24 somites). By E11.5 (27–28 somites), TH-positive cells are evident in sensory ganglion primordia of the glossopharyngeal and vagal cranial nerves. By E12.0 (35–36 somites), only rare TH-positive cells are detectable in the rostral trigeminal, whereas TH is observed in the more caudal nodose and petrosal ganglia. At this stage, isolated bipolar, TH-containing cells are visible in the rostral DRGs. By E13.0 (46–48 somites), TH is no longer detectable in any of these structures; disappearance follows the rostrocaudal pattern of initial appearance. Even at this early stage of investigation, several conclusions may be warranted. Since some trigeminal neurons are derived from the crest, whereas nodose neurons are epibranchial placode derivatives, heterogeneous populations, differing embryologically, anatomically, and functionally, may exhibit transient phenotypic expression during development. Consequently, transient expression may represent a widespread phenomenon, occurring throughout the embryo. This contention is supported by the recent description of transient TH neurons in the spinal cord (Teitelman *et al.*, 1981a) and of transient CA expression in islet cells of the pancreas, which later express glucagon and insulin (Teitelman *et al.*, 1981b). Finally, the phenomenon of transient phenotypic expression is not restricted to CA characters, since (1) transient *cholinergic* expression occurs in crest derivatives (see above) and (2) transient serotonergic cells have been described in developing spinal cord (Cabana *et al.*, 1981).

A number of questions remain outstanding: What mechanisms underlie initial expression and subsequent disappearance of transmitter characters? What are the transmitter fates of these cells and, in fact, do all these populations survive to maturity? What are the functional implications, if any, of transient expression? Does transmitter plasticity persist into the postnatal period?

3. TRANSMITTER PLASTICITY IN THE NEONATE

Recent studies suggest that transmitter plasticity is not restricted to the prenatal period. Sympathetic neurons innervating the rat eccrine sweat glands convert from noradrenergic to cholinergic in the neonate, retaining, however, the U_1 system for α-methyl-NE (Landis, 1983; Landis and Keefe, 1980). During postnatal ontogeny, small granular vesicles, which store CAs, disappear in these sweat gland terminals, while TH and DBH immunoreactivity decreases. Further, administration of 6-hydroxydopamine to neonates, which selectively destroys CA neurons, renders TH, DBH, and the putative peptide transmitter vasoactive intestinal peptide undetectable in adult fibers. It may be concluded that these neurons convert from catecholaminergic to cholinergic and peptidergic phenotypes during postnatal development.

The foregoing studies suggest that transmitter phenotypic plasticity is a property of neurons throughout ontogeny, extending from the embryo to the fetus and neonate. It is logical to ask, consequently, whether phenotypic plasticity is strictly a developmental phenomenon or, alternatively, extends into maturity. In fact, is transmitter plasticity a lifelong property of at least some neuronal populations *in vivo?*

4. TRANSMITTER PLASTICITY DURING MATURITY

To examine transmitter plasticity in the mature rat *in vivo,* we employed the sympathetic superior cervical ganglion (SCG) as a model system. Conventional wisdom maintains that sympathetic neurons use only NE or ACh as transmitters and that fibers that innervate ganglia are cholinergic (Mayer, 1980). The recent discovery of SP in ganglion nerve fibers (Hökfelt *et al.,* 1977a,b), however, raised the possibility that transmitter organization is considerably more complicated. Nevertheless, these studies failed to detect SP in ganglion perikarya, suggesting that the positive fibers emanated from extraganglionic sources. SP does, however, appear to play a physiological role in sympathetic ganglia: the peptide is released by a high potassium pulse in a calcium-dependent manner, and application of SP to sympathetic neurons evokes membrane depolarization and neuronal discharge (Dun and Karczmar, 1979; Konishi *et al.,* 1979).

We initiated studies of SP in the SCG to address a number of specific issues. Initially, and most fundamentally, what is the source of SP-positive fibers in the adult rat ganglion? Do classic noradrenergic sympathetic neurons express peptidergic phenotypic characters during maturity? If so, what factors regulate expression of SP? In fact, do mature, postmitotic sympathetic neurons exhibit transmitter plasticity, varying the expression of peptidergic and CA characters quantitatively or even qualitatively?

To begin defining the source of SP in the SCG, ganglia were unilaterally decentralized (denervated) in adult rats, to determine whether peptidergic fibers simply entered the SCG with the preganglionic cholinergic trunk. Paradoxically, decentralization elicited a dramatic *increase* in SP after 12 days (Fig. 2) (Kessler and Black, 1982), suggesting that preganglionic fibers normally suppress SP. In

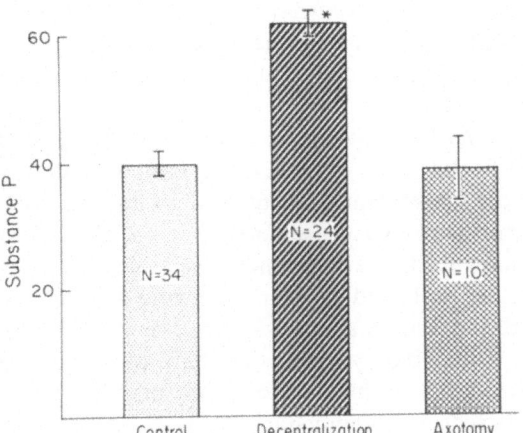

Figure 2. Effects of decentralization or axotomy on SP in the adult rat SCG. Unilateral decentralization or axotomy of the SCG was performed with the contralateral unoperated ganglion serving as control. At 12 days later, the ganglia were examined, with the sample size for each group indicated (N). SP is expressed as mean pg/ganglion ± S.E.M. (⊥⊤). *Differs from other groups at $p < 0.005$. Reprinted from Kessler and Black (1982) with permission.

contrast, postganglionic axotomy did not alter SP (Fig. 2), suggesting that (1) peptidergic fibers do not simply enter the SCG via this route and (2) retrograde factors are less important than orthograde effects in SP regulation.

To determine whether the effects of preganglionic nerves are mediated by transsynaptic transmission, rats were treated with chlorisondamine, a specific ganglionic-blocking agent. The drug prevents postsynaptic depolarization by competing with ACh for postsynaptic nicotinic receptor sites (Crimson *et al.,* 1955). Treatment significantly increased ganglion SP, reproducing the effects of surgical decentralization. Our observations suggest that presynaptic nerves normally decrease ganglion SP through the mediation of transsynaptic ACh and postsynaptic depolarization (Kessler and Black, 1982).

To ascertain whether increased presynaptic nerve impulse activity, conversely, decreases SP, rats were treated with phenoxybenzamine, an α-receptor-blocking agent that reflexly increases sympathetic impulse flow (Dontas and Nickerson, 1957). Treatment significantly decreased SP, suggesting that ganglion SP content is inversely related to impulse activity (Kessler and Black, 1982).

In summary, our studies suggest that presynaptic nerves decrease SP in postsynaptic sympathetic neurons of the adult SCG through the transsynaptic release of ACh and postsynaptic nicotinic stimulation. Transsynaptic factors also regulate ganglion SP in neonates. Decentralization of the neonatal SCG increases SP 36 hr postoperatively (Kessler *et al.,* 1981). It may be concluded that transsynaptic nicotinic stimulation decreases ganglion SP throughout life. This is directly relevant to the issue of transmitter plasticity during development and maturity, since it has long been recognized that the same, or similar, transsynaptic mechanisms *increase* the CA characters TH and DBH in sympathetic neurons of the SCG in the neonate and adult (for a review, see Black, 1982). Consequently, transsynaptic impulse flow may have opposite effects on catecholaminergic and peptidergic expression during development and maturity.

To define molecular mechanisms in greater detail, and to definitively localize SP in the SCG, we turned to a tissue-culture system. Since neonatal ganglia are more conveniently maintained in culture, this model system was employed. Short-term explant cultures were initially examined, obviating the need for addition of NGF, a confounding variable, to the medium. SP increased dramatically in the explanted ganglia, achieving a 5-fold rise by 12 hr a 20-fold rise by 24 hr, and a more than 50-fold rise by 48 hr (Fig. 3) (Kessler *et al.*, 1981).

Importantly, more than 85% of the immunoreactive SP coeluted with authentic SP on high-performance liquid chromatography (Kessler *et al.*, 1983). Consequently, authentic SP, and not some other peptide, was indeed being measured by our radioimmunoassay.

To determine whether the increase in SP was dependent on protein, RNA, or DNA synthesis, a number of metabolic inhibitors were added to the culture for a 12-hr period (Kessler *et al.*, 1981). Cycloheximide completely prevents the increase in SP, suggesting that protein synthesis is necessary. In contrast, arabinosylcytosine, an inhibitor of DNA synthesis, had no effect; actinomycin D and camptothecin, inhibitors of RNA synthesis, partially blocked the increase. To summarize, ongoing protein synthesis, and perhaps RNA synthesis, are required for the rise in SP.

The marked increase in SP in culture afforded an optimal opportunity to localize SP immunocytochemically. In fact, after 48 hr in culture, the principal (postsynaptic) ganglion neurons exhibited intense SP immunofluorescence (Fig.

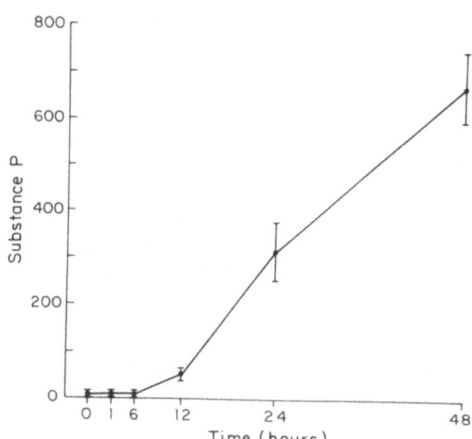

Figure 3. Time–course of ganglion SP accumulation in culture. Ganglia were placed on filter-paper rafts in Ham's nutrient mixture F12 with 10% fetal calf serum, penicillin (50 U/ml), and streptomycin (50 µg/ml). Cultures were maintained in an atmosphere of 95% air–5% CO_2 at nearly 100% relative humidity. Ganglia were examined after varying times in culture. SP is expressed as mean pg/ganglion ± S.E.M. (I) for 8 animals. Reprinted from Kessler *et al.* (1981) with permission.

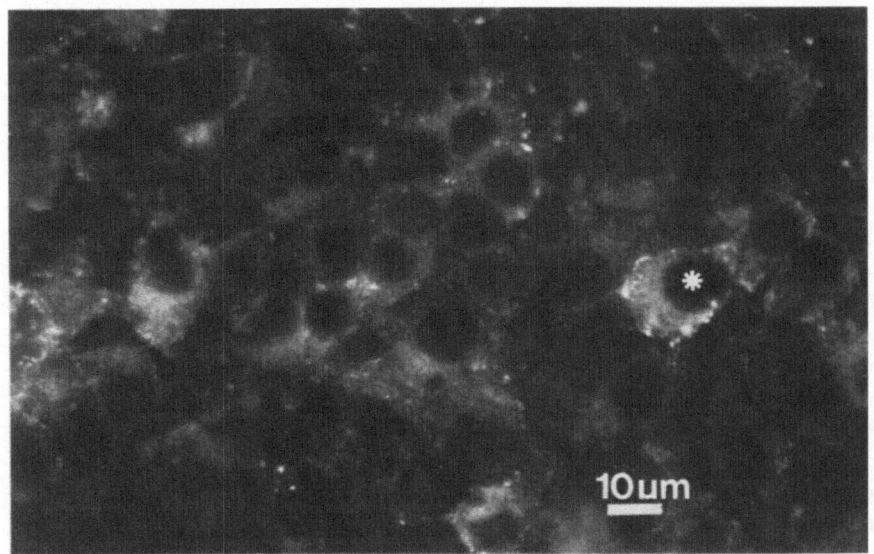

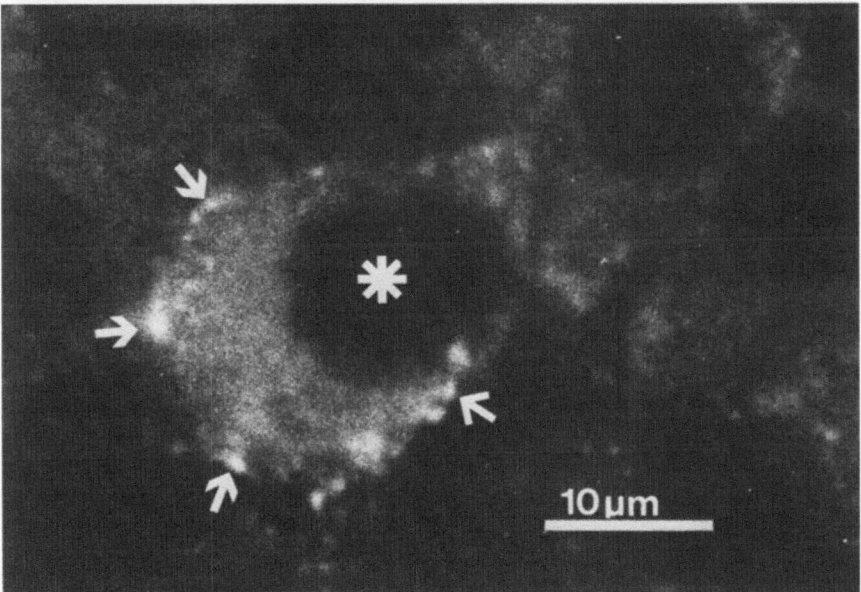

Figure 4. Demonstration of SP in principal neurons of the SCG. Ganglia were cultured as in Fig. 3 for 48 hr, but with added NGF (100 ng/ml). Sections (10 μm) were stained with antiserum to SP; control ganglia incubated with serum from nonimmunized rabbits were negative. (A) SP immunoreactivity was observed in most perikarya and in many beaded fibers coursing through the ganglia. The staining intensity varied from cell to cell. (B) Cell in (A) (✱) at higher magnification. Note that SP immunofluorescence is granular, filling the cytoplasm and leaving the nucleus clear. Apparent fluorescent boutons were observed impinging on somata of many SP-positive neurons (◊). Photomicrograph by M.C. Bohn.

4) (Kessler *et al.*, 1981). Consequently, classic sympathetic noradrenergic neurons do indeed express SP.

The effects of membrane depolarization were examined in culture, to define mechanisms by which impulse activity decreases SP *in vivo*(see above). Veratridine, which increases membrane sodium flux by binding to sodium channels (Ulbricht, 1969), completely blocked the rise in SP (Fig. 5) (Kessler *et al.*, 1981). Tetrodotoxin (TTX), which specifically antagonizes the ion effects of veratridine (Evans, 1972; Catterall and Nirenberg, 1973) prevented the influence of veratridine on SP (Fig. 5), suggesting that veratridine inhibits the SP increase by increaing sodium influx. On the other hand, membrane depolarization by high potassium also prevents the increase in SP (Kessler *et al.*, 1983), suggesting that membrane depolarization rather than sodium influx itself may mediate the effects of veratridine. Viewed in conjunction with the *in vivo* studies, these results suggest that transsynaptic impulses, mediated by (1) ACh release (2) nicotinic receptor stimulation, (3) postsynaptic membrane depolarization, and (4) sodium influx, decrease SP in principal (noradrenergic) ganglion neurons. Moreover, the inhibitory effects of depolarization are not simply due to increased *release* of SP from cellular stores, since the amount of peptide released by depolarization is only a fraction of the total increase in SP that normally occurs *in vitro* (Kessler *et al.*, 1983). Consequently, depolarization appears to inhibit *net synthesis* (synthesis less catabolism) of the peptide by SCG neurons.

Our studies indicate that classic postsynaptic sympathetic neurons, which were formerly thought to use only NE or ACh as transmitters also express the

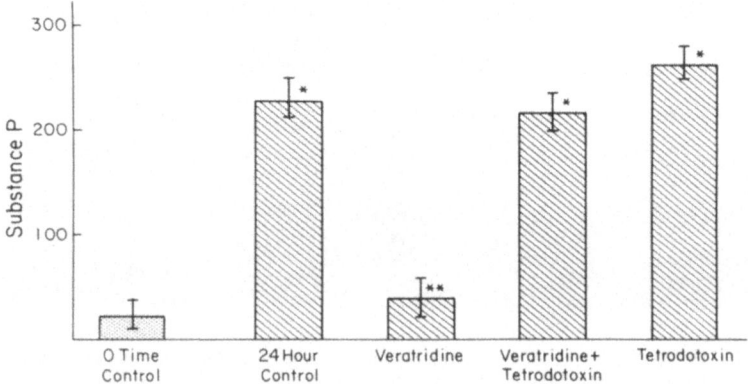

Figure 5. Effects of membrane depolarizaton on SP in the SCG. Ganglia were cultured as in Fig. 3, in the presence of veratridine (2×10^{-4} M), TTX (10^{-7} M), or both, and compared to uncultured ganglia (0 Time Control) and those cultured without additions (24 Hour Control). After 24 hr, groups of 8 ganglia were assayed for SP. *Differs from 0 Time Control at $p < 0.001$. **Differs from 24 Hour Control at $p < 0.001$. Reprinted from Kessler *et al.* (1981) with permission.

putative transmitter SP. Moreover, these postmitotic, principal-ganglion neurons in adults exhibit marked transmitter plasticity: their apparent putative transmitter, revealed immunocytochemically, is regulated by transsynaptic impulse flow.

Sympathetic activity itself is thus one stimulus that appears to govern transmitter metabolism in mature ganglion neurons. Transsynaptic stimulation, mediated by ACh, nicotinic receptor stimulation, and postsynaptic depolarization, decreases the net synthesis of SP. However, transsynaptic stimulation has long been known to biochemically induce TH and DBH in mature sympathetic neurons, thereby *increasing* NE biosynthesis (Molinoff *et al.,* 1970; Thoenen *et al.,* 1971; Otten *et al.,* 1973). Consequently, transsynaptic activation and postsynaptic depolarization exert diametrically opposite effects on CA and SP metabolism, and perhaps expression. We have not yet determined whether the effects on SP are qualitative as well as quantitative. Although SP is immunocytochemically undetectable in sympathetic perikarya *in vivo,* with normal impulse flow, and is detectable only in the denervated ganglion in culture, the appropriate methods are subject to severe threshold effects.

While it is possible that at least some sympathetic neurons do not express SP unless there is a marked reduction in transsynaptic stimulation and postsynaptic depolarization, additional experimentation is required. Nevertheless, it is abundantly clear that extracellular signals, such as ACh, may profoundly alter the transmitter status of the neuron. Consequently, mature neurons exhibit marked transmitter plasticity, analogous to developing neuroblasts and neurons.

5. OTHER PEPTIDES, OTHER POPULATIONS

We examined the putative peptide transmitter somatostatin (SS) in the SCG *in vitro,* to determine whether similar mechanisms regulate the expression and metabolism of different neuropeptides. SS exhibits a dramatic 6-fold increase in culture over 24 hr, and the rise is prevented by veratridine, reproducing the results with SP (Kessler *et al.,* 1983). TTX blocks the effects of veratridine on SS, suggesting that sodium influx and depolarization decrease SS as well as SP in the ganglion.

To determine, conversely, whether SP is similarly regulated in an entirely different neuronal population, we examined primary sensory neurons of the nodose cranial ganglion. In contrast to visceral motor neurons of the SCG, SP-containing, sensory neurons of the nodose are derived from the epibranchial placode (van Campenot, 1937), not from the neural crest. In summary, veratridine blocked the increase of nodose SP in culture, in a TTX-sensitive manner, entirely reproducing results obtained with the SCG.

Our observations indicate that multiple peptides, in neuronal populations that differ embryologically, anatomically, and functionally, may be subject to the same or similar regulatory mechanisms. More specifically, the state of neuronal depolarization, and transmembrane sodium flux, appear to regulate SP and SS content in diverse neuronal populations. Depolarization induces TH and DBH, and increases NE biosynthesis, while it decreases SP and SS in sympathetic neurons. It is apparent that expression of transmitter characters by mature neurons represents a dynamic, mutable process, dependent on extracellular signals and the physiological state of the neuron.

6. CONCLUSIONS AND PROSPECTS

Transmitter phenotypic characters such as TH, DBH, and PNMT, as well as the peptides SP and SS, vary over a wide range in developing and mature neurons *in vivo* and in culture. Extracellular signals that elicit depolarization and transmembrane sodium flux profoundly alter transmitter metabolism of mature, postmitotic neurons; the precise signals that govern plasticity in neuroblasts and developing neurons *in vivo* have yet to be defined. Moreover, underlying intracellular mechanisms that regulate transmitter plasticity remain to be characterized. Do individual neurons normally synthesize entirely different species of mRNA, coding for different transmitter characters at different times *in vivo?* Alternatively, do apparent changes in phenotypic characters simply reflect marked quantitative shifts in metabolism? Do neurons transcribe a variety of mRNAs continuously, but translate only selected species, elaborating different transmitters under the influence of different extracellular signals? In light of these and related unanswered questions, issues of quantitative vs. qualitative shifts in transmitter characters have yet to be resolved. Consequently, it is not yet clear whether the same or similar processes underlie apparent phenotypic plasticity of neurons in the embryo, neonate, and adult. Nevertheless, it is now evident that transmitter metabolism and perhaps expression are mutable processes, regulated by the physiological state of the neuron. The aforementioned studies suggest, moreover, that plasticity persists into maturity and may contribute to normal function of the adult nervous system *in vivo*.

ACKNOWLEDGMENTS. This work was supported by NIH Grants NS 10259, HD 12108, and NS 17814 and aided by grants from the Cerebral Palsy Association, the National Foundation–March of Dimes, and the Dysautonomia Foundation.

REFERENCES

Andrew, A., 1969, The origin of intramural ganglia. II. The trunk neural crest as a source of enteric ganglion cells, *J. Anat.* **105:**89–101.
Andrew, A., 1970, The origin of intramural ganglia. III. The "vagal" source of enteric ganglion cells, *J. Anat.* **107:**327–336.

Black, I.B., 1982, Stages of neurotransmitter development in autonomic neurons, *Science* **215:**1198–1204.

Black, I.B., and Patterson, P.H., 1980, Developmental regulation of neurotransmitter phenotype, *Curr. Top. Dev. Biol.* **15:**27–40.

Bohn, M.C., Goldstein, M., and Black, I.B., 1981, Role of glucocorticoids in expression of the adrenergic phenotype in rate embryonic adrenal gland, *Dev. Biol.* **82:**1–10.

Cabana, T., DiTirro, F., Ho, R., and Martin, G.G., 1981, The development of serotonergic pathways within the spinal cord: Studies using the North American opossum as an experimental model, *Soc. Neurosci. Abstr* **7:**181.

Catterall, W.A., and Nirenberg, M., 1973, Sodium uptake associated with activation of action potential ionophores of cultured neuroblastoma and muscle cells, *Proc. Natl. Acad. Sci. U.S.A.* **70:**3759–3763.

Cochard, P., Goldstein, M., and Black, I.B., 1978, Ontogenetic appearance and disappearance of tyrosine hydroxylase and catecholamines in the rat embryo, *Proc. Natl. Acad. Sci. U.S.A.* **75:**2986–2990.

Cochard, P., Goldstein, M., and Black, I.B., 1979, Initial development of the noradrenergic phenotype in autonomic neuroblasts of the rat embryo, *Dev. Biol.* **71:**100–114.

Coulombre, A.J., Johnston, M.C., and Weston, J.A., 1974, Conference on neural crest in normal and abnormal embryogenesis, *Dev. Biol.* **36:**fl–5.

Crimson, K., Tarazi, A., and Frazer, J., 1955, A new orally active quatenary ammonium ganglion blocking drug capable of reducing blood pressure, SV-3088, *Circulation* **11:**733–741.

Dontas, A.S., and Nickerson, M., 1957, Central and peripheral components of the action of "ganglionic" blocking agents, *J. Pharmacol. Exp. Ther.* **120:**147–159.

Dun, N.J., and Karczmar, A.G., 1979, Actions of substance P on sympathetic neurons, *Neuropharmacology* **18:**215–218.

Evans, H., 1972, Tetrodotoxin, saxitoxin, and related substances: Their applications in neurobiology, *Int. Rev. Neurobiol.* **15:**83–166.

Hökfelt, T., Elfvin, L.-G., Schultzberg, M., Goldstein, M., and Lift, R., 1977a, Occurrence of somatostatin-like immunoreactivity in some peripheral sympathetic noradrenergic neurons, *Proc. Natl. Acad. Sci. U.S.A.* **74:**3587–3591.

Hökfelt, T., Elfvin, L.-G., Schultzberg, M., Goldstein, M., and Nilsson, G., 1977b, On the occurrence of substance P fibers in sympathetic ganglia: Immunohistochemical evidence, *Brain Res.* **132:**2941.

Horstadius, S., 1950, *The Neural Crest, Oxford University Press, London,* p.111

Jonakait, G.M., Wolf, J., Cochard, P., Goldstein, M., and Black, I.B., 1979, Selective loss of noradrenergic phenotypic characters in neuroblasts of the rat embryo, *Proc. Natl. Acad. Sci. U.S.A.* **76:**4683–4686.

Jonakait, G.M., Bohn, M.C., and Black, I.B., 1980, Maternal glucocorticoid hormones influence neurotransmitter phenotypic expression in embryos, *Science* **210:**51–553.

Jonakait, G.M., Bohn, M.C., Markey, M., Goldstein, M., and Black, I.B., 1981, Elevation of maternal glucocorticoid hormones alters neurotransmitter phenotypic expression in embryos, *Dev. Biol.* **88:**288–296.

Jonakait, G.M., Markey, K.A., Goldstein, M., and Black, I.B., 1982, *Transient* expression of catecholaminergic traits in cranial nerve ganglia of the embryonic rat, *Soc. Neurosci. Abstr.* **8:**754.

Kessler, J.A., and Black, I.B., 1982, Regulation of substance P in adult rat sympathetic ganglia, *Brain Res.* **234:**182–187.

Kessler, J.A., Cochard, P., and Black, I.B., 1979, Nerve growth factor alters the fate of embryonic neuroblasts, *Nature (London)* **280:**141–142.

Kessler, J.A., Adler, J.E., Bohn, M.C., and Black, I.B., 1981, Substance P in principal sympathetic neurons: regulation by impulse activity, *Science* pp. 335–336.

Kessler, J.A., Adler, J. E., Bell, W. O., and Black, I.B., 1983, Substance P and somatostatin metabolism in sympathetic and special sensory ganglia *in vitro, Neuroscience* **9**:309–318.

Konishi, S., Isunov, A., and Otsuka, M., 1979, Substance P and noncholinergic excitatory synaptic transmission in guinea pig sympathetic ganglia, *Proc. Jpn. Acad. Ser. B.* **55**:525–530.

Landis, S.C., 1983, Development of cholinergic sympathetic neurons: Evidence for transmitter plasticity *in vivo, Fed. Proc. Fed. Am. Soc. Exp. Biol.* **42**:1633–1638.

Landis, S.C., and Keefe, D., 1980, Development of cholinergic sympathetic innervation of eccrine sweat glands in rat footpad, *Soc. Neurosci. Abstr.* **6**:379.

Le Douarin, N.M., 1980, The ontogeny of the neural crest in avian embryo chimaeras, *Nature (London)* **286**:663–669.

Le Douarin, N., and Teillet, M.A., 1973, The migration of neural crest cells to the wall of the digestive tract in avian embryo, *J. Embryol. Exp. Morphol.* **30**:31–48.

LeDouarin, N., and Teillet, M.A., 1974, Experimental analysis of the migration and differentiation of neuroblasts of the autonomic nervous system and of neurectodermal mesenchymal derivatives, using a biological cell marking technique, *Dev. Biol.* **41**:162–184.

Mayer, S.E., 1980, Neurohumoral transmission and the autonomic nervous system, in: *The Pharmacologic Basis of Therapeutics,* 6th ed. (A.C.Gilman, L.A. Gilman, eds.), Macmillan, New York. pp. 56–90.

Molinoff, P.B., Brimijoin, S., Weinshiboum, R., and Axelrod, J., 1970, Neurally mediated increase in dopamine-β-hydroxylase activity, *Proc. Natl. Acad. Sci. U.S.A.* **66**:453–458.

Otten, U., Paravicini, U., Oesch, F., and Thoenen, H., 1973, Time requirement for the single steps of trans-synaptic induction of tyrosine hydroxylase in the peripheral sympathetic nervous system, *Naunyn-Schmiedeberg's Arch. Pharmacol.* **280**:117–127.

Patterson, P.H., 1978, Environmental determination of autonomic neurotransmitter functions, *Annu. Rev. Neurosci.* **1**:1–18.

Sabban, E., Goldstein, M., Bohn, M.C., and Black, I.B., 1982, Development of the adrenergic phenotype: Increase in adrenal messenger RNA coding for phenylethanolamine-*N*-methyltransferase, *Proc. Natl. Sci. U.S.A.* **79**:4823–4827.

Teitelman, G., Gershon, M.D., Rothman, T.P., Joh, T.H., and Reis, D.J., 1981a, Proliferation and distribution of cells that transiently express a catecholaminergic phenotype during development in mice and rats, *Dev. Biol.* **86**:348–355.

Teitelman, G., Joh, T., and Reis, D.J., 1981b, Transformation of catecholaminergic precursors into glucagon (A) cells in mouse embryonic pancreas, *Proc. Natl. Acad. Sci. U.S.A.* **78**:5225–5229.

Thoenen, H., Kettler, R., Burkard, W., and Saner, A., 1971, Neurally mediated control of enzymes involved in the synthesis of norepinephrine: Are they regulated as an operational unit?, *Naunyn-Schmiedeberg's Arch. Pharmacol.* **270**:146–160.

Ulbricht, W., 1969, The effect of veratridine on excitable membranes of nerve and muscle, *Ergeb. Physiol.* **61**:18–70.

Van Campenot, E., 1937, Le rôle des placodes epiblastiques au cours du development embryonnaire du porc et du poulet, *Bull. Acad. R. Med. Belg.* **2**:169–184.

Varon, S.S., and Bunge, R.P., 1978, Trophic mechanisms in the peripheral nervous system, *Annu. Rev. Neurosci.* **1**:327–362.

Weston, J.A., 1963, A radioautographic analysis of the migration and localization of trunk neural crest cells in the chick, *Dev. Biol.* **6**:279–310.

Weston, J.A., 1970, The migration and differentiation of neural crest cells, *Adv. Morphog.* **8**:41–114.

Weston, J.A., and Butler, S.L., 1966, Temporal factors affecting localization of neural crest cells in the chicken embryo, *Dev. Biol.* **14**:246–266.

Yntema, C.L., and Hammond, W.S., 1945, depletions and abnormalities in the cervical sympathetic system of the chick following extirpation of the neural crest, *J. Exp. Zool.* **100**:237–263.

Nerve Growth Factor as a Model Growth Factor

Session Chairman: Eric M. Shooter

Chapter 9

Mechanisms of the Promotion of Neurite Outgrowth by Nerve Growth Factor

Lloyd A. Greene, David E. Burstein, James L. Connolly, Steven H. Green, P. John Seeley, and Michael L. Shelanski

1. INTRODUCTION

Nerve growth factor (NGF) has a variety of actions on its physiological targets, sympathetic and sensory neurons (Levi-Montalcini and Angeletti, 1968; Greene and Shooter, 1980; Thoenen and Barde, 1980). Among the most striking of these actions is the promotion of neurite outgrowth. This chapter will focus on the mechanisms by which NGF causes neurites to be initiated and maintained.

The neurite is a particularly singular structure. On one hand, it must be capable of exploratory and locomotive movements so that it can elongate and reach its targets. It is the neuronal growth cone that carries out these activities. On the other hand, for the neurite to maintain its greatly elongated shape, it must attain a degree of rigidity and stability along the neuronal shaft between the growth cone and the cell body. We shall develop herein the concept that NGF stimulates neurite outgrowth by regulating both growth-cone movement and neuritic stability. We shall also raise the issue that NGF carries out these two types of actions via distinct and separable mechanisms.

LLOYD A. GREENE, DAVID E. BURSTEIN, STEVEN H. GREEN, P. JOHN SEELEY, and MICHAEL L. SHELANSKI • Department of Pharmacology, New York University School of Medicine, New York, New York 10016.   JAMES L. CONNOLLY • Department of Pathology, Harvard Medical School, Boston, Massachusetts 02115.

2. Actions of Nerve Growth Factor on Growth Cones

Campenot (1977, 1982a,b) was among the first to demonstrate that NGF has local actions on the growth and maintenance of neurites. In these studies, sympathetic neurons were cultured under conditions in which the distal portion of neurites and their growth cones could be exposed to medium different from that bathing the rest of the cell. Among the findings made with this system was that neurites will not grow or be maintained within a local environment lacking NGF and that local withdrawal of NGF from the distal ends of neurites results in their degeneration. The notion that such local actions could occur rapidly and could affect movement of the growth cone derives from the work of Gundersen and Barrett (1980) and Griffin and Letourneau (1980). The former workers showed that growth cones of cultured chick sensory neurons would turn toward and follow a gradient of NGF concentration created by allowing the factor to leak slowly from a nearby micropipette. This tropic guidance occurred within 20 min and was suggested to be associated with an increase in filopodial growth on the side of the growth cone nearest the NGF source. The experiments of Griffin and Letourneau (1980) showed that a large, stepwise increment in NGF concentration from a minimal level caused cultured chick sensory neurons to retract the ends of their processes.

Work in our own laboratories has complemented and extended the aforecited studies by demonstrating rapid, local actions of NGF on the shape, surface organization, motility, and locomotion of growth cones. In these experiments, rat sympathetic neurons and PC12 rat pheochromocytoma cells were first cultured with NGF so that they extended neurites with growth cones. NGF was then withdrawn from the cultures for several hours and subsequently readded. The growth cones and neurites of cells maintained in each condition (continuous NGF exposure, NGF withdrawal, and NGF readdition) were observed by either time-lapse video recording (TLV recording) or scanning electron microscopy (SEM). As monitored by TLV recording (Seeley and Greene, 1983), the growth cones, in accordance with previous studies, were flattened, spread structures that were characterized by the continuous extension and withdrawal of motile fingerlike projections that probed the environment. The growth cones also exhibited locomotive movements that resulted in neurite elongation. Withdrawal of NGF led to progressive changes in growth cones and neurites. Within a few hours of NGF deprivation, the growth cones tended to lose their spread, flattened shape and to round up. The motile fingerlike projections ceased formation, and motility and locomotive elongation stopped. By 3–5 hr after removal of NGF, the growth cones appeared to be essentially "frozen." Readdition of NGF elicited a rapidly onsetting series of responses. Within several minutes, motile fingerlike projections reappeared at the neurite tip as well as along processes. With a latency of about 20 min, the growth cones reflattened and resumed locomotive extension. The response to NGF readdition could be generated locally and independently of the cellular synthetic machinery. This was established in PC12 cultures by demonstrating the occurence of the aforementioned responses in the presence of inhibitors of RNA and protein synthesis as well as in growth cones of processes that had been completely severed from their cell bodies with glass knives.

SEM studies have provided further information (Connolly *et al.*, 1980, 1982, and unpublished results). In the continuous presence of NGF, the growth cones of rat sympathetic neurons and PC12 cells were again seen to have a flat, spread-out configuration. In addition, well-defined ruffles were present on the surfaces of growth cones and, occasionally, along neurites. This contrasts with the absence of these structures on cell bodies. Removal of NGF for several hours led to the rounding up of growth cones and the loss of projections as noted in TLV recordings; in addition, withdrawal of NGF also resulted in the disappearance of ruffles. Readdition of NGF triggered the morphological changes noted above: the growth cones returned to a flattened shape and re-formed fingerlike projections. Furthermore, within 30 sec after NGF was readded, ruffles were again formed on the growth cones and neurites as well as on cell bodies. By 15 min of readdition and thereafter, ruffles, while present on growth cones, were diminished in number along neurites and again absent from cell bodies. Thus, within a short time, the growth cones and other areas of the neuron resumed the general appearance that they had prior to NGF withdrawal. Recent SEM studies have also been carried out with cultured chick embryo dorsal-root-ganglion neurons (J.L. Connolly, P.J. Seeley, and L.A. Greene, unpublished results). In the presence of NGF, these cells possess many more clearly defined filopodia than PC12 cells and rat sympathetic neurons, but do not display ruffles. Withdrawal of NGF results in the retraction of the filopodia, and NGF readdition rapidly triggers their reappearance.

To summarize the findings discussed above, NGF has rapid and local actions on growth cones and neurites that affect their shape, motility, and locomotion. There are several possible functional consequences of these responses. Membrane ruffles and filopodia have been associated with cell locomotive movements, and regulation of these may therefore, in turn, at least partially underlie the action of NGF on neurite elongation. By controlling the formation and motility of projections at growth cones, NGF may also regulate the capacity of neurites to probe and interact with their immediate environment. Furthermore, local changes of NGF concentration may not only rapidly promote or repress fiber growth, but also, by inhomogeneous presentation to neurite growth cones, affect chemotropic guidance.

3. STABILIZATION OF NEURITES AND EFFECTS OF NERVE GROWTH FACTOR

As suggested above, locomotion of cytoplasm or axoplasm to create cellular extensions seems necessary, but not sufficient, to account for the growth of neurites. The growing process must also be stabilized so as to maintain its form and rigidity. One major element of the cell cytoskeleton that is frequently suggested to play a major role in supporting and stabilizing the neurite is the microtubule (for a review, see Lasek and Shelanski, 1981). Neurites characteristically contain extensive parallel arrays of microtubules (Peters et al., 1976). Treatments that interfere with the formation of microtubules from soluble tubulin or that promote disassembly of microtubules also inhibit the capacity of neurons to initiate or maintain neurites (Daniels 1972, 1975; Yamada et al., 1970).

The PC12 clonal line of rat pheochromocytoma cells is a model system that has been used to study the mechanisms of NGF-promoted neurite outgrowth (Greene and Tischler, 1976, 1982) and, in particular, the role of microtubules in this phenomenon. When grown in culture medium without exogenously supplied NGF, PC12 cells resemble chromaffinlike cells in that they lack neurites, contain chromaffin granules, and synthesize, store, and release catecholamines. Addition of physiological levels of NGF to the cultures causes the cells to slowly extend (i.e., over a time–course of several days) processes that become varicose and branching, possess growth cones, and reach lengths of at least 1 mm. The PC12 line is particularly advantageous for examining the role of microtubules and other cytoskeletal elements in neurite outgrowth because the cells can be compared at various times before and after NGF treatment and neurite outgrowth. Experiments on this system discussed below suggest that NGF does indeed affect the properties of microtubules and that it does so in a manner that may confer neurite stability. A possible molecular basis for these effects is also discussed.

Ultrastructural examination of "naïve" PC12 cells (i.e., cells unexposed to NGF) reveals the presence of microtubules that are in a more or less random arrangement (Luckenbill-Edds et al., 1979; Tischler and Greene, 1978). During the first several days of NGF treatment, the processes may reach up to several hundred microns in length, but still possess few if any parallel arrays of microtubules (Luckenbill-Edds et al., 1979). It is only after several days to a week of NGF exposure that the processes begin to exhibit the packed arrays of microtubules that characterize mature neurites (Tischler and Greene, 1978; Luckenbill-Edds et al., 1979). Light-microscopic observations and TLV recordings provide the impression that these alterations in ultrastructure are paralleled by changes in behavior of the processes (P.J. Seeley and L.A. Greene, unpublished results). During the first days of NGF treatment, the processes are often likely to withdraw and to be resorbed by their cell bodies. This occurs either spontaneously or, to a greater degree, in response to a fall in temperature (from 37°C to ambient) or a rise in pH. In contrast, at later times of treatment when the neurites have reached lengths of greater than a few hundred microns, they appear to be substantially more stable with respect to withdrawal and resorption.

Biochemical studies have also indicated that long-term NGF treatment brings about changes in the properties of PC12 microtubules. Black and Greene (1982) exposed PC12 cultures to colchicine after various times (0, 1, and 21 days) of pretreatment with NGF. Colchicine binds to soluble tubulin and promotes disassembly and inhibits formation of microtubules (Sternlicht and Ringel, 1979). Sister cultures were assayed before and after colchicine treatment for their contents of polymerized tubulin (the vast bulk of which is most likely to be in the form of microtubules). Analysis of the data revealed that in cultures grown without NGF or for only 1 day with NGF, colchicine treatment led to the rapid (<30 min) loss of polymerized tubulin within the cells. In contrast, in cultures exposed to NGF for 3 weeks, nearly two thirds of the original content of polymerized tubulin was present, even after 6 hr of colchicine treatment. EM examination confirmed that parallel arrays of microtubules were still present in the neurites of the latter colchicine-treated cells. A plausible explanation for these observations is that the microtubules in long-term NGF-treated cultures have a much slower rate of depolymerization than those in cultures with no or short-term NGF treatment. That is, in cells without or with a short time of NGF exposure, the microtubules may have a sufficiently high rate of disassembly such that they are lost relatively rapidly when assembly is blocked with colchicine. In the cells treated with NGF for longer times, in contrast, the microtubules may be more stable and therefore more resistant to colchicine.

If the formation and stabilization of neurites may be mediated, at least in part, via changes in microtubules, then one might next inquire as to the molecular mechanism by which NGF causes such changes to occur. Isolation of microtubules by various means has revealed that in addition to tubulin, these structures also contain a discrete group of proteins known as microtubule-associated proteins (MAPs) (for a review, see Sloboda and Rosenbaum, 1982). A number of different types of experiments have established that MAPs greatly enhance the formation and stability of microtubules (cf. Murphy et al., 1977; Sloboda and Rosenbaum, 1979) and that they may also serve to cross-link microtubules to other elements of the cytoskeleton such as neurofilaments (Leterrier et al., 1982). On these bases, it has been suggested that MAPs may therefore play a critical role in regulating microtubule properties and, consequently, neurite outgrowth (cf. Olmsted and Lyon, 1981).

Of particular relevance to the aforestated point, recent studies have demonstrated that NGF brings about specific changes in the MAPs of PC12 cells (Greene et al., 1983). Examination of PC12 cell phosphoproteins by sodium dodecylsulfate (SDS)–polyacrylamide gel electrophoresis revealed a high-molecular-weight (>300,000-dalton) species, the level of which relative to that of other phosphoproteins significantly increases after 1–2 days of NGF exposure. Biochemical experiments showed that this phosphoprotein has the behavior of a MAP, and biochemical and immunological data indicated that it is identical or closely related to a specific high-molecular-weight MAP (designated MAP 1.2) that is also present in brain. The change in this MAP appears to be due entirely, or in large part, to increases in actual levels of protein rather than to enhancement of phosphorylation, and inhibitor studies suggested that this effect requires transcription. The NGF-induced increase in the MAP also occurred when the cells were blocked from extending neurites by being cultured in suspension. Hence, this response is not merely a consequence of neurite outgrowth. A search for additional MAPs that might be affected by NGF has been initiated (S. Green, unpublished data). These studies indicate that NGF causes few other changes in MAPs, with one exception being a species of apparent molecular weight in SDS of 34,000. Experiments are ongoing to determine whether this peptide represents an independent MAP or whether it is a fragment or subunit of MAP 1.2. (Green and Greene, 1983).

To summarize the findings discussed above, experiments with PC12 cells have been consistent with the notion that changes in the properties of microtubules play an important role in the NGF-promoted establishment and maintenance of neurites. Specific and restricted changes in the MAP content of target cells may, at least in part, underlie these actions.

4. PRIMING MODEL AND MULTIPLE ACTIONS OF NERVE GROWTH FACTOR

Comparison of the characteristics of NGF-promoted neurite outgrowth by naïve (i.e., without prior NGF exposure) and primed (i.e., pretreated with NGF for at least 5 days) PC12 cells has led to proposal of the "priming" model (Burstein and Greene, 1978; Greene et al., 1982). Initiation of outgrowth by naïve cells has a lag of about 18 hr, occurs slowly so that the average time for a cell to produce neurites is 2–3 days, is characterized by a slow, average net rate of process elongation (30–50 μm/day), and is suppressed by inhibition of RNA synthesis. Outgrowth in primed cultures is also NGF-dependent but, in contrast, has little if any lag, occurs rapidly so that 80–95% of the cells produce neurites within 24 hr, is characterized by a high initial rate of process elongation (~200 μm over the first day), and is not blocked by inhibitors of RNA synthesis. The model based on these data can be summarized as follows: (1) Initiation of outgrowth by NGF requires a transcription-dependent pathway that leads to the increased synthesis and accumulation of material that is necessary for formation of neurites. This accounts for the delayed, transcription-dependent nature of outgrowth by naïve cells. Also, the capacity of primed cells to produce neurites rapidly even when RNA synthesis is blocked is explained by the presence of material that was synthesized and accumulated during pretreatment with NGF. (2) In addition to the transcriptional action of NGF, neurite outgrowth also requires that the factor be present in a rapidly expressed nontranscriptional capacity. This accounts for the rapidity with which primed cells produce neurites and for their inability to undergo outgrowth when NGF is absent.

In the preceding discussion, it has been argued that NGF promotes neurite outgrowth by regulating both the movement and the stability of processes. How do these actions correlate with the priming model? The effects of NGF on growth cones were shown to occur with short latency and independently of the synthetic machinery for RNA and protein. In contrast, the effects of the factor on the properties of microtubules and on MAP levels occur with a delay of at least a day and, at least in the latter case, require RNA synthesis. It is therefore reasonable to suggest that the transcription-independent pathway of the priming model includes or corresponds to the growth-cone actions of NGF and that the transcription-dependent pathway includes or corresponds to the neurite-stabilizing actions of the factor.

5. CONCLUSIONS AND POSSIBLE RELEVANCE TO GENETIC AUTONOMIC DYSFUNCTION

This chapter has presented the viewpoint that the mechanism by which NGF promotes neurite outgrowth comprises two separable and distinct components. One involves rapid, local transcription-independent regulation of growth-cone shape, motility, and locomotion. The other involves a delayed, transcription-dependent stabilization of neurites that may be mediated via regulation of the properties of microtubules.

Might these mechanisms be related in some manner to inherited dysfunction of the peripheral nervous system? Evidence has been presented that peripheral neurons are maintained by trophic factors and that these factors may be supplied by target tissues (for reviews, see Varon and Bunge, 1978; Thoenen and Barde, 1980). A deficit in the neurite-generating mechanism would hence result not only in malformation of target innervation, but also in a reduced access to trophic factors. Ganglionic degeneration could occur as a consequence of the latter. It is therefore possible that the symptoms associated with familial autonomic dysfunction could result from a deficiency in neurite outgrowth that is distal to either the production or the primary response to trophic factors.

ACKNOWLEDGMENTS. Portions of the work described herein were supported by grants from the National Foundation–March of Dimes and the USPHS (NS 16036, NS 17888, and AM 26920. L.A.G. is the recipient of an Irma T. Hirschl Foundation Career Development Award. We thank Yvel Calderon and Julia Cohen for their aid in preparation of this manuscript.

REFERENCES

Black, M.M., and Greene, L.A., 1982, Changes in colchicine susceptibility of microtubules associated with neurite outgrowth: Studies with nerve growth factor-responsive PC12 pheochromocytoma cells, *J. Cell Biol.* **95:**379–386.

Burstein, D.E., and Greene, L.A., 1978, Evidence for both RNA-synthesis-dependent and -independent pathways in stimulation of neurite outgrowth by nerve growth factor, *Proc. Natl. Acad. Sci. U.S.A.* **75**:6059–6063.

Campenot, R.B., 1977, Local control of neurite development by nerve growth factor, *Proc. Natl. Acad. Sci. U.S.A.* **74**:4516–4519.

Campenot, R.B., 1982a, Development of sympathetic neurons in compartmentalized cultures. I. Local control of neurite growth by nerve growth factor, *Dev. Biol.* **93**:1–12.

Campenot, R.B., 1982b, Development of sympathetic neurons in compartmentalized cultures. II. Local control of neurite survival by nerve growth factor, *Dev. Biol.* **93**:13–21.

Connolly, J.L., Green, S.A., and Greene, L.A., 1980, Pit formation and rapid changes in surface morphology of sympathetic neurons in response to nerve growth factor, *J. Cell Biol.* **90**:176–180.

Connolly, J.L., Seeley, P.J., Green, S.A., and Greene, L.A., 1982, Effects of nerve growth factor on the growth cone, *J. Cell Biol.* **95**:204a.

Daniels, M.P., 1972, Colchicine inhibition of nerve fiber formation *in vitro, J. Cell Biol.* **53**:164–176.

Daniels, M.P., 1975, The role of microtubules in the growth and stabilization of nerve fibers, *Ann. N. Y. Acad. Sci.* **253**:535–544.

Green, S.H., and Greene, L.A., 1983, Increased low molecular weight microtubule-associated protein (MAP) in PC12 cells following long term exposure to NGF, *Soc. Neurosci. Abstr.* **9**:206.

Greene, L.A., and Shooter, E.M., 1980, The nerve growth factor: Biochemistry, synthesis and mechanism of action, *Annu. Rev. Neurosci.* **3**:353–402.

Greene, L.A., and Tischler, A.S., 1976, Establishment of a noradrenergic clonal line of rat adrenal pheochromocytoma cells which respond to nerve growth factor *Proc. Natl. Acad. Sci. U.S.A.* **73**:2424–2428.

Greene, L.A., and Tischler, A.S., 1982, PC12 pheochromocytoma cultures in neurobiological research, *Adv. Cell. Neurobiol.* **3**:273–414.

Greene, L.A., Burstein, D.E., and Black, M.M., 1982, The role of transcription-dependent priming in nerve growth factor promoted neurite outgrowth, *Dev. Biol.* **91**:305–316.

Greene, L.A., Liem, R.K.H., and Shelanski, M.L., 1983, Regulation of a high molecular weight microtubule-associated protein in PC12 cells by nerve growth factor, *J. Cell Biol.* **96**:76–83.

Griffin, C.G., and Letourneau, P.C., 1980, Rapid retraction of neurites by sensory neurons in response to increased concentrations of nerve growth factor, *J. Cell Biol.* **86**:156–161.

Gundersen, R.W., and Barrett, J.N., 1980, Characterization of the turning response of dorsal root neurites toward nerve growth factor, *J. Cell Biol.* **87**:546–554.

Lasek, R.J., and Shelanski, M.L. (eds.), 1981, Cytoskeletons and the architecture of nervous systems, *Neurosci. Res. Program Bull.* **19**:1–153.

Letterier, J.F., Liem, R.K.H., and Shelanski, M.L., 1982, Interactions between neurofilaments and microtubule-associated proteins: A possible mechanism for intraorganellar bridging, *J. Cell Biol.* **95**:982–986.

Levi-Montalcini, R., and Angeletti, P.U., 1968, Nerve growth factor, *Physiol. Rev.* **48**:534–569.

Luckenbill-Edds, L., Van Horn, C., and Greene, L.A., 1979, Fine structure of initial outgrowth of processes induced in a pheochromocytoma cell line (PC12) by nerve growth factor, *J. Neurocytol.* **8**:493–511.

Murphy, D.B., Johnson, K.A., and Borisy, G.G., 1977, Role of tubulin-associated proteins in microtubule nucleation and stabilization, *J. Mol. Biol.* **117**:33–52.

Olmsted, J.B., and Lyon, H.D., 1981, A microtubule-associated protein specific to differentiated neuroblastoma cells, *J. Biol. Chem.* **256**:3507–3511.

Peters, A., Palay, S.L., and Webster, H.deF., 1976, *The Fine Structure of the Nervous System: The Neurons and Supporting Cells,* W.B. Saunders, Philadelphia, 406 pp.

Seeley, P.J., and Greene, L.A., 1983, Short-latency local actions of nerve growth factor at the growth cone, *Proc. Natl. Acad. Sci. U.S.A.* **80**:2789–2793.

Sloboda, R.D., and Rosenbaum, J.L., 1979, Decoration and stabilization of intact, smooth-walled microtubules with microtubule-associated proteins, *Biochemistry* **18**:48–55.

Sloboda, R.D., and Rosenbaum, J.L., 1982, Purification and assay of microtubule-associated proteins (MAPs), in: *Methods in Enzymology, Vol. 85* (D.W. Fredriksen and L.W. Cunningham, eds.), Academic Press, New York, pp. 409–433.

Sternlicht, H., and Ringel, I., 1979, Colchicine inhibition of microtubule assembly via copolymer formation, *J. Biol. Chem.* **254**:10,540–10,550.

Thoenen, H., and Barde, Y.A., 1980, *Physiol. Rev.* **60**:1284–1385.

Tischler, A.S., and Greene, L.A., 1978, Morphological and cytochemical properties of a clonal line of rat adrenal pheochromocytoma cells which respond to nerve growth factor, *Lab. Invest.* **39**:77–89.

Varon, S.S., and Bunge, R.P., 1978, Trophic mechanisms in the peripheral nervous system, *Annu. Rev. Neurosci.* **1**:327–361.

Yamada, K.M., Spooner, B.S., and Wessells, N.K., 1970, Axon growth: Roles of microfilaments and microtubules, *Proc. Natl. Acad. Sci. U.S.A.* **66**:1206–1212.

Chapter 10

Cultured Sympathetic Neurons in the Study of Nerve Growth Factor Action

Edward Hawrot

1. LONG-TERM PRIMARY CULTURE OF DISSOCIATED RAT SYMPATHETIC NEURONS

The ability to maintain long-term cultures of rat sympathetic neurons in the absence of nonneuronal cells has facilitated a large number of studies dealing with factors important in neuronal growth and development (Patterson, 1978). The work of Mains and Patterson (1973a) demonstrated that the neurons in such cultures exhibited many of the properties of sympathetic neurons *in vivo*. Furthermore, the time–course of development of neurotransmitter functions in culture paralleled that seen *in vivo* (Mains and Patterson, 1973b). In the complete absence of nonneuronal influences, the isolated neurons differentiate along the expected adrenergic pathway, showing a developmental increase in the ability to synthesize, store, release, and transport norepinephrine (Patterson *et al.*, 1975; Burton and Bunge, 1975). The numerous axonal processes elaborated in these neuronal cultures are electrically active, and the cell bodies are sensitive to iontophoretically applied acetylcholine (O'Lague *et al.*, 1975, 1978a).

When dissociated sympathetic neurons are cocultured with nonneuronal cells (Patterson and Chun, 1974) or incubated with medium that has been conditioned by nonneuronal cells, the neurons develop cholinergic properties (Patterson and Chun, 1977). Furthermore, the neurons form cholinergic synaptic interactions

EDWARD HAWROT ● Department of Pharmacology, Yale University School of Medicine, New Haven, Connecticut 06510.

between themselves and with cocultured skeletal-muscle or cardiac myocytes (Furshpan *et al.*, 1976; Nurse, 1981; O'Lague *et al.*, 1978b). Because of these extremely interesting cellular and developmental interactions, cultured sympathetic neurons have formed the basis for a wide variety of biochemical, electrophysiological, and morphological investigations.

As is the case for their *in vivo* counterparts (Levi-Montalcini, 1976), rat sympathetic neurons in culture require nerve growth factor (NGF) for survival and development of their differentiated phenotype. Chun and Patterson (1977a) have shown that NGF induces neuronal survival, growth, and differentiation in a dose-dependent fashion. In the absence of nonneuronal cells, the dissociated sympathetic neurons are usually absolutely dependent on exogenously added NGF. Recent studies have suggested that certain components of extracellular matrix may be capable, under some conditions, of eliciting neurite outgrowth from sympathetic neurons even in the absence of NGF (Lander *et al.*, 1982). The neuronal response to the extracellular matrix is not sustained, however, since the neurons do not survive beyond 24 hr in culture in the continued absence of NGF. Besides influencing neurite growth, the culture substrate can also have dramatic effects on the phenotypic characteristics of sympathetic neurons (Hawrot, 1980). With respect to neurotransmitter-synthetic capability, NGF appears to play a permissive rather than an instructive role, since both cholinergic and adrenergic sympathetic cultures respond similarly to NGF supplementation or withdrawal (Chun and Patterson, 1977b).

Because of the enhanced accessibility possible in dissociated cultures of sympathetic neurons, a number of studies dealing with NGF action in sympathetic-neuron development and function can be more readily interpreted using dissociated cultures as opposed to organ culture. Not only can the concentrations of various agents be more readily correlated with biological response, but also the possible influence of nonneuronal cells can be minimized. Recent work on NGF-mediated tyrosine hydroxylase (TH) induction in dissociated sympathetic neurons has shown that the concentration of NGF required for maximal TH induction (100 ng/ml) is higher than that required for maximal neuronal survival (10 ng/ml) (Hefti *et al.*, 1982). This work was in general agreement with the study of Chun and Patterson (1977a), which also showed a distinction between the dose–response curves for NGF-dependent neuronal survival and stimulation of catecholamine synthesis. A similar distinction had been observed using sympathetic-ganglion explants, but higher concentrations of NGF were required to achieve maximal TH activity (Hill and Hendry, 1976). Using dissociated cultures, Hefti *et al.* (1982) further demonstrated that the NGF-mediated induction of TH was not due to the action of cyclic AMP or Ca^{2+} as second messenger. In addition, inhibitor studies suggested that the increase in enzyme molecules was being regulated at the post-transcriptional level.

2. USE OF ^{125}I-LABELED NERVE GROWTH FACTOR TO CHARACTERIZE THE NERVE GROWTH FACTOR RECEPTOR IN SYMPATHETIC-NEURON CULTURES

It is generally agreed that a complete understanding of the action of polypeptide hormones, such as NGF, will require the identification and characterization of the cell-surface receptor that mediates the hormone action. A number of laboratories are involved in studies on the NGF receptor in the attempt to elucidate thereby the molecular signals produced by NGF binding. One convenient cell system for studying the NGF receptor, and its involvement in mediating the multiple biological responses to NGF, is the PC12 pheochromocytoma rat cell line developed by Greene and Tischler (1976). The PC12 cell line is a tumor-derived line of sympathoblastlike cells that, on exposure to NGF, develop several neuronal features. Further descriptions of the use of this interesting cell line and its mutant variants are presented in Chapters 9 and 13. The NGF receptor in the PC12 cell line has been studied using ^{125}I-labeled NGF ($[^{125}I]$-NGF) (Herrup and Thoenen, 1979; Landreth and Shooter, 1980; Schechter and Bothwell, 1981).

The properties of the NGF receptor are also being investigated in embryonic ganglionic nerve cells and in membranes prepared from adult sympathetic ganglia (Sutter *et al.*, 1979; Olender and Stach, 1980; Massague *et al.*, 1981; for recent reviews, see Greene and Shooter, 1980; Harper and Thoenen, 1981; Yankner and Shooter, 1982). $[^{125}I]$-NGF has been affinity-cross-linked to membranes from adult rabbit superior cervical ganglia (SCGs) using a photoreactive agent (Massague *et al.*, 1981). Two membrane proteins with apparent molecular weights of 143,000 and 112,000 were thus specifically labeled and are believed to represent components of the NGF receptor. Since isolated membrane fractions were used for these studies, the possibility exists that these labeled proteins are derived from intracellular membranes. Determination of whether similar cell-surface-localized proteins can be labeled with $[^{125}I]$-NGF will require the appropriate studies with intact cells.

Since rat sympathetic neurons in culture are dependent on NGF for survival and full expression of their differentiated phenotype, they provide an attractive system for attempting to correlate biological responses to NGF with a characterization of the properties and distribution of specific NGF-binding sites. A further advantage of the culture system lies in the ability to maintain these cultures in the absence of nonneuronal cells. Thus, not only can the concentrations of NGF and other additives be carefully controlled, but also any observations can be confidently ascribed to the neurons alone. In addition, by the application of specialized culture chambers (Campenot, 1977) (see Section 3), the retrograde axonal transport of $[^{125}I]$-NGF can be studied under defined and fully manipulatable conditions. For these reasons, in collaboration with Dr. P. Claude (University of Wisconsin) and Dr. R.B. Campenot (Cornell), we have used [^{125}I]-NGF to study the binding and subsequent intracellular accumulation of NGF in dissociated cultures of rat sympathetic neurons (Claude *et al.*, 1982, 1983; Hawrot, 1982). In these studies, we have attempted to pursue a combined biochemical and morphological investigation of the cellular events that follow the binding of NGF to its receptor on the cell surface of rat sympathetic neurons. Since these primary neuron cultures exhibit many of the developmental characteristics normally seen in sympathetic neurons *in vivo*, it is anticipated that such studies will effectively complement other investigations utilizing the sympathoblastlike PC12 cell line.

We have used [^{125}I]-NGF of high specific activity to quantitate and visualize the presence of NGF receptors on rat sympathetic neurons in culture. Other laboratories have used indirect immunofluorescent, immunocytochemical, and autoradiographic techniques to localize NGF binding in cultures of mouse and chick sympathetic neurons (Kim *et al.*, 1979; Marchisio *et al.*, 1981). NGF (2.5 S) was prepared from the salivary glands of adult male mice according to the procedure of Bocchini and Angeletti (1969) with minor modifications (Claude *et al.*, 1982). The purified NGF was labeled with ^{125}I using a lactoperoxidase–glucose oxidase coupled system, as previously described (Claude *et al.*, 1982). Greater than 95% of the final [^{125}I]-NGF (specific activity 50–75 cpm/pg) could be precipitated by trichloroacetic acid (TCA) or by rabbit antiserum to NGF, and more than 95% of the label was associated with authentic NGF on sodium dodecyl sulfate–polyacrylamide gel electrophoresis (SDS-PAGE). The [^{125}I]-NGF was fully active in a biological assay measuring survival of rat sympathetic neurons in culture for 4–7 days. Under our culture conditions, half-maximal neuron survival was obtained with an NGF concentration of 10 ng/ml (4×10^{-10} M).

Sympathetic-neuron suspensions were prepared by mechanically dissociating the SCGs obtained from newborn rat pups (CD strain, Charles River). The dissociated cells were grown as a monolayer on modified culture dishes coated with a substrate of air-dried collagen. L15–CO_2 medium was supplemented with 5% rat serum and 7 S NGF (1 µg/ml), as previously described (Hawrot and Patterson, 1979). The growth of nonneuronal cells was suppressed by the use of the antimitotic agent cytosine arabinoside. In general, 1000–3000 neurons were plated per culture dish (approximately 0.8 cm^2 in area).

Initially, binding studies were performed with neurons that had been maintained in culture for 4–7 weeks. At this time, the neurons have elaborated a massive axonal network, and thus the amount of cellular material available for binding of NGF was maximal. Under these conditions, steady-state binding studies performed at 25 or at 1°C indicated that each neuron had approximately 2×10^7 receptors with an apparent affinity constant of 2–5 $\times 10^{-9}$ M (Claude *et al.*, 1982). Since it is estimated that in such neuron cultures, the cell-body plasma membrane is only 0.2–1% of the total cellular, primarily axonal, plasma membrane, the final surface density of receptors (15–70/μm^2 of membrane) is comparable to that of other membrane proteins.

Specificity of [^{125}I]-NGF binding was assessed by competition with excess unlabeled NGF, insulin, and epidermal growth factor (EGF). Excess unlabeled NGF alone was capable of reducing [^{125}I]-NGF binding by 80%. The collagen culture substrate alone was also capable of binding significant amounts of [^{125}I]-NGF. In fact, a component of the collagen binding was saturable with an apparent affinity constant of 10^{-6} M. The binding due to the collagen substrate could be substantially reduced, however, by a wash of suitable length (typically 1 hr at 25°C) that served to eliminate a large part of the binding to collagen without affecting the binding of [^{125}I]-NGF to the neurons (Hawrot, 1982). A comparison of the apparent kinetic constants for association and dissociation in neuronal cultures of similar age produced an apparent K_D of 1 × 10^{-9} M, in reasonably good agreement with the steady-state measurements (Hawrot, 1982). Autoradiographic studies at the light-microscopic (LM) level and at the electron-microscopic (EM) level (see below) supported the conclusion that [^{125}I]-NGF was binding to all regions of the neuronal cell surface. Figure 1 illustrates an experiment in which 7-day neuron cultures were incubated with 100 ng/ml [^{125}I]-NGF for 2 hr at 25°C in the presence or absence of 20 μg/ml unlabeled NGF, washed for 1 hr at 25°C, and then fixed with glutaraldehyde. The coverslips were coated with Kodak NTB2 emulsion, exposed for 5 days at 4°C, and then developed. Photographic grains were visualized by bright-field optics (Fig. 1B, D), while cellular details were observed with phase-contrast optics (Fig. 1A, C). In the presence of excess NGF, only background binding of [^{125}I]-NGF to the collagen was observed. In contrast, saturable [^{125}I]-NGF binding to all the axonal elements in the cultures was observed. Binding to cell bodies could not be confidently determined, since the photographic emulsion did not evenly coat the relatively thick cell bodies. Using similarly labeled neuronal cultures, Dr. P. Claude has carried out a localization of [^{125}I]-NGF using EM autoradiographic techniques (Claude *et al.*, 1982). These studies demonstrated that the density of binding sites on the axonal plasma membranes was approximately twice that on the plasma membrane of the cell bodies.

Recently, we have performed additional steady-state binding studies with sympathetic neurons that have been cultured for shorter periods of time. Much older cultures were used previously, and it is known that cultured sympathetic neurons become less dependent on NGF under continued culture (Lazarus *et al.*, 1976). Figure 2 is a Scatchard plot of one such study with sympathetic neurons after 12 days in culture. In this experiment, binding was carried out at 1°C, and an apparent affinity constant of 6 × 10^{-10} M (14 ng/ml) was obtained. Compared to older cultures, only about 10% as many specific binding sites were observed.

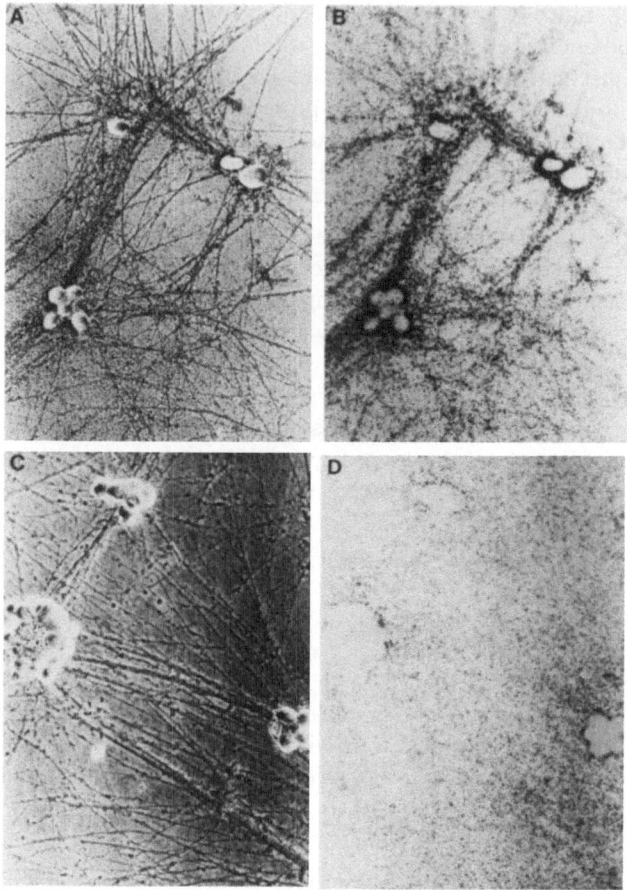

Figure 1. Autoradiographic analysis of NGF receptors in rat sympathetic-neuron cultures. Neurons were maintained in culture for 7 days. (A, B) Set of replicate cultures incubated with [^{125}I]-NGF (100 ng/ml) for 2 hr at 25°C. (C, D) Another set of cultures incubated with the same amount of [^{125}I]-NGF, but in the presence of excess unlabeled NGF (20 μg/ml). The cultures were washed for 1 hr in NGF-free medium, fixed with glutaraldehyde, and coated with photographic emulsion. After 5 days at 4°C, the emulsion was developed. (A, C) Cellular morphology observed under phase-contrast optics using a Zeiss ICM 405 photomicroscope. (B, D) Photographic grains observed with bright-field optics.

This is consistent, however, with the overall increase in the amount of cellular protein between 1 and 6 weeks in culture. It is unclear at present whether the higher-affinity binding observed with younger cultures represents a separate receptor entity or whether developmental changes in membrane configuration and composition could account for an approximately 10-fold variation in affinity

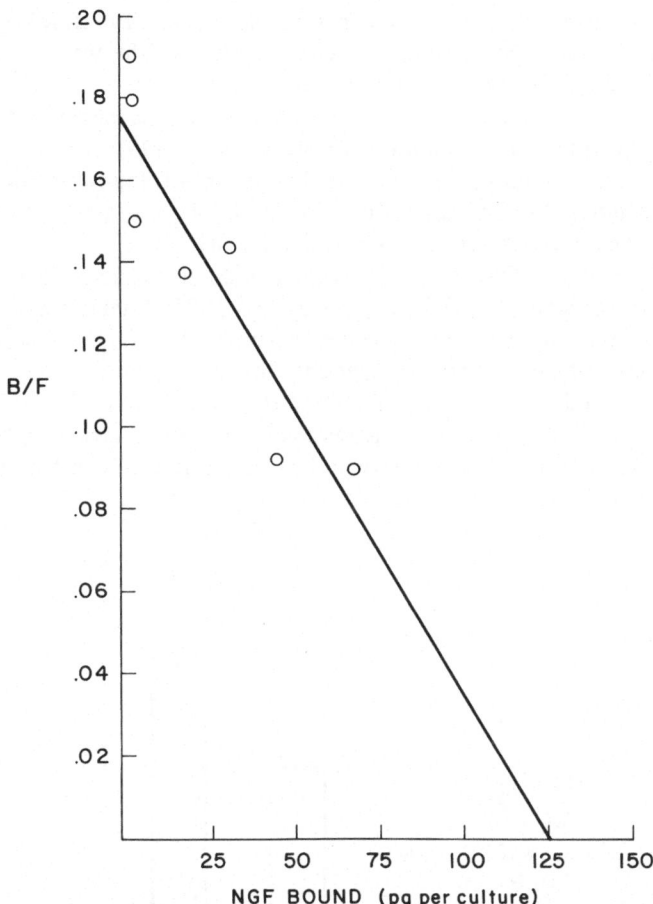

Figure 2.   Steady-state binding at 1°C of [^{125}I]-NGF to 12-day sympathetic cultures. After 12 days in culture, sympathetic neurons were washed with NGF-free medium and incubated with various concentrations of [^{125}I]-NGF as previously described (Claude *et al.*, 1982). After 24 hr at 1°C, the cultures were washed and the cell-bound radioactivity was determined. Nonspecific binding was determined in parallel cultures treated with 20 μg/ml of unlabeled NGF. The nonspecific binding was subtracted from the total binding observed to arrive at the specific binding (B) component. Aliquots of the incubation mixture were taken for a direct determination of the free (F) concentration of [^{125}I]-NGF. The binding data were presented as a Scatchard plot (Bylund, 1980).

constant over the course of several weeks in culture. If two receptor classes persist in older cultures, it would be difficult to detect them reliably on Scatchard plots if indeed the higher-affinity receptor accounts for only 10% of the maximal binding. In any case, the affinity constants that we have obtained are within the range of NGF concentrations known to be important for survival of sympathetic neurons and for the mediation of TH induction (Hefti *et al.*, 1982). It is important to note that all our binding measurements involve a 1-hr wash to decrease nonspecific binding. Presumably, any potential "fast"-dissociating binding sites, as are seen in chick embryonic sensory and sympathetic nerve cells, would dissociate under our wash conditions (Sutter *et al.*, 1979). Thus, we cannot strictly rule out the additional presence, in these neuron cultures, of low-affinity, rapidly dissociating NGF-binding sites. The physiological significance of the low-affinity sites in chick sensory nerve cells is unresolved.

There is evidence that the NGF receptor is down-regulated in PC12 cells (Calissano and Shelanski, 1980; Shooter *et al.*, 1981; Yankner and Shooter, 1982). We wanted to determine whether a similar phenomenon occurs with primary cultures of rat sympathetic neurons. Since the neuron cultures require NGF for continued survival, we examined this question by determining the number of NGF-binding sites on cultures that had been incubated in NGF-free medium ("starved") to allow any rebound expression of cell-surface receptors down-regulated during growth in the normal growth medium. The binding of

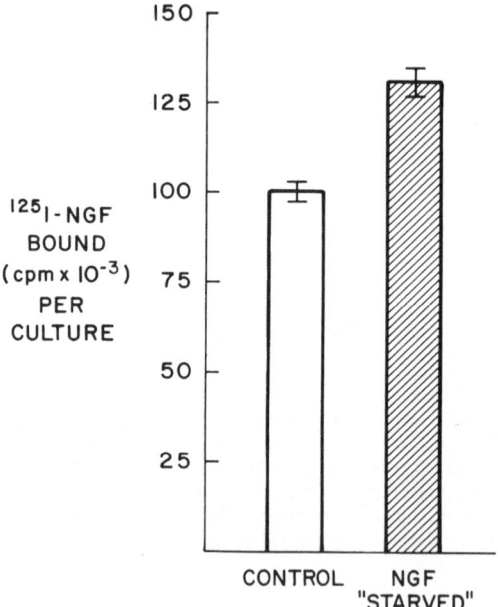

Figure 3. Regulation of NGF receptors in sympathetic-neuron cultures. After 20 days of culture, one set of cultures was washed free of NGF and incubated for 26 hr in NGF-free medium ("starved"). A replicate set of control cultures was incubated with 2 μg/ml of NGF (2.5 S) for 21 hr to maximize any possible down-regulation of NGF receptors. After the NGF was washed out, the specific binding of [^{125}I]-NGF was determined ([^{125}I]-NGF at 200 ng/ml, 10-hr incubation at 1°C).

[^{125}I]-NGF to "starved" cultures was compared to that of cultures rapidly washed free of NGF just prior to the binding assay. Figure 3 shows that although in many cases a slight (25–30%) increase in [^{125}I]-NGF binding is observed in "NGF-starved" cultures, this increase would correspond to an apparent down-regulation of only 23%. This level of down-regulation is rather meager compared to the very dramatic decrease (80%) seen in the EGF system. (Aharonov *et al.*, 1978). The observations made with these primary cultures thus differ significantly from those reported for PC12 cells. The lack of a marked down-regulation in the NGF receptor of sympathetic neurons would suggest either that the receptor is rapidly recycled or replenished from an intracellular reservoir or, possibly, that the receptor is in fact not internalized along with the NGF. In several other receptor-mediated endocytotic systems, there is evidence for extremely rapid ($t_{1/2}$ = 2–4 min) recycling of endocytosed receptors (Ciechanover *et al.*, 1983).

3. RETROGRADE TRANSPORT OF ^{125}I-LABELED NERVE GROWTH FACTOR IN RAT SYMPATHETIC-NEURON CULTURES

A number of studies have demonstrated that sympathetic-nerve terminals *in vivo* are capable of taking up specific macromolecules, including NGF, and retrogradely transporting these agents to the neuronal cell body (Hendry *et al.*, 1974a,b; Iversen *et al.*, 1975; Johnson *et al.*, 1978; Schwab and Thoenen, 1977; Schwab, 1977; Schwab *et al.*, 1979; Dumas *et al.*, 1979). The specificity of retrograde transport relies on the presence of specific binding sites on the nerve-terminal membrane. On internalization, NGF is retrogradely transported within the axon in membrane-limited organelles (Schwab, 1977; Schwab and Thoenen, 1977; Schwab *et al.*, 1979). It is currently believed that sympathetically inner-vated target organs produce endogenous NGF that is then taken up by nerve terminals and transported to the neuronal cell bodies to serve a neuronotrophic function (Harper and Thoenen, 1981; Schwab *et al.*, 1982). The numerous *in vivo* studies have been of great value in elucidating the biological significance of the retrograde transport of NGF.

Since an accurate determination of the concentration–response character-istics of retrograde transport is difficult *in vivo*, we decided to examine the retrograde transport of NGF using rat sympathetic-neuron cultures. This analysis was made possible by the inventive work of Campenot (1977), who developed a compartmentalized culture system for sympathetic neurons. In such compart-mentalized cultures, dissociated neurons were plated onto a collagen substrate in the central chamber of a three-compartment culture dish (Campenot, 1977). On continued culture, the neuronal processes extend under the partitions into two adjacent independent compartments. In this system, the medium that bathed each of the three compartments could be manipulated independently of the other compartments (Campenot, 1979). Campenot has shown that the neuronal cell bodies in the central chamber will survive as long as the nerve endings in a side chamber are exposed to NGF. In contrast, nerve endings deprived of NGF will deteriorate even though the cell bodies and nerve endings in the opposite side chamber are continuously exposed to NGF (Campenot, 1979). Thus, NGF plays a local function in maintaining axonal integrity, and the presence of NGF in one region of a neuronal arborization does not rescue axonal regions deprived of NGF.

Using rat sympathetic neurons grown in compartmentalized cultures, we were able to demonstrate retrograde axonal transport of $[^{125}I]$-NGF in agreement with the *in vivo* studies (Claude *et al.*, 1982). $[^{125}I]$-NGF at various concentrations, in the presence or absence of excess unlabeled NGF, was added to the outer well of a three-compartment culture dish containing neurites emanating from the neuronal cell bodies in the central compartment. At various times, individual cultures were washed to remove the $[^{125}I]$-NGF and dismantled, and then the radioactivity associated with the cell bodies was determined. The rate of appearance of $[^{125}I]$-NGF in the cell bodies was consistent with an axoplasmic transport rate of 3 mm/hr, similar to that observed *in vivo*. Transport was inhibited by colchicine and appeared to be due to a specific interaction with saturable receptors, since excess unlabeled NGF reduced $[^{125}I]$-NGF transport by 90%. In addition, excess insulin or EGF had no effect on the retrograde transport of $[^{125}I]$-NGF. Bathing the neuronal cell bodies in excess unlabeled NGF also did not reduce the amount of $[^{125}I]$-NGF retrogradely transported from the outer chamber.

After about 8 hr of continued exposure of nerve endings to $[^{125}I]$-NGF, a steady state was reached in which retrograde transport was balanced by metabolism and release of radioactivity into the bathing medium. Under these steady-state conditions, approximately 60% of cell-body-associated radioactivity was TCA-precipitable and migrated along with authentic NGF on SDS-PAGE. The specific, retrograde transport of $[^{125}I]$-NGF could be detected with as little as 0.5 ng/ml (2×10^{-11} M) in the outer chamber and appeared to plateau at a concentration of approximately 100 ng/ml (4×10^{-9} M). In contrast, the nonsaturable accumulation of $[^{125}I]$-NGF in the central compartment increased linearly with concentration (Hawrot, 1982). Even after extended periods of continuous incubation of nerve endings with $[^{125}I]$-NGF, no radioactivity was ever translocated past the neuronal cell bodies into the neurites in the chamber directly opposite the incubation chamber. This lack of anterograde transport of $[^{125}I]$-NGF from the cell bodies is consistent with the *in vivo* studies of Brunso-Bechtold and Hamburger (1979), who showed that in embryonic chick, NGF was retrogradely transported to the neuronal cell bodies in the dorsal-root ganglia, but did not move beyond the sensory ganglia into the spinal cord. Using similar compartmentalized culture systems, other laboratories have been able to study the retrograde transport of horseradish peroxidase (HRP) isoenzymes (Chan *et al.*, 1981) or of [³H]adrenaline within the axons of sympathetic neurons (Schwab and Thoenen, 1982).

Increasing neuronal activity did not affect the amount of NGF retrogradely transported. Compartmentalized cultures were chronically stimulated electrically under conditions that produce marked effects on neurotransmitter choice and development in sympathetic neurons (Walicke *et al.*, 1977). Such cultures did not accumulate any more [¹²⁵I]-NGF than the unstimulated controls (Hawrot, 1982). The addition of the lectin wheat germ agglutinin (WGA) (10 μg/ml) together with [¹²⁵I]-NGF did, however, reduce the retrograde transport of [¹²⁵I]-NGF consistent with the observed reduction of [¹²⁵I]-NGF binding to sympathetic membranes or solubilized receptor by WGA (Costrini and Kogan, 1981). Furthermore, the retrograde transport of [¹²⁵I]-NGF appeared to be regulated by the external NGF concentration. When nerve endings were incubated in NGF-free medium for 24 hr, a test dose of 2 ng/ml of [¹²⁵I]-NGF resulted in a 5-fold increase in the amount of [¹²⁵I]-NGF transported in 12 hr as compared to suitable washed, but unstarved, controls (Hawrot, 1982). This effect did not appear to be due to an increase in NGF-binding sites, since direct binding studies show only a 30% increase in receptor binding (Fig. 3).

4. INTRACELLULAR FATE OF ¹²⁵I-LABELED NERVE GROWTH FACTOR IN PRIMARY NEURON CULTURES

Having prepared a fluorescent conjugate of NGF, Levi *et al.* (1980) demonstrated that the NGF–receptor complex appeared to cluster on PC12 or embryonic chick sensory ganglia cells. At 37°C, these patches are subsequently endocytosed. In other receptor-mediated endocytosis systems, such as EGF, α-macroglobulin, or low-density lipoprotein, the internalized material is localized first within coated pits and then in coated vesicles (Goldstein *et al.*, 1979). No direct information is currently available on the possible role of coated vesicles or pits in the internalization of NGF. *In vivo*, during retrograde axonal transport by rat sympathetic neurons, NGF is localized within membrane-delimited compartments, such as smooth vesicles, cisternae, and tubules (Schwab, 1977; Schwab and Thoenen, 1977; Schwab *et al.*, 1979). Once the NGF arrives in the cell body, it is localized to lysosomal structures. No labeling of nuclei or Golgi was observed *in vivo*. In contrast to these EM morphological studies, a number of other studies utilizing LM or subcellular fractionation have suggested that a large fraction of internalized NGF is localized to the nucleus or the nuclear membrane (Andres *et al.*, 1977; Johnson *et al.*, 1978; Yankner and Shooter, 1979; Marchisio *et al.*, 1980).

The issue of intracellular localization was addressed further using primary cultures of rat sympathetic neurons. Dr. P. Claude, using EM autoradiographic techniques, calculated the labeling density of various intracellular organelles to compare the accumulation of label in various intracellular sites (Claude *et al.*, 1982). The tabulated labeling density represents the percentage of total grains assigned to a particular organelle category divided by the fractional area of that organelle type as determined in cross sections. Of the [^{125}I]-NGF internalized into neuronal cell bodies following a 1-hr incubation at 36°C, the highest concentration of grains was associated with multivesicular bodies and lysosomal organelles. The labeling densities of cytoplasm, nucleus, and Golgi apparatus were uniformly low.

The distribution of intracellular [^{125}I]-NGF was also examined in compartmentalized sympathetic-neuron cultures after 5, 8, or 24 hr of retrograde transport (Claude *et al.*, 1982, 1983). Within the proximal neurites of such compartmentalized cultures, not directly exposed to exogenous NGF but involved in the retrograde axonal transport of [^{125}I]-NGF, grains were predominantly associated with lysosomes and multivesicular bodies. In addition, a large number of grains were found in association with elongated, membrane-limited profiles. These structures included tubular bodies and elongated multivesicular bodies and were in some cases very similar in appearance to smooth endoplasmic reticulum (Claude *et al.*, 1983). In chick optic nerve, elongated membrane-limited structures, similar to but distinct from smooth endoplasmic reticulum, appear to be involved in the retrograde transport of HRP (LaVail *et al.*, 1980). Although vesicles made up a large component of the axonal fractional volume, they were not preferentially labeled.

At the level of the cell body, most of the retrogradely transported [^{125}I]-NGF was concentrated in lysosomes and multivesicular bodies. Lysosomes contained 45–60% of the cell-body-associated grains, while the multivesicular bodies accounted for 5–10% of the grains but had the greatest labeling density (Claude *et al.*, 1983). Even after 24 hr of retrograde transport, no evidence was obtained for a nuclear association of internalized [^{125}I]-NGF. Because of the suggestion that lysosomotropic agents such as chloroquine or methylamine might enhance the level of association of [^{125}I]-NGF with nuclei within PC12 cells (Shooter *et al.*, 1981), the effect of these agents was examined in sympathetic-neuron cultures. Neurons, incubated for 22 hr with [^{125}I]-NGF and 0.05 mM chloroquine or 10 mM methylamine, were processed for EM autoradiography (Claude *et al.*, 1983). With each of the two drugs, lysosomes appeared swollen with membranous material, but were still highly labeled with [^{125}I]-NGF. No further increase in the labeling of multivesicular bodies was observed, nor was there any evidence for nuclear accumulation of NGF. Furthermore, even when dissociated sympathetic neurons from newborn rats were plated directly into [^{125}I]-NGF-supplemented medium and grown for 4 days, no accumulation of [^{125}I]-NGF into nuclei was apparent on EM autoradiographic analysis.

The autoradiographic studies on rat sympathetic neurons in culture thus support, in general, the earlier observations made with neuronal retrograde transport of [125I]-NGF *in vivo* (Schwab, 1977; Schwab and Thoenen, 1977; Schwab *et al.*, 1979). [125I]-NGF was contained primarily within membrane-limited compartments, and no significant accumulation of label was observed within the cytoplasm or with the cell nucleus. Consistent with these observations are the reports that the direct introduction of NGF into the cytoplasm does not induce neurite outgrowth in PC12 cells (Heumann *et al.*, 1981; Huttner and O'Lague, 1981). Similarly, introduction of antibodies to NGF into the cytoplasm does not prevent the normal biological action of exogenously added NGF (Heumann *et al.*, 1981). Furthermore, using both EM autoradiography and subcellular fractionation techniques, Rohrer *et al.* (1982) have shown that under normal growth conditions, [125I]-NGF is taken up by PC12 cells, but is not transferred to the nuclear membrane or the nuclear chromatin.

5. BINDING OF 125I-LABELED NERVE GROWTH FACTOR TO CULTURES OF VARIOUS NEURAL-CREST AND PLACODE DERIVATIVES

In those cases in which NGF elicits a cellular response, the effects can generally be shown to be mediated by specific receptors on the cell surface. Thus, the identification of specific binding sites for NGF on cells suggests the potential for a biological response. Furthermore, the cellular expression of NGF receptors might provide a valuable phenotypic characteristic to be used in the study of neuronal development. Along these lines, we have been interested in the presence of NGF-binding sites on various cellular derivatives of the embryonic neural crest and ectodermal placode. These embryonic structures are largely responsible for the formation of the peripheral nervous system (PNS).

As in the case of sympathetic neurons, we have used autoradiographic techniques to examine the specific association of [125I]-NGF with primary cultures of neurons derived from neural crest or placode. In these studies, we examined cultures of rat sensory neurons obtained from dorsal-root ganglion, trigeminal ganglion, and nodose ganglion. The sensory-neuron cultures were kindly provided by Dr. P.I. Baccaglini and Dr. E. Cooper (Harvard Medical School). The neural-crest-derived dorsal-root-ganglion neurons bind [125I]-NGF, as expected from the fact that NGF is required for neuronal survival (Fig. 4A, B). Trigeminal-neuron cultures also require NGF, and, similarly, nearly all the neuronal processes in these cultures could be shown to bind [125I]-NGF (Fig. 4C, D). Since the neurons of the trigeminal ganglion are believed to be derived from both the neural crest and the placode (Noden, 1978; Le Douarin *et al.*, 1981; Ayer-Le Lièvre and Le Douarin, 1982), the possibility could not be ruled out that only the neural-crest-derived neurons survived the culture conditions (Baccaglini and Hogan, 1983).

At least in avian species, the neurons of the nodose ganglion appear to originate exclusively from the placodes (Ayer-Le Lièvre and Le Douarin, 1982). Dissociated cultures of nodose neurons obtained from newborn rats require NGF for survival (Baccaglini and Cooper, 1982). When such cultures were incubated with [^{125}I]-NGF, it appeared that essentially all the neuronal processes contained NGF-binding sites (Fig. 4E, F). The presence of NGF receptors on the nodose neurons is consistent with the observations of Hedlund and Ebendal (1980) that murine NGF promotes survival and neurite outgrowth in chick embryo nodose-ganglion explants. In contrast, however, Lindsay (1979) has indicated that dissociated embryonic rat nodose neurons can survive in culture for 6 days in the absence of NGF and in the presence of antibodies to NGF when the neurons are cocultured with astrocytes. In addition, in the autoimmune model of Johnson *et al.* (1980), newborn guinea pigs exposed to antibodies to NGF *in utero* exhibited no change in nodose-ganglion cell number, whereas both dorsal-root ganglia and superior cervical ganglia were severely affected.

It is possible that nodose neurons are only transiently dependent on NGF *in vivo* and that other trophic factors can replace the NGF requirement. For example, the loss, in culture, of NGF receptors by chick dorsal-root-ganglion sensory neurons grown in the presence of rat brain extract has recently been demonstrated with autoradiographic techniques (Rohrer and Barde, 1982). The observation, described here, that cultured rat nodose neurons exhibit NGF-binding sites suggests that NGF may play an important role in the development of some placode-derived neurons.

It has been suggested that the molecular defect responsible for the abnormal development of the PNS in familial dysautonomia may lie at the level of NGF or the NGF receptor (Pearson, 1979). To gain more information concerning the characteristics of the human NGF receptor, we have been studying the binding of [^{125}I]-NGF to human melanocytes and melanoma cells. Melanocytes are derived from laterally migrating neural-crest cells (Weston, 1970). Furthermore, several human melanoma cell lines contain high levels of NGF receptors (Fabricant *et al.*, 1977). One such identified line, A875, has approximately 7×10^5 receptors per cell with an apparent affinity constant of 10^{-9} M. In addition, two of the human melanoma cell lines with the greatest density of NGF receptors were isolated from tumors that had metastasized to the central nervous system (Fabricant *et al.*, 1977), suggesting a possible link between NGF receptor expression and melanoma tropism. NGF receptors have also been identified on several human neuroblastoma cell lines (Perez-Polo *et al.*, 1979; Sonnenfeld and Ishii, 1982). NGF can induce neurite outgrowth in responsive human neuroblastoma cell lines, but is, in general, incapable of altering growth rate (Sonnenfeld and Ishii, 1982).

To determine the binding of [^{125}I]-NGF to human melanoma cells, cultures were treated with 1 mM ethylenediamine tetraacetic acid (EDTA) to dissociate the cells. Binding studies were then carried out, at either 1 or 25°C, using a

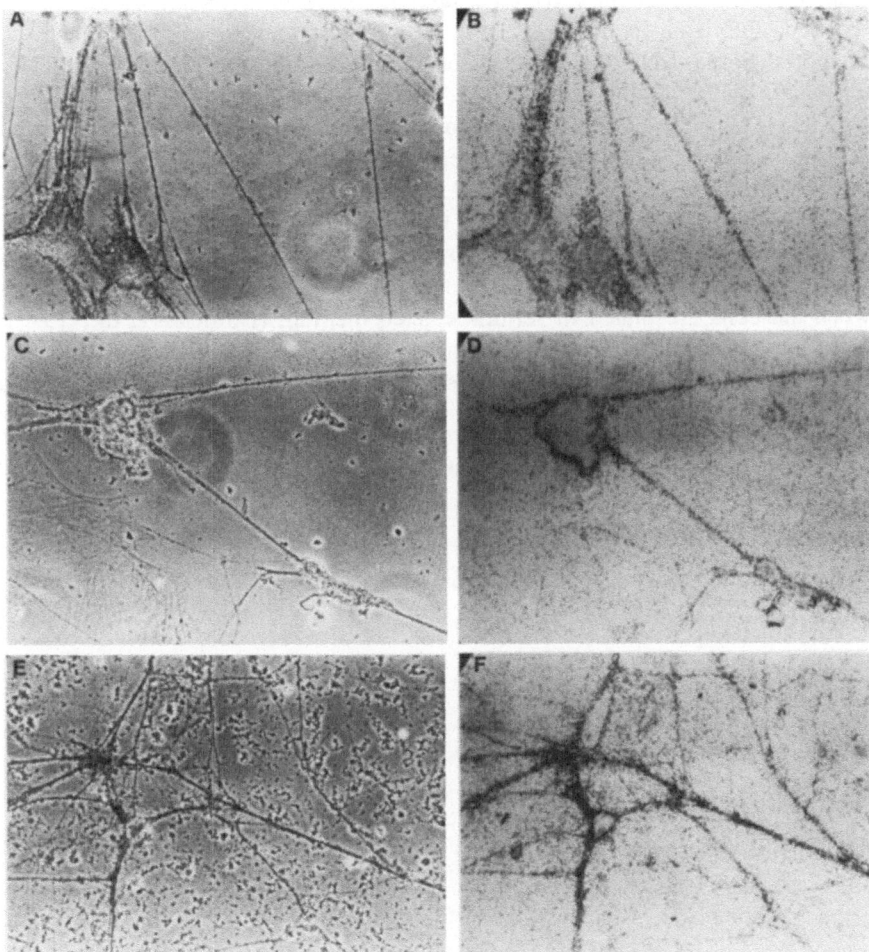

Figure 4.   Autoradiographic analysis of [^{125}I]-NGF binding to rat dorsal-root-, trigeminal-, and nodose-ganglion neurons. Neuron cultures were prepared as previously described (Baccaglini and Cooper, 1982). Incubation with [^{125}I]-NGF, and subsequent autoradiography, were carried out as described in the Fig. 1 caption. The neuronal binding of [^{125}I]-NGF was saturable, since excess unlabeled NGF markedly reduced the grain density associated with neuronal processes, as in Fig. 1. (A, C, E) Phase-contrast views; (B, D, F) same fields under bright-field optics. (A, B) Rat dorsal-root-ganglion sensory neurons; (C, D) rat trigeminal-ganglion sensory neurons; (E, F) rat nodose-ganglion sensory neurons.

centrifugation assay (Sutter *et al.*, 1979). Nonspecific binding was determined in the presence of a 100-fold excess of unlabeled NGF. As shown in Fig. 5, human A875 melanoma cells bind high levels of [^{125}I]-NGF, as previously reported (Fabricant *et al.*, 1977). Of five human melanoma lines isolated at Memorial Sloan–Kettering Cancer Center, New York, and made available for this study by Dr. R. Halaban (Yale), one line, designated SKM-37, also had high levels of [^{125}I]-NGF binding. Of an additional five human cell lines isolated at Yale by Dr. J. Kirkwood, two were shown to have binding characteristics comparable to those of PC12 cells. Several commonly used murine melanoma cell lines (B16 and various subclones, Cloudman S91 and various variants) were also examined, but were found to lack high-affinity binding.

It has been reported that NGF has a survival-enhancing effect on A875 cells when the cells are plated in serum-depleted medium (Fabricant *et al.*, 1977). No other biological effects of NGF have been reported in the case of human melanoma cells. In preliminary studies with human melanoma cell lines isolated

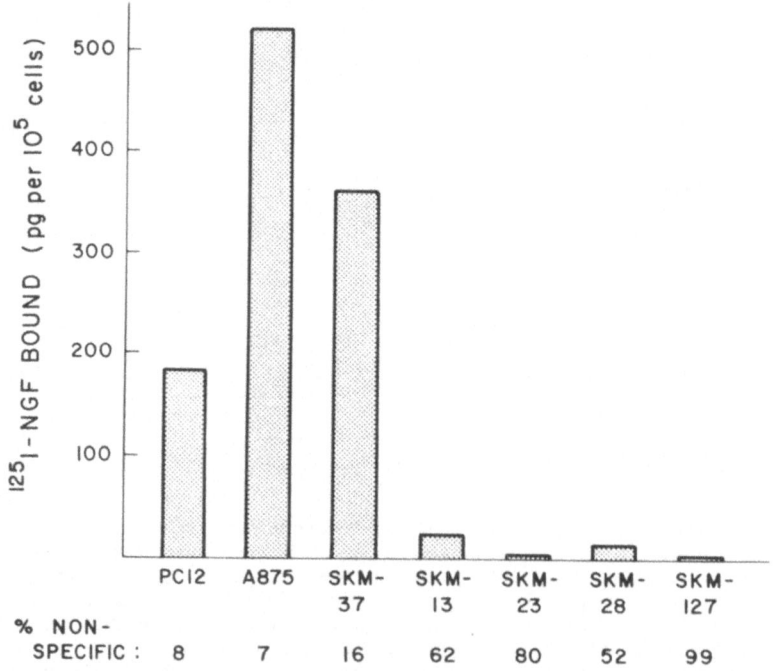

Figure 5. NGF receptors on human melanoma cells. Cultured cells were suspended by treatment with 1 mM EDTA. Binding of [^{125}I]-NGF (56 ng/ml) was determined by centrifugation after 90-min incubation at 25°C in a phosphate buffer containing Ca^{2+}, Mg^{2+}, and 1 mg/ml albumin (Sutter *et al.*, 1979). The amount of [^{125}I]-NGF bound in the presence of 4 μg/ml unlabeled NGF was also determined. Cell counts were made on a Coulter cell-counting device.

Table I. Binding of [^{125}I]-NGF to Human Melanocytes and Melanoma Cells

Cells	Specific binding (pg [^{125}I]-NGF bound/ 10^5 cells)
Rat PC12	128
Ymel 81-180 (human melanoma)	64
Human	
Caucasian melanocytes (Expt. 1)	2
Pigmented melanocytes (Expt. 2)	2

at Yale, we also have failed to observe any effects either on morphology or on growth rate by the addition of NGF. The addition of NGF to hormone-supplemented, serum-free medium was similarly ineffectual in altering morphology or growth characteristics. Thus, although the melanoma NGF receptor has not yet been strongly correlated with a biological response, the human melanoma system may provide further access to the investigation of human NGF receptors.

Because of the presence of NGF receptors on melanoma cells, we were interested in determining whether normal melanocytes contain NGF receptors. Little is known concerning the possible role of NGF in normal melanocyte development. Normal melanocytes, obtained from newborn human foreskins, were propagated in culture by Dr. Halaban using the procedure of Eisinger and Marko (1982). Cells were suspended with 1 mM EDTA, and binding of [^{125}I]-NGF was determined on the cell suspension. As indicated in Table I, the normal human melanocytes did not bind [^{125}I]-NGF. Thus, in melanocytes, as in the parasympathetic derivatives of the neural crest (Helfand et al., 1978), the absence of NGF receptors suggests a lack of involvement of NGF in the development of these particular neural-crest derivatives. Our studies used melanocytes that had been cultured for several months in the presence of phorbol ester and cholera toxin (Eisinger and Marko, 1982). Although melanocytes washed free of phorbol ester for 3 days prior to NGF binding showed no increase in [^{125}I]-NGF binding, it is possible that under these growth conditions, the expression of NGF receptors could be repressed. Therefore, additional studies with earlier-passage cells are required.

ACKNOWLEDGMENTS. Much of the original work described herein was initiated in the laboratory of Dr. Paul H. Patterson. I thank Drs. P. Baccaglini, E. Cooper, R. Halaban, and J. Kirkwood for providing cell lines and cultures, A. Doupe for help with autoradiography, and M. Ahern for assistance with the manuscript. Support was obtained from the NINCDS, NSF (BNS 81-18777), Swebilius Foundation, and the Biomedical Research Support Grant Program (Grant RRO5358), Division of Research Resources, NIH.

REFERENCES

Aharonov, A., Pruss, R.M., and Herschman, H.R., 1978, Epidermal growth factor: Relationship between receptor regulation and mitogenesis in 3T3 cells, *J. Biol. Chem.* **253**:3970–3977.

Andres, R.Y., Jeng, I., and Bradshaw, R.A., 1977, Nerve growth factor receptors: Identification of distinct classes in plasma membranes and nuclei of embryonic dorsal root neurons, *Proc. Natl. Acad. Sci. U.S.A.* **74**:2785–2789.

Ayer-Le Lièvre, C.S., and Le Douarin, N.M., 1982, The early development of cranial sensory ganglia and the potentialities of their component cells studied in quail–chick chimeras, *Dev. Biol.* **94**:291–310.

Baccaglini, P.I., and Cooper, E., 1982, Electrophysiological studies of newborn rat nodose neurones in cell culture, *J. Physiol.* **324**:429–439.

Baccaglini, P.I., and Hogan, P.G., 1983, Some rat sensory neurons in culture express characteristics of differentiated pain sensory cells, *Proc. Natl. Acad. Sci. U.S.A.* **80**:594–598.

Bocchini, V., and Angeletti, P.U., 1969, The nerve growth factor: Purification as a 30,000-molecular-weight protein, *Proc. Natl. Acad. Sci. U.S.A.* **64**:787–794.

Brunso-Bechtold, J., and Hamburger, V., 1979, Retrograde transport of nerve growth factor in chicken embryo, *Proc. Natl. Acad. Sci. U.S.A.* **76**:1494–1496.

Burton, H., and Bunge, R.P., 1975, A comparison of the uptake and release of [^{3}]norepinephrine in rat autonomic and sensory ganglia in tissue culture, *Brain Res.* **97**:157–162.

Bylund, D.B., 1980, Analysis of receptor binding data, in: 1980 *Short Course Syllabus: Receptor Binding Techniques*, Bethesda, MD, Society for Neuroscience, pp. 70–99.

Calissano, P., and Shelanski, M.L., 1980, Interaction of nerve growth factor with pheochromocytoma cells: Evidence for tight binding and sequestration, *Neuroscience* **5**:1033–1039.

Campenot, R.B., 1977, Local control of neurite development by nerve growth factor, *Proc. Natl. Acad. Sci. U.S.A.* **74**:4516–4519.

Campenot, R.B., 1979, Independent control of the local environment of somas and neurites, *Methods Enzymol.* **58**: 302–307.

Chan, K.Y., Bunt, A.H., and Haschke, R.H., 1981, *In vitro* retrograde neuritic transport of horseradish peroxidase isoenzymes by sympathetic neurons, *Neuroscience* **6**:59–69.

Chun, L.L.Y., and Patterson, P.H., 1977a, Role of nerve growth factor in the development of rat sympathetic neurons *in vitro*. I. Survival, growth, and differentiation of catecholamine production, *J. Cell Biol.* **75**:694–704.

Chun, L.L.Y., and Patterson, P.H., 1977b, Role of nerve growth factor in the development of rat sympathetic neurons *in vitro*. III. Effect on acetylcholine production, *J. Cell Biol.* **75**:712–718.

Ciechanover, A., Schwartz, A.L., and Lodish, H.F., 1983, The asialoglycoprotein receptor internalizes and recycles independently of the transferrin and insulin receptors, *Cell* **32**:267–275.

Claude, P., Hawrot, E., Dunis, D.A., and Campenot, R.B., 1982, Binding, internalization, and retrograde transport of ^{125}I-nerve growth factor in cultured rat sympathetic neurons, *J. Neurosci.* **2**:431–442.

Claude, P., Hawrot, E., and Parada, I., 1982, Ultrastructural studies on the intracellular fate of ^{125}I-nerve growth factor in cultured rat sympathetic neurons, *J. Cell. Biochem.* **20**:1–13.

Costrini, N.V., and Kogan, M., 1981, Lectin-induced inhibition of nerve growth factor binding by receptors of sympathetic ganglia, *J. Neurochem.* **36**:1175–1180.

Dumas, M., Schwab, M.E., and Thoenen, H., 1979, Retrograde axonal transport of specific macromolecules as a tool for characterizing nerve terminal membranes, *J. Neurobiol.* **10**:179–197.

Eisinger, M., and Marko, O., 1982, Selective proliferation of normal human melanocytes *in vitro* in the presence of phorbol ester and cholera toxin, *Proc. Natl. Acad. Sci. U.S.A.* **79**:2018–2022.

Fabricant, R.N., DeLarco, J.E., and Todaro, G.J., 1977, Nerve growth factor receptors on human melanoma cells in culture, *Proc. Natl. Acad. Sci. U.S.A.* **74**:565–569.

Furshpan, E.J., MacLeish, P.R., O'Lague, P.H., and Potter, D.D., 1976, Chemical transmission between rat sympathetic neurons and cardiac myocytes developing in microcultures: Evidence for cholinergic, adrenergic, and dual-function neurons, *Proc. Natl. Acad. Sci. U.S.A.* **73**:4225–4229.

Goldstein, J.L., Anderson, R.G.W., and Brown, M.S., 1979, Coated pits, coated vesicles, and receptor-mediated endocytosis, *Nature (London)* **279**:679–685.

Greene, L.A., and Shooter, E.M., 1980, The nerve growth factor: Biochemistry, synthesis, and mechanism of action, *Annu. Rev. Neurosci.* **3**:353–402.

Greene, L.A., and Tischler, A.S., 1976, Establishment of a noradrenergic clonal line of rat adrenal pheochromocytoma cells which respond to nerve growth factor, *Proc. Natl. Acad. Sci. U.S.A.* **73**:2424–2428.

Harper, G.P., and Thoenen, H., 1981, Target cells, biological effects, and mechanism of action of nerve growth factor and its antibodies, *Annu. Rev. Pharmacol. Toxicol.* **21**:205–229.

Hawrot, E., 1980, Cultured sympathetic neurons: Effects of cell-derived and synthetic substrata on survival and development, *Dev. Biol.* **74**:136–151.

Hawrot, E., 1982, Characteristics of the association of nerve growth factor with primary cultures of rat sympathetic neurons, *J. Neurosci. Res.* **8**:213–224.

Hawrot, E., and Patterson, P.H., 1979, Long-term culture of dissociated sympathetic neurons, *Methods Enzymol.* **58**:574–584.

Hedlund, K.-O., and Ebendal, T., 1980, The chick embryo nodose ganglion: Effects of nerve growth factor in culture, *J. Neurocytol.* **9**:665–682.

Hefti, F., Gnahn, H., Schwab, M.E., and Thoenen, H., 1982, Induction of tyrosine hydroxylase by NGF and by elevated K^+ concentrations in cultures of dissociated sympathetic neurons, *J. Neurosci.* **2**:1554–1566.

Helfand, S.L., Riopelle, R.J., and Wessells, N.K., 1978, Non-equivalence of conditioned medium and nerve growth factor for sympathetic, parasympathetic and sensory neurons, *Exp. Cell Res.* **113**:39–45.

Hendry, I.A., Stach, R., and Herrup, K., 1974a, Characteristics of the retrograde axonal transport system for nerve growth factor in the sympathetic nervous system, *Brain Res.* **82**:117–128.

Hendry, I.A., Stoeckel, K., Thoenen, H., and Iversen, L.L., 1974b, The retrograde axonal transport of nerve growth factor, *Brain Res.* **68**:103–121.

Herrup, K., and Thoenen, H., 1979, Properties of the nerve growth factor receptor of a clonal line of rat pheochromocytoma (PC12) cells, *Exp. Cell Res.* **121**:71–78.

Heumann, R., Schwab, M., and Thoenen, H., 1981, A second messenger required for NGF biological activity?, *Nature (London)* **292**:838–840.

Hill, C.E., and Hendry, I.A., 1976, Differences in sensitivity to nerve growth factor of axon formation and tyrosine hydroxylase induction in cultured sympathetic neurons, *Neuroscience* **1**:489–496.

Huttner, S., and O'Lague, P., 1981, Absence of morphological responses to intracellular injection of nerve growth factor into cells of the neuron-like clone PC-12, *Soc. Neurosci.* Abstr. **7**:551.

Iversen, L.L., Stoeckel, K., and Thoenen, H., 1975, Autoradiographic studies of the retrograde axonal transport of nerve growth factor in mouse sympathetic neurons, *Brain Res.* **88**:37–43.

Johnson, E.M., Jr., Andres, R.Y., and Bradshaw, R.A., 1978, Characterization of the retrograde transport of nerve growth factor (NGF) using high specific activity [^{125}I]NGF, *Brain Res.* **150**:319–333.

Johnson, E.M., Jr., Gorin, P.D., Brandeis, L.D., and Pearson, J., 1980, Dorsal root ganglion neurons are destroyed by exposure *in utero* to maternal antibody to nerve growth factor, *Science* **210**:916–918.

Kim, S.U., Hogue-Angeletti, R., and Gonatas, N.K., 1979, Localization of nerve growth factor receptors in sympathetic neurons cultured *in vitro*, *Brain Res.* **168**:602–608.

Lander, A.D., Fujii, D.K., Gospodarowicz, D., and Reichardt, L.F., 1982, Characterization of a factor that promotes neurite outgrowth: Evidence linking activity to a heparan sulfate proteoglycan, *J. Cell Biol.* **94**:574–585.

Landreth, G.E., and Shooter, E.M., 1980, Nerve growth factor receptors on PC12 cells: Ligand-induced conversion from low- to high-affinity states, *Proc. Natl. Acad. Sci. U.S.A.* **77**:4751–4755.

LaVail, J.H., Rapisardi, S., and Sugino, I.K., 1980, Evidence against the smooth endoplasmic reticulum as a continuous channel for the retrograde axonal transport of horseradish peroxidase, *Brain Res.* **191**:3–20.

Lazarus, K.J., Bradshaw, R.A., West, N.R., and Bunge, R.P., 1976, Adaptive survival of rat sympathetic neurons cultured without supporting cells or exogenous nerve growth factor, *Brain Res.* **113**:159–164.

Le Douarin, N.M., Smith, J., and Le Lièvre, C.S., 1981, From the neural crest to the ganglia of the peripheral nervous system, *Annu. Rev. Physiol.* **43**:653–671.

Levi, A., Shechter, Y., Neufeld, E.J., and Schlessinger, J., 1980, Mobility, clustering, and transport of nerve growth factor in embryonal sensory cells and in a sympathetic neuronal cell line, *Proc. Natl. Acad Sci. U.S.A.* **77**:3469–3473.

Levi-Montalcini, R., 1976, The nerve growth factor: Its role in growth, differentiation and function of the sympathetic adrenergic neuron, *Prog. Brain Res.* **45**:235–258.

Lindsay, R.M., 1979, Adult rat brain astrocytes support survival of both NGF-dependent and NGF-insensitive neurons, *Nature (London)* **282**:80–82.

Mains, R.E., and Patterson, P.H., 1973a, Primary cultures of dissociated sympathetic neurons. I. Establishment of long-term growth in culture and studies of differentiated properties, *J. Cell Biol.* **59**:329–345.

Mains, R.E., and Patterson, P.H., 1973b, Primary cultures of dissociated sympathetic neurons. III. Changes in metabolism with age in culture, *J. Cell Biol.* **59**:361–366.

Marchisio, P.C., Naldini, L., and Calissano, P., 1980, Intracellular distribution of nerve growth factor in rat pheochromocytoma PC12 cells: Evidence for a perinuclear and intranuclear location, *Proc. Natl. Acad. Sci. U.S.A.* **77**:1656–1660.

Marchisio, P.C., Cirillo, D., Naldini, L., and Calissano, P., 1981, Distribution of nerve growth factor in chick embryo sympathetic neurons *in vitro*, *J. Neurocytol.* **10**:45–55.

Massague, J., Guillette, B.J., Czech, M.P., Morgan, C.J., and Bradshaw, R.A., 1981, Identification of a nerve growth factor receptor protein in sympathetic ganglia membranes by affinity labeling, *J. Biol. Chem.* **256**:9419–9424.

Noden, D., 1978, The control of avian cephalic neural crest cytodifferentiation. II. Neural tissues, *Dev. Biol.* **67**:313–329.

Nurse, C.A., 1981, Interactions between dissociated rat sympathetic neurons and skeletal muscle cells developing in cell culture. I. Cholinergic transmission, *Dev. Biol.* **88**:55–70.

O'Lague, P.H., MacLeish, P.R., Nurse, C.A., Claude, P., Furshpan, E.J., and Potter, D.D., 1975, Physiological and morphological studies on developing sympathetic neurons in dissociated cell culture, *Cold Spring Harbor Symp. Quant. Biol.* **40**:399–407.

O'Lague, P.H., Potter, D.D., and Furshpan, E.J., 1978a, Studies on rat sympathetic neurons developing in cell culture. I. Growth characteristics and electrophysiological properties, *Dev. Biol.* **67**:384–403.

O'Lague, P.H., Potter, D.D., and Furshpan, E.J., 1978b, Studies on rat sympathetic neurons developing in cell culture. III. Cholinergic transmission, *Dev. Biol.* **67**:424–443.

Olender, E.J., and Stach, R.W., 1980, Sequestration of ^{125}I-labeled beta nerve growth factor by sympathetic neurons, *J. Biol. Chem.* **255**:9338–9343.

Patterson, P.H., 1978, Environmental determination of autonomic neurotransmitter functions, *Annu. Rev. Neurosci.* **1**:1–17.

Patterson, P.H., and Chun, L.L.Y., 1974, The influence of nonneuronal cells on catecholamine and acetylcholine synthesis and accumulation in cultures of dissociated sympathetic neurons, *Proc. Natl. Acad. Sci. U.S.A.* **71:**3607–3610.

Patterson, P.H., and Chun, L.L.Y., 1977, The induction of acetylcholine synthesis in primary cultures of dissociated rat sympathetic neurons. I. Effects of conditioned medium, *Dev. Biol.* **56:**263–280.

Patterson, P.H., Reichardt, L.F., and Chun, L.L.Y., 1975, Biochemical studies on the development of primary sympathetic neurons in cell culture, *Cold Spring Harbor Symp. Quant. Biol.* **40:** 389–397.

Pearson, J., 1979, Familial dysautonomia (a brief review), *J. Autonomic Nervous System* **1:**119–126.

Perez-Polo, J.R., Werrbach-Perez, K., and Tiffany-Castiglioni, E., 1979, A human clonal cell line model of differentiating neurons, *Dev. Biol.* **71:**341–355.

Rohrer, H., and Barde, Y.-A., 1982, Presence and disappearance of NGF receptors on sensory neurons in culture, *Dev. Biol.* **89:**309–315.

Rohrer, H., Schäfer, T., Korsching, S., and Thoenen, H., 1982, Internalization of NGF by pheochromocytoma PC12 cells: Absence of transfer to the nucleus, *J. Neurosci.* **2:**687–697.

Schechter, A.L., and Bothwell, M.A., 1981, Nerve growth factor receptors on PC12 cells: Evidence for two receptor classes with differing cytoskeletal association, *Cell* **24:**867–874.

Schwab, M.E., 1977, Ultrastructural localization of a nerve growth factor-horseradish peroxidase coupling product after retrograde axonal transport in adrenergic neurons, *Brain Res.* **130:**190–196.

Schwab, M.E., and Thoenen, 1977, Selective trans-synaptic migration of tetanus toxin after retrograde axonal transport in peripheral sympathetic nerves: A comparison with nerve growth factor, *Brain Res.* **122:**459–474.

Schwab, M.E., and Thoenen, H., 1982, Retrograde axonal transport of H^3-noradrenaline (NA) in dissociated sympathetic neurons *in vitro*, *Soc. Neurosci. Abstr.*, **8:**786.

Schwab, M.E., Suda, K., and Thoenen, H., 1979, Selective retrograde transsynaptic transfer of a protein, tetanus toxin, subsequent to its retrograde axonal transport, *J. Cell Biol.* **82:** 798–809.

Schwab, M.E., Heumann, R., and Thoenen, H., 1982, Communication between target organs and nerve cells: Retrograde axonal transport and site of action of NGF, *Cold Spring Harbor Symp. Quant. Biol.* **46:**125–154.

Shooter, E.M., Yankner, B. A., Landreth, G.E., and Sutter, A., 1981, Biosynthesis and mechanism of action of nerve growth factor, *Rec. Progr. Horm. Res.* **37:**417–446.

Sonnenfeld, K.H., and Ishii, D.N., 1982, Nerve growth factor effects and receptors in cultured human neuroblastoma cell lines, *J. Neurosci. Res.* **8:**371–391.

Sutter, A., Riopelle, R.J., Harris-Warrick, R.M., and Shooter, E.M., 1979, Nerve growth factor receptors: Characterization of two distinct classes of binding sites on chick embryo sensory ganglia cells, *J. Biol. Chem.* **254:**5972–5982.

Walicke, P., Campenot, R., and Patterson, P., 1977, Determination of transmitter function by neuronal activity, *Proc. Natl. Acad. Sci. U.S.A.* **74:**5767–5771.

Weston, J. A., 1970, The migration and differentiation of neural crest cells, *Advan. Morphogen.* **8:**41–114.

Yankner, B.A., and Shooter, E.M., 1979, Nerve growth factor in the nucleus: Interaction with receptors on the nuclear membrane, *Proc. Natl. Acad. Sci. U.S.A.* **76:**1269–1273.

Yankner, B.A., and Shooter, E.M., 1982, The biology and mechanism of action of NGF, *Annu. Rev. Biochem.* **51:**845–868.

Guanethidine-Induced Destruction of Sympathetic Neurons: An Autoimmune "Disease" Prevented by Nerve Growth Factor

Eugene M. Johnson, Jr., and Pamela Toy Manning

1. INTRODUCTION

Guanethidine, the structure of which is shown in Fig. 1, is a guanidinium adrenergic-neuron-blocking agent. It is the prototype of this class of drugs that act to reduce the efficacy of sympathetic neurotransmission primarily by dissociating the action potential from the secretion of the neurotransmitter norepinephrine (NE). By virtue of its ability to block NE release, it lowers blood pressure, and has thus been used clinically as an antihypertensive agent since the 1960s. In 1971, it was demonstrated by Burnstock et al. (1971), Eränkö and Eränkö (1971), and Jensen-Holm and Juul (1971) that chronic treatment of rats with high doses of guanethidine (about 10 times that needed to produce adrenergic-neuron blockade) resulted in the destruction of sympathetic neurons in both neonatal and adult animals. The destruction was selective for sympathetic neurons; other neurons such as sensory or parasympathetic neurons were not affected. The basis for this selectivity arises from the ability of guanethidine to accumulate to high concentrations in sympathetic neurons via the catecholamine (CA)-uptake pump as the neuronal destruction is prevented by an inhibitor of this uptake mechanism, desmethylimipramine (Juul and Sand, 1973). It is ex-

EUGENE M. JOHNSON, JR. and PAMELA TOY MANNING ● Department of Pharmacology, Washington University School of Medicine, St. Louis, Missouri 63110.

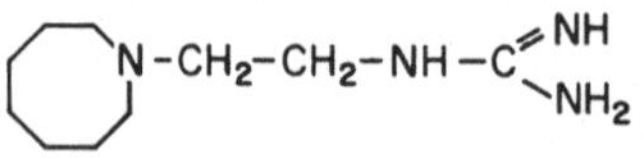

Figure 1.   Structure of guanethidine.

cluded from neurons that lack this active-uptake mechanism because it is a highly charged guanidinium compound.

Our interest in the mechanism that underlies the cytotoxic effects of guanethidine on sympathetic neurons has been heightened by an observation made ten years ago by one of us while working with Luigi Aloe in the laboratory of Dr. Levi-Montalcini. We (Johnson and Aloe, 1974) observed that concomitant treatment of neonatal rats with 10 mg/kg per day of mouse nerve growth factor (NGF) completely prevented the neuronal destruction produced by guanethidine. We have recently determined that doses of 1 mg/kg per day of NGF prevent neuronal destruction (Fig. 2). The prevention of the cytotoxicity by NGF is not due to a block of the accumulation of guanethidine within the neurons. Therefore, in the presence of exogenous NGF, concentrations of guanethidine that normally kill sympathetic neurons lack the cytotoxic effect that occurs in the absence of NGF. We feel that an understanding of the mechanism by which guanethidine destroys neurons might provide useful information on the mechanism by which NGF normally sustains these neurons.

Several mechanisms have been proposed to explain the cytotoxic effects of guanethidine. Most frequently suggested was a mechanism based on the demonstrated ability of guanidinium compounds to inhibit oxidative phosphorylation (Malmquist and Oates, 1968). However, structure/activity-relationship studies, in which we systematically examined the accumulation of several analogues, their cytotoxicity, and their ability to inhibit oxidative phosphorylation in isolated mitochondria, indicate that it is unlikely that this mechanism is responsible for the observed cytotoxicity. These studies showed that only guanethidine and a few close analogues are capable of destroying neurons; cytotoxicity is not a general property of adrenergic-neuron-blocking agents (Johnson and Hunter, 1979.)

The possibility that the neuronal destruction produced by guanethidine might be immunologically mediated was suggested to us by several observations in the literature, some of which are listed in Table I. First was the observation that the neuronal destruction observed following guanethidine treatment is accompanied by an intense small-cell infiltration similar to that seen in cell-mediated, delayed-type hypersensitivity reactions. This dense infiltration was noted in the early papers of Jensen-Holm and Juul and of Eränkö and was described as "lymphocytic-like" in nature. The similarity of this small-cell infiltrate to that which occurs in immune reactions was, in fact, noted by both these workers. We have recently examined the nature of this infiltrate at the ultrastructural and light-

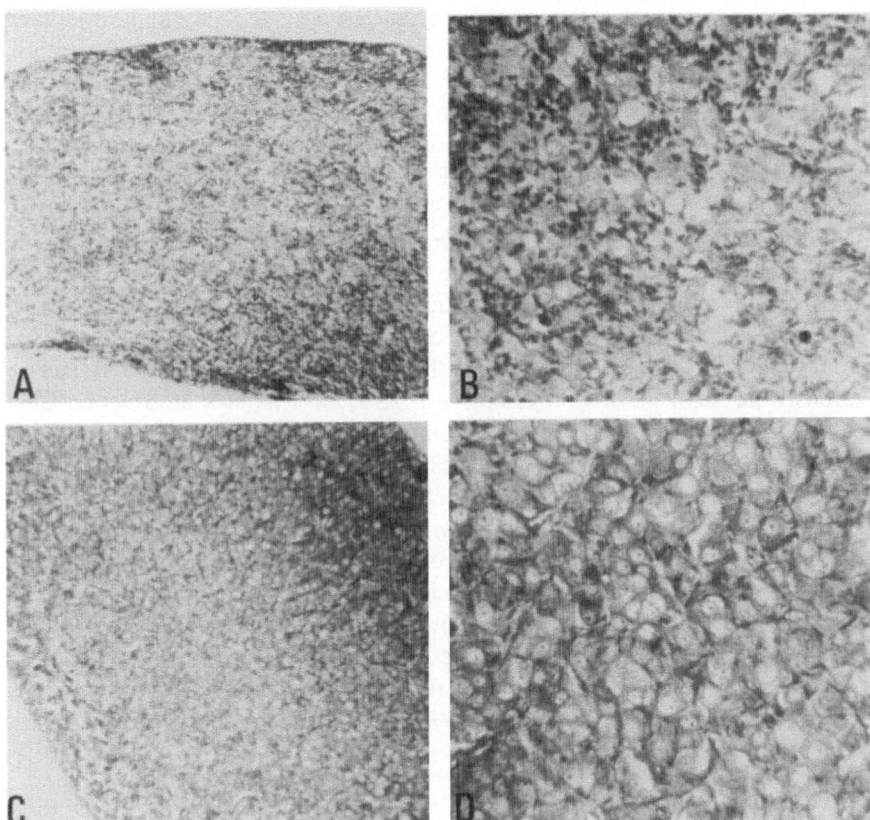

Figure 2.   Protection by NGF against guanethidine-induced sympathetic-neuronal destruction. Newborn Sprague–Dawley rats were treated with guanethidine sulfate (50 mg/kg per day) alone on days 2–7 (A, B), or with guanethidine plus 1 mg/kg of NGF (C, D). Note that ganglia from animals treated with NGF + guanethidine are hypertrophied and that neuronal degeneration and small-cell infiltration are not present. Tissues shown are superior cervical ganglia (8-μm section; toluidine blue stain). (A, C) ×70; (B, D) ×175.

Table I.   Observations in the Literature Consistent with an Immune-Mediated Mechanism

1. Presence of small-cell infiltrate concurrent with neuronal destruction.
2. Lack of toxicity of guanethidine *in vitro* under normal tissue-culture conditions.
3. Inability of guanethidine to destroy CNS neurons.
4. Species specificity—guanethidine causes destruction of sympathetic neurons only in the rat.

microscopic (LM) level at the time when neurons are beginning to die and have found that it consists primarily of lymphocytes and macrophages and very few neutrophils, a picture consistent with an immune-mediated infiltrate rather than a nonspecific response to tissue damage. A second observation consistent with an immune mechanism is the *failure* of guanethidine to destroy sympathetic neurons *in vitro* in either explants (Eränkö *et al.*, 1972; Johnson and Aloe, 1974) or dissociated cells (Wakshull *et al.*, 1978; Manning, unpublished) under normal tissue-culture conditions. This lack of *in vitro* toxicity would be expected if the killing of neurons required the presence of immune effector cells that enter the ganglion from the circulation. Third, attempts to destroy noradrenergic neurons in the central nervous system (CNS) have failed even though the drug was administered chronically directly into the CNS and the central noradrenergic neurons were depleted of CA (Evans *et al.*, 1975). This lack of effect on CNS neurons might be explained by the immunologically privileged nature of the brain. The fourth observation is the striking species specificity of this phenomenon. Guanethidine administration results in a permanent sympathectomy *only* in the rat. Treatment of mice, hamsters, cats, rabbits, gerbils, chickens, toads, or guinea pigs with chronic high doses of guanethidine does not cause neuronal destruction (O'Donnell and Saar, 1974; Johnson *et al.*, 1977; Evans *et al.*, 1979 and unpublished observations). Where it has been examined, guanethidine does accumulate in the sympathetic neurons of these species and does cause ultrastructural alterations, but, again, does not cause neuronal death (Evans *et al.*, 1979; Révész and van der Zypen, 1979). Autoimmune diseases often show species and strain specificities.

2. EFFECTS OF IMMUNOSUPPRESSIVE AGENTS ON GUANETHIDINE-INDUCED NEURONAL DESTRUCTION

Because all these observations suggested, but certainly did not prove, that guanethidine-induced sympathectomy might be a drug-induced autoimmune "disease," we carried out a series of experiments aimed at testing this idea. Initial experiments were carried out to test the ability of various immunosuppressive agents to prevent guanethidine-induced neuronal destruction (Manning *et al.*, 1982). The treatment protocol that we employed in these experiments involved treating neonatal rats with 50 mg/kg per day of guanethidine sulfate for 5 days beginning at 1 week of age. Animals were usually killed at 14 days of age. This protocol was chosen because the time–course of the small-cell infiltration and neuronal destruction is very reproducible. Sympathetic ganglia were examined by LM, and tyrosine hydroxylase (TH) activity was measured as a biochemical correlate of neuronal destruction. At 4 days after initiation of guanethidine treatment, the superior cervical ganglion (SCG) appears normal by LM examination.

Usually on the 5th day, the small-cell infiltrate becomes apparent and degenerating "ghost" cells appear. Over the next 2 days, the small-cell infiltrate becomes dense and covers the entire ganglion, and virtually all the neurons are destroyed. Decreases in TH activity paralleled these histological changes. Pretreatment of neonatal rats with either the alkylating agent cyclophosphamide or whole-body γ-irradiation a few hours before initiation of guanethidine treatment prevented guanethidine-induced sympathectomy in a dose-related manner. For example, shown in Fig. 3 are SCGs from 14-day-old rats treated with either

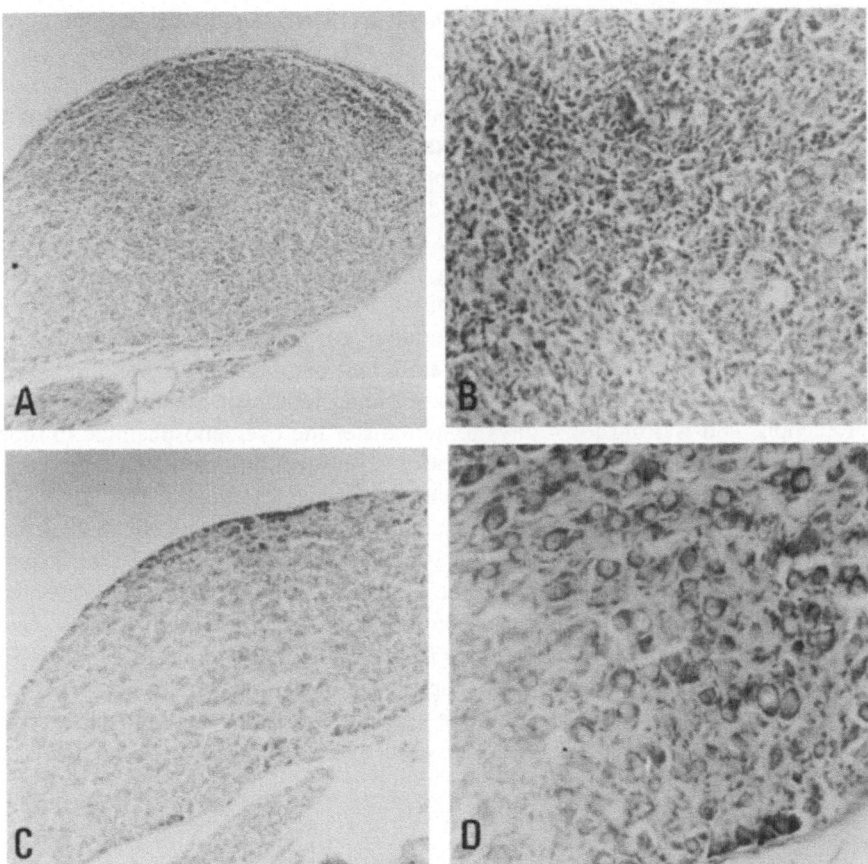

Figure 3. Protection by γ-irradiation against guanethidine-induced sympathetic-neuronal destruction. Sprague–Dawley rats, 1 week old, received guanethidine sulfate (50 mg/kg per day) for 5 days and were killed 2 days after the last injection (A, B). Another group of rats (C, D) was pretreated with 900 rads of whole-body γ-irradiation prior to guanethidine treatment. Tissues shown are SCGs (8-μm section; toluidine blue stain). (A, C) ×70; (B, D) ×175.

guanethidine alone (Fig. 3A, B) or with guanethidine $+$ 900 rads of γ-irradiation (Fig. 3C, D). Note the massive small-cell infiltrate and the virtual absence of viable neurons in the SCG from the animal with an intact immune system following guanethidine treatment. In contrast, the SCG from the immunosuppressed animal given guanethidine was indistinguishable from that of an untreated control animal (not shown). Lesser but clear degrees of protection were afforded by maximally tolerated doses of two other immunosuppressants that act by different mechanisms, azathioprine and dexamethazone.

Several experiments were carried out to determine whether other possible effects of these admittedly relatively nonspecific drugs could account for the protection against guanethidine sympathectomy. For example, the accumulation of guanethidine in the sympathetic ganglia was determined in normal, cyclophosphamide-treated, and irradiated rats. Little or no inhibition of guanethidine accumulation was caused by either immunosuppressant. The time dependence of the immunosuppression was also examined. If cyclophosphamide or γ-irradiation was acting as an immunosuppressant, i.e., blocking the proliferative phase of an immune response, it would be necessary to treat the animal at the time of exposure to antigen or shortly thereafter. Treating the animal at later times would reduce tissue damage only if the drug was acting by a nonspecific antiinflammatory effect. When newborn rats were treated with a single injection of cyclophosphamide at the time of initiation of guanethidine treatment, complete protection against the small-cell infiltration and neuronal death was produced. Waiting until 2 days after starting guanethidine treatment greatly reduced the protection, and waiting for 4 days to administer the cyclophosphamide (a time still prior to small-cell invasion) provided no protection against the small-cell infiltration and neuronal destruction that ensued over the next 3 days.

We also investigated whether immunosuppressants were specific in the sense that other methods of destroying the sympathetic nervous system that do not depend on an immune response by the treated animal should not be prevented by immunosuppressive agents. To examine this question, newborn rats were treated with either 6-hydroxydopamine (6-OH-DA) or antibodies to mouse NGF (anti-NGF). Doses of either cyclophosphamide or gamma irradiation that completely protected against guanethidine did not alter the neuronal destruction produced by these agents as assessed by either LM examination or the decrease in TH activity in the SCG.

3. NATURE OF THE SMALL-CELL INFILTRATE IN SYMPATHETIC GANGLIA OF GUANETHIDINE-TREATED RATS

The experiments described above strongly suggested that an immunological, presumably cellularly mediated, destruction of the rat sympathetic nervous system occurred following guanethidine treatment. As previously discussed, the small-cell infiltrate in 14-day-old rats treated with guanethidine consisted primarily of lymphocytes and macrophages (Manning *et al.*, 1983). To further characterize this infiltrate and to test the hypothesis that the neuronal destruction was cellularly mediated, we looked for T lymphocytes in the small-cell infiltrate. An anti-rat T-cell antiserum was prepared by previously published procedures (Ishii *et al.*, 1976). Briefly, an enriched population of rat T cells was prepared by passing cells obtained from mesenteric lymph nodes over nylon wool. The nonadherent cells were injected intravenously into a rabbit that was boosted at monthly intervals with similarly prepared cells. This antithymocyte serum (ATS) and normal rabbit serum (NRS) were both extensively adsorbed with rat liver and skeletal muscle to eliminate nonspecific binding to rat tissue and thus enhance the specificity of the antiserum for rat T lymphocytes. The specificity of this serum was demonstrated using conventional indirect immunohistochemistry with peroxidase-conjugated goat anti-rabbit IgG. Binding of ATS to rat spleen is shown in Fig. 4A. Note that the reaction product appears to mark the plasma membranes of small cells (approximately 7 μm in diameter) that are densely packed around the arterioles. Occasional staining occurs in cells of the same size in other regions of the spleen. These are the size and distribution within the spleen that are expected of T lymphocytes. No staining of these cells occurs

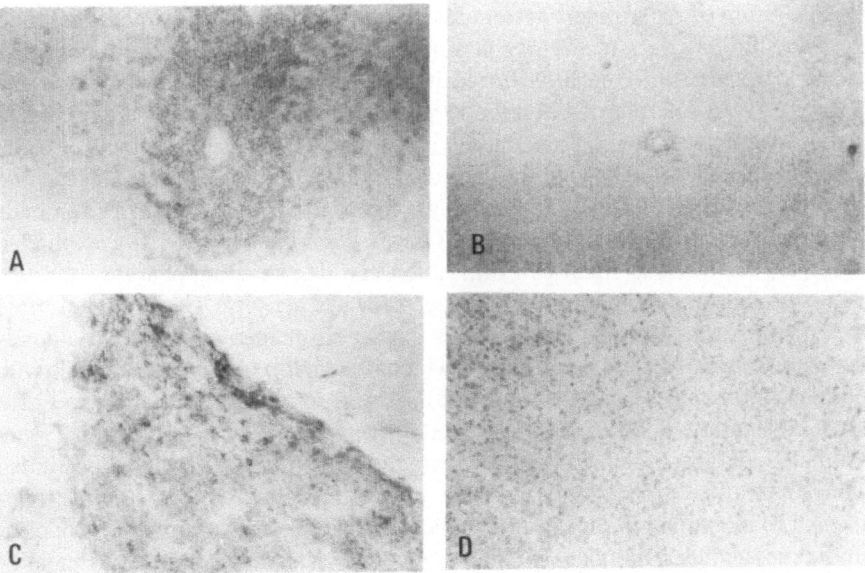

Figure 4. Immunohistochemical identification of T lymphocytes in SCGs of guanethidine-treated rats. (A) Normal rat spleen stained with ATS; (B) as in (A), stained with NRS; (C) SCG from guanethidine-treated rat stained with ATS; (D) SCG from guanethidine-treated rat stained with NRS. (A–D) ×175. See the description in the text.

with NRS (Fig. 4B). Staining of the same-size cells was seen in the thymus (not shown). No staining or only very occasional staining was seen in other tissues including liver and skeletal muscle. Figure 4C shows the staining in ganglia of 14-day-old rats treated with guanethidine using ATS. Numerous individual or, more frequently, clusters of cells are stained by the ATS. These cells appear identical to those stained in the spleen: the cells are about 7 μm in diameter and the reaction product appears on the plasma membrane. NRS does not stain sections from the same ganglia (Fig. 4D). Similarly, neither ATS nor NRS stains ganglia obtained from untreated animals. Preliminary experiments in 6-OH-DA-treated animals likewise show no evidence of the presence of T lymphocytes.

4. IMMUNE-RECONSTITUTION EXPERIMENTS

The aforecited experiments in which the protective effects of immunosuppressants and the small-cell infiltrate were examined provided a strong case for an immune-mediated destruction of sympathetic neurons in guanethidine-treated rats. The traditional way to prove that the immune system mediates tissue destruction is to perform adoptive-transfer (reconstitution) experiments. We administered immunocompetent cells to an irradiated animal (which therefore cannot destroy neurons after guanethidine treatment) and demonstrated that reconstitution of the immune system restored the ability of the animal to respond to guanethidine (i.e., to destroy neurons). The immune-reconstitution experiments were performed in 3-week-old inbred Lewis rats. Initially, some of the rats were given 850 rads of whole-body γ-irradiation. Irradiated and nonirradiated animals were given daily guanethidine injections. The immune systems of some of the irradiated animals were reconstituted by injection of syngeneic spleen and bone-marrow cells on the 1st and 2nd days of guanethidine treatment. The animals were killed after 9–10 days; one SCG was examined by LM (by a blinded observer) for obvious small-cell infiltration and degenerating neurons, and one SCG was frozen and subsequently assayed for TH activity. The results of these experiments are shown in Table II. As expected, guanethidine-treated animals that were irradiated were not affected, whereas guanethidine-treated animals with normal immune systems (not irradiated) showed small-cell infiltration and neuronal degeneration. The critical observation was that 14 of the 15 rats that were irradiated and the immune systems of which were at least partially reconstituted showed small-cell infiltration and neuronal destruction after guanethidine treatment. LM examination (not shown) indicated that although the effects were clear in the reconstituted animals, the magnitude of the response was not as great as that seen in nonirradiated animals that received guanethidine. The decrease in TH activity in the reconstituted animals, which was statistically significant but less that that in nonirradiated guanethidine-treated animals, was consistent with

Table II. Dependence of Guanethidine-Induced Destruction of Sympathetic Neurons on Immune Reconstitution

Treatment[a]	Effect (affected/total)[b,c]	Tyrosine hydroxylase activity [% of control (B)][c]
A. Nonirradiated + guanethidine	10/10	$45.5 \pm 4.9^{d,f}$
B. Irradiated control	0/13	100 ± 3.9
C. Irradiated + guanethidine	0/15	$88.9 + 4.5$
D. Irradiated + spleen and bone-marrow cells + guanethidine	14/15	$74.8 \pm 3.2^{d,e,g}$

[a] Lewis rats (groups B–D), 3 weeks old, were treated with 850 rads of γ-irradiation using a Gamma-Cell 40 immediately prior to initial reconstitution. Initially (day 1), recipients (group D) received intravenously a combination of 3.1×10^8 (Expt. I) or 6×10^8 (Expt. II) bone-marrow and spleen cells obtained from adult syngeneic, naïve donors. At 24 hr later (day 2), recipients also received 2×10^8 syngeneic spleen cells (Expt. I) or 7×10^8 spleen and bone-marrow cells (Expt.II). All groups except the control group (B) received daily injections of 50 mg/kg guanethidine sulfate for 8 days. Animals were killed on day 9 or 10, and the SCGs were quickly removed for assay of TH activity and for histological evaluation.
[b] Presence of obvious small-cell infiltration and neuronal destruction. Scored blindly by a neuropathologist.
[c] Numbers represent the pooled data for Expts. I and II.
[d] Differs from irradiated control (group B) at the $P < 0.01$ level of significance, by Student's t test.
[e] Differs from irradiated + guanethidine (group C) at the $P < 0.05$ level of significance.
[f] Differs from irradiated + guanethidine (group C) at the $P < 0.01$ level of significance.
[g] Differs from nonirradiated + guanethidine (group A) at the $P < 0.01$ level of significance.

the histological changes. Photomicrographs of this experiment and a discussion of the reasons for the intermediate response in the immune-reconstituted animals are presented in Manning *et al.* (1983).

5. VALUE OF GUANETHIDINE-INDUCED SYMPATHECTOMY AS A MODEL OF AUTOIMMUNE DISEASE OF THE NERVOUS SYSTEM AND OF DRUG-INDUCED AUTOIMMUNE DISEASE

The experiments described herein demonstrate that guanethidine-induced destruction of the sympathetic nervous system is immune (presumably cell)-mediated. As an autoimmune disease of the nervous system, guanethidine-induced sympathectomy offers several advantages over existing experimental models as a possible system in which to gain a better understanding of these phenomena. These potential advantages are listed in Table III. Most of these advantages have been mentioned previously. With respect to the second advantage listed in Table III, we have observed clear strain specificities in the sensitivity to guanethidine, including strains (Wistar–Fürth) in which the small-cell infiltration is transient despite continued drug treatment, a pattern consistent with activation of potent suppressor-cell mechanisms. This strain specificity may allow a detailed analysis of the genetic component(s) that regulate(s) the immune response.

Table III. Advantages of Guanethidine-Induced Sympathectomy as a Model for the Study of Drug-Induced Autoimmune Disease

1. Highly reproducible response.
2. Species and strain differences.
3. Cell attacked may be isolated in culture, free of other cell types.
4. Trigger (guanethidine) is a single defined chemical entity that is readily attainable. Some data on structure–activity relationships are available.
5. Animals in which the immunological attack is proceeding are not sick. The sympathetic neuron is not necessary for survival.

A point of particular interest, as discussed in Section 1, is the ability of NGF to prevent guanethidine-induced neuronal destruction. We feel that three general mechanisms exist by which NGF may exert its protective effect. The first possibility is that NGF is a general immunosuppressant when administered in high doses (i.e., greater than 1 mg/kg per day) to neonatal animals. There are no data in the literature on the effects of NGF *in vitro* or *in vivo* that would support such a suggestion. In addition, experiments that we have carried out to directly test this possibility have been negative. For example, treatment with high doses of NGF does not produce decreases in spleen weight of neonatal animals (gross measure of immunosuppression), nor do the animals appear to be more susceptible to infection. Also, *in vitro* killing or proliferation of cytotoxic lymphocytes directed against major histocompatibility antigens (on tumor target cells) is unaffected by concentrations of NGF of 10 μg/ml (Manning and Russell, unpublished observations).

More specific mechanisms include an ability of NGF to prevent the expression of the antigen on sympathetic neurons, which is the object of the immunological attack; an ability of NGF to block immune recognition of the expressed antigen at the proliferative or effector stage, or both, of the immune response; or an ability to block the killing activity of the immune effectors on sympathetic neurons without blocking recognition. The latter of these three mechanisms would seem unlikely, since NGF treatment prevents not only neuronal destruction, but also the appearance of the small-cell infiltration. Distinguishing between the first two of these specific mechanisms will be difficult, if not impossible, until an identification of the antigen that is the object of the immune attack is made. Experiments aimed at determining the nature of the antigen(s) are currently under way in our laboratory.

In conclusion, our studies have shown that guanethidine-induced sympathectomy represents an unusual, perhaps unique, drug-induced autoimmune "disease." For reasons outlined above, we feel that study of this phenomenon may provide useful insights into mechanisms of drug-induced and nervous system autoimmune disorders. More speculatively, we propose that an understanding of the means by which NGF prevents guanethidine-induced destruction may provide insights into the mechanism(s) by which NGF promotes the survival of NGF-dependent cell types.

ACKNOWLEDGMENTS. We would like to thank our colleagues Luigi Aloe, John Russell, Robert Schmidt, and Christine Powers for their collaboration in the experiments described in this chapter. P.T.M. is a Fellow of the Missouri and American Heart Associations. E.M.J. is an Established Investigator of the American Heart Association. This work was supported by NIH grant HL20604.

REFERENCES

Burnstock, G., Evans, B., Gannon, B.J., Heath, J.D., and James, V., 1971, A new method of destroying adrenergic nerves in adult animals using guanethidine, *Br. J. Pharmacol.* **43**:195–301.

Eränkö, L., and Eränkö, O., 1971, Effect of guanethidine on nerve cells and small intensively fluorescent cells in sympathetic ganglia of newborn and adult rats, *Acta Pharmacol. Toxicol.* **30**:403–416.

Eränkö, L., Hill, C., Eränkö, O., and Burnstock, G., 1972, Lack of toxic effects of guanethidine on nerve cells and small intensely fluorescent cells in cultures of sympathetic ganglia of newborn rats, *Brain Res.* **43**:501–513.

Evans, B.K., Singer, G., Armstrong, S., Saunders, P.E., and Burnstock, G., 1975, Effects of chronic intracranial injection of low and high concentrations of guanethidine on the rat, *Pharmacol. Biochem. Behav.* **3**:219–228.

Evans, B.K., Heath, J.W., and Burnstock, G., 1979, Effects of chronic guanethidine on the sympathetic nervous system of mouse and toad, *Comp. Biochem. Physiol.* **63c**:81–92.

Ishii, Y., Koshiba, H., Yamanaka, H., and Kikachi, K., 1976, Rat T lymphocyte specific antigens and their cross reactivity with mouse T cells, *J. Immunol.* **117**:497–503.

Jensen-Holm, J., and Juul, P., 1971, Ultrastructural changes in the rat superior cervical ganglion following prolonged guanethidine administration, *Acta Phamacol. Toxicol.* **30**:308–320.

Johnson, E.M., and Aloe, L., 1974, Suppression of the *in vitro* and *in vivo* cytotoxic effects of guanethidine in sympathetic neurons by nerve growth factor, *Brain Res.* **81**:519–532.

Johnson, E.M., and Hunter, F.E., 1979, Chemical sympathectomy by guanidinium adrenergic neuron blocking agents, *Biochem. Pharmacol.* **28**:1525–1531.

Johnson, E.M., Macia, R.A., and Yellin, T.O., 1977, Marked difference in the susceptibility of several species to guanethidine-induced chemical sympathectomy, *Life Sci.* **20**:107–112.

Juul, P., and Sand, O., 1973, Determination of guanethidine in sympathetic ganglia, *Acta Pharmacol. Toxicol.* **2**:487–499.

Malmquist, J., and Oates, J.A., 1968, Effects of adrenergic neuron-blocking guanidine derivatives on mitochondrial metabolism, *Biochem. Pharmacol.* **17**:1845–1854.

Manning, P.T., Russell, J.H., and Johnson, E.M., 1982, Immunosuppressive agents prevent guanethidine-induced destruction of rat sympathetic neurons, *Brain Res.* **214**:131–143.

Manning, P.T., Schmidt, R.E., and Johnson, E.M., 1983, Guanethidine-induced destruction of peripheral sympathetic neurons occurs by an immune-mediated mechanism, *J. Neurosci.* **3**:214–224.

O'Donnell, S.R., and Saar, N., 1974, The effects of 6-hydroxydopamine and guanethidine on peripheral adrenergic nerves in the guinea pig, *Eur. J. Pharmacol.* **28:**251–256.

Révész, E., and van der Zypen, E., 1979, Ultrastructural changes in adrenergic neurons in following chemical sympathectomy, *Acta Anat.* **105:**198–208.

Wakshull, E., Johnson, M.I., and Burton, H., 1978, Persistence of an amine uptake system in cultured rat sympathetic neurons which use acetylcholine as their transmitter, *J. Cell Biol.* **79:**121–131.

Enhanced Dependence of Fetal Mouse Neurons on Trophic Factors after Taxol Exposure in Organotypic Cultures

Stanley M. Crain and Edith R. Peterson

1. INTRODUCTION

The factors that lead to selective degeneration of specific types of neurons in many neurological disorders of the embryonic and mature organism are unknown. Appel (1981) has suggested that degenerative disorders such as amyotrophic lateral sclerosis (ALS), Parkinson's disease, and Alzheimer's disease may all result from the "lack of a neurotrophic hormone which is elaborated or stored in the synaptic target of the affected neurons and exerts a specific effect by acting in a retrograde fashion. These neurotrophic hormones would be of vital importance to survival of neurons during development, to their sprouting during regeneration, and to their maintenance throughout the life cycle" (see also reviews in Varon and Bunge, 1978; Thoenen and Barde, 1980). Removal of a limb during embryonic development enhances the extent of "programmed" motoneuron death (Hamburger, 1975; Prestige and Wilson, 1974; Oppenheim *et al.*, 1978; Lamb, 1981), whereas addition of supernumerary limbs increases motoneuron survival (Hollyday and Hamburger, 1976). Tissue-culture analyses have provided further evidence of specific neurotrophic factors derived from

STANLEY M. CRAIN and EDITH R. PETERSON ● Departments of Neuroscience and Physiology, and the Rose F. Kennedy Center for Research in Mental Retardation and Human Development, Albert Einstein College of Medicine, Bronx, New York 10461.

target tissues that enhance survival, neurite extension, and neurotransmitter-synthetic enzymes in presynaptic neurons, e.g., skeletal-muscle factors on cultured ventral spinal cord cholinergic neurons (Giller *et al.*, 1973, 1977; Bennett *et al.*, 1980; Smith and Appel, 1981, 1982, 1983; Appel, 1981), striatal factors on cultured substantia nigra dopaminergic neurons (Hemmendinger *et al.*, 1981; Prochiantz *et al.*, 1979; Appel, 1981), hippocampal factors on cultured septal cholinergic neurons (Appel, 1981), and tectal factors on cultured retinal ganglion cells (Nurcombe and Bennett, 1981; Smalheiser *et al.*, 1981; McCaffery *et al.*, 1982) (see also Chapters 16 and 17). On the basis of these developmental studies *in situ* and *in vitro*, Appel (1981) suggested that "the primary manifestation of ALS, Parkinson disease and Alzheimer disease may be failure of the target tissue to supply the necessary [specific] neurotrophic hormone" to the presynaptic ventral-horn motoneurons, substantia nigra cells, and septal cells, respectively, resulting in selective degeneration of these neurons.

Similarly, it has been suggested that a deficiency in nerve growth factor (NGF) may account for the selective degeneration of peripheral sensory- and sympathetic-ganglion neurons that occurs in familial dysautonomia (FD) *in utero* (e.g., Pearson *et al.*, 1974, 1978; Pearson, 1979; Crain *et al.*, 1980) (see also Chapter 22). This hypothesis is based on the selective degeneration of sympathetic ganglion cells in neonatal rats following anti-NGF antibody injections (Levi-Montalcini and Angeletti, 1966) and the failure of chick, rodent, and human embryonic dorsal-root ganglion (DRG) neurons to survive in cultures deficient in NGF (Levi-Montalcini and Angeletti, 1963; Crain and Peterson, 1974a; Crain *et al.*, 1980). Further support for this view derives from studies by E.M. Johnson *et al.* (1980) and Aloe *et al.* (1981) demonstrating that prenatal exposure of pregnant rats and guinea pigs to anti-NGF antibodies results in the selective destruction of up to 85% of DRG neurons as well as destruction of sympathetic neurons. E.M. Johnson *et al.* (1980) noted that their "data *in utero* are quantitatively consistent with experiments showing that most embryonic DRG neurons will survive *in vitro* in the presence, but not in the absence, of NGF" (Crain and Peterson, 1974a). However, the severe depletion of parasympathetic sphenopalatine ganglion neurons in FD (Pearson and Pytel, 1978)—which does not occur in embryos treated with anti-NGF antibodies (E.M. Johnson *et al.*, 1980)—suggests that the neuronal degeneration in FD may involve other types of neural-crest cells in addition to NGF-dependent sensory- and sympathetic-ganglion neurons (Pearson, 1979 and personal communication).

Although the selective neuronal degeneration that occurs in FD, as well as in the central nervous system (CNS) diseases noted above, may be due to a failure of target tissues to supply specific neurotrophic hormones [or to molecular defects in the synthesis of the trophic factors (e.g., Schwartz and Breakefield 1980)], more systematic attempts are needed to determine whether these neurological disorders could alternatively result from impairments in the ability of certain neurons to utilize or respond to the neuronotrophic survival factors. As noted by Appel (1981), "any process such as trauma, toxins, ischemia, or infections that added to the age-dependent loss of neurotrophic hormone or impaired the neuron uptake or retrograde transport process would accentuate the difficulty and increase the likelihood of neuronal degeneration and central nervous system symptomatology."

We have recently used taxol, an antitumor plant alkaloid, as a probe to selectively disrupt the neuronal dependencies on trophic factors that are expressed in organotypic cultures of fetal mouse DRGs explanted together with attached spinal cord. Our *in vitro* studies draw attention to the possibility that the selective degeneration of sensory- and autonomic-ganglion cells as occurs in FD—and perhaps of CNS neurons in some types of degenerative disorders—may be due to an abnormal requirement of these neurons for much higher levels of NGF, or of other specific hormones and trophic factors (or of both) than are normally available *in utero* or in the adult organism.

2. PRIMARY EFFECTS OF TAXOL

Taxol, an antitumor alkaloid (Wani *et al.*, 1971) isolated from the plant *Taxus brevifola*, promotes the assembly of microtubules (MTs) and binds to them in cell-free systems; such MTs resist depolymerization by cold or calcium (Schiff *et al.*, 1979; Parness and Horwitz, 1981). In fibroblasts and other proliferating cells exposed to taxol, abnormal bundles of MTs form throughout the cytoplasm and mitosis is blocked (Schiff and Horwitz, 1980; Masurovsky *et al.*, 1982). Many of the taxol-induced MTs appeared to form at aberrant sites in the cytoplasm, rather than in orderly spatial relationship to cellular MT-organizing centers (MTOCs) (DeBrabander *et al.*, 1981a; Simone *et al.*, 1981). In contrast, cytoplasmic MTs attached to MTOCs were not stabilized by taxol, and they gradually disappeared during exposure to the drug (DeBrabander *et al.*, 1981a). Taxol exposure also leads to the formation of unusual numbers and arrays of MTs in neurons and supporting cells in explants of spinal-cord tissue with attached DRGs (Fig. 1) (Masurovsky *et al.*, 1981a,b, 1983). The primary effects of taxol appear to be on the tubulin–MT system; no other specific mode of action has been reported. In addition to the dramatic effects of taxol on MTs in neurons and other cells, this drug may also have significant effects on tubulin associated with certain other cellular components. Recent studies have provided evidence that specific forms of tubulin are associated with endoplasmic reticulum, plasma and synaptic membranes in neurons (e.g., Soifer and Czosnek, 1980; Strocchi *et al.*, 1981; Gozes and Littauer, 1979; Zisapel *et al.*, 1980; Burke and De-Lorenzo, 1982; see also Jacobs, 1982). Some of the changes produced by taxol, and by other drugs that interact with tubulin complexes, may therefore be me-

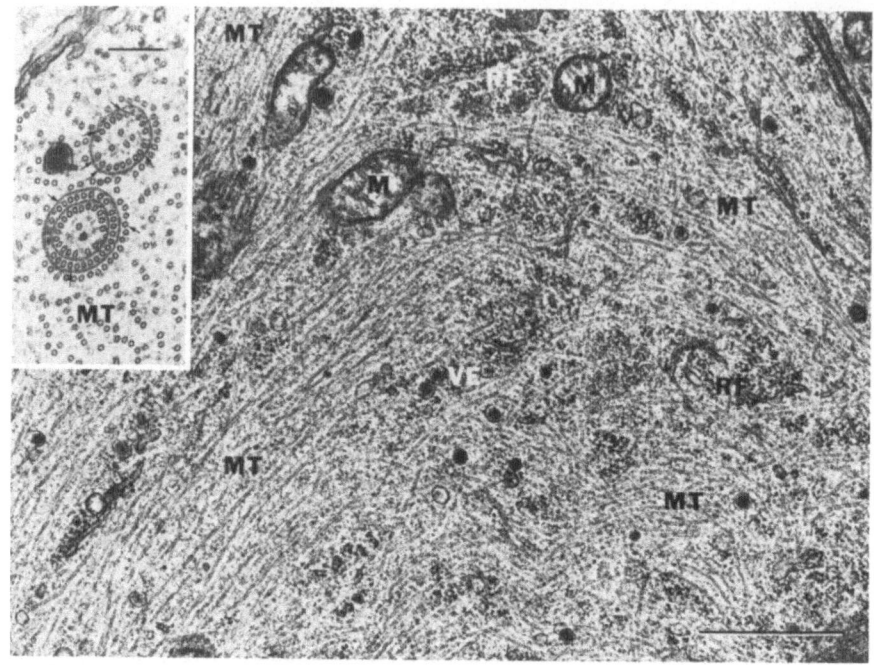

Figure 1. Electron micrograph illustrating an unusual abundance of microtubules (MT) coursing through the cytoplasm near an exiting process of a neuron in a 13-day fetal mouse DRG explant exposed to taxol (1 μM) for 6 days (+NGF), after an initial period of development for more than 2 weeks in control culture medium. The MTs appear in various orientations interspersed with foci of vesicles (VE), mitochondria (M), and ribosomal formations (RF). $\times$40,000. Scale bar: 1 μm. *Inset:* Transverse section through concentric ordered arrays of MTs alternating with layers of macromolecular material in a portion of a neuritic extension near the soma. Connections between some MTs and these nonmembranous lamellae appear at various points in these complexes (e.g., arrows). Some nearby MTs appear to be deployed in various linear and other groupings. $\times$80,000. Scale bar: 0.2 μm. From Masurovsky *et al.* (1983), courtesy of IBRO and Pergamon Press (see also Masurovsky *et al.*, 1981a).

diated by direct effects on membrane tubulin systems, rather than by alterations of the polymerized tubulin in MTs (e.g., Stephens, 1983).

Analyses of the effects of the plant alkaloid taxol on neurons are of further interest in view of the possibility raised by Schiff *et al.* (1979) that "there may be low molecular weight cell constituents with activity similar to that of taxol which may be involved in the regulation of MT assembly in the cell" (see also Schiff and Horwitz, 1981). Furthermore, DeBrabander *et al.* (1981a–c) have proposed that a factor that locally lowers the critical concentrations of tubulin required for polymerization and assembly of MTs may normally be released or concentrated at MTOCs in mammalian cells. The implications of these analyses in fibroblasts, together with studies of the effects of taxol on cultures of DRG and cord tissues (see below), have led us to speculate that a defect that underlies the selective degeneration of nerve cells in some types of neurological disorders might involve failure of the cellular system regulating the level of such a postulated endogenous factor with activity similar to that of taxol (see Section 6).

3. REVERSIBLE BLOCKADE OF NEURITIC OUTGROWTH DURING EXPOSURE TO TAXOL

In view of the abundance of MTs that form in relatively mature mouse DRG neurons, and in at least some cord neurons, exposed to taxol after 2 weeks in culture (Masurovsky *et al.*, 1981a, 1983), tests were made to determine the effects of this drug when introduced during the first few days after explantation of fetal DRG–cord tissues.

For these experiments, spinal cord from 13-day-old fetal mice was transversely cross-sectioned between each pair of DRGs. Cord cross-sections with attached DRGs were placed on collagen-coated coverslips and incubated in Maximow depression-slide chambers at 35°C in a medium containing 30% human placental serum and 10% chick embryo extract (see details of culture technique in Crain and Peterson, 1974a; Crain *et al.*, 1980). NGF (300 U/ml; ≈3 μg/ml) was added to the medium during the first week *in vitro* to ensure optimal survival and growth of a large fraction of the DRG neurons (>1000 per ganglion) (Fig. 2). In the absence of exogenous NGF, more than 90% of the DRG neurons in these 13-day fetal explants degenerated during the first week *in vitro*, even in our enriched nutrient media (Fig. 2B) (Crain and Peterson, 1974a).

In cultures exposed to 1–2 μM taxol during the first few days after explantation, survival of DRG neurons was comparable to that in control cultures as long as NGF (300 U/ml) was present. The DRG perikarya showed some cytoplasmic aberrations and nuclear distortions, but these were reversible after taxol was withdrawn (Peterson *et al.*, 1980). Neuritic outgrowth, on the other hand, was sharply attenuated or completely blocked* (Peterson *et al.*, 1980;

---

* In cultures of cord–DRG explants exposed to lower concentrations of taxol (0.1 μM) in the presence of NGF, abundant outgrowth occurred, but the neurites showed abnormal contours, e.g., large, tapered proximal processes instead of characteristic cylindrical fibers of fine diameter; wide variations in neurite diameter all the way to their terminations, with unusual tortuous branching and arborizations (Peterson and Crain, unpublished observations). Other types of abnormal neurite growth patterns have been observed in cultures of PC12 pheochromocytoma cells exposed to 0.2 μM taxol + 50 ng/ml NGF (Corvaja *et al.*, 1982, 1983). Electron microscopy of taxol-treated PC12 cells showed that their cytoplasm and neuritic processes were filled with an extremely large number of MTs (Corvaja *et al.*, 1982, 1983).

Peterson and Crain, 1982a), in contrast to the thousands of neurites that normally grow out from NGF-enhanced fetal DRGs during the first few days in control medium. Similar inhibition of neurite elongation occurred if taxol was introduced after several days of initial neuritic outgrowth *in vitro* (Fig. 3). Withdrawal of taxol after 3–4 days of exposure led to resumption of abundant neurite outgrowth within 1–2 days, but this recovery occurred only in media containing high levels of NGF (see below). The reversible blockade of neurite growth by taxol may

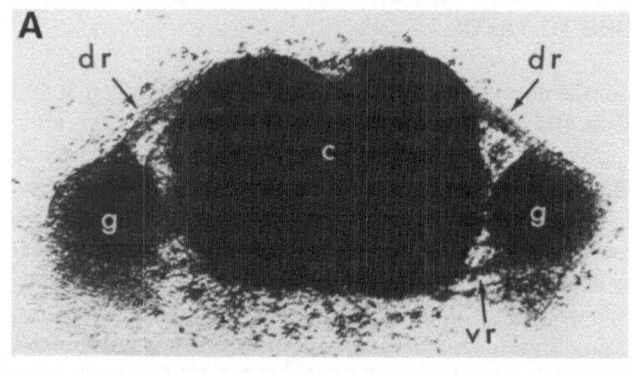

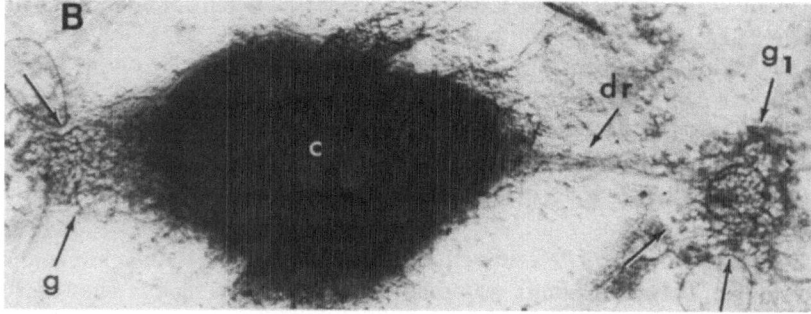

Figure 2. Photomicrographs of 13-day fetal mouse spinal cord explants (transverse cross sections, (≈0.5 mm thick) with attached DRGs. These are living, unstained cultures. (A) Shortly after explantation (1 day *in vitro*). Note the size of the DRGs (g) relative to cord tissue (c). (dr) Dorsal roots; (vr) ventral roots. (B) After 1 month in normal culture medium. Most of the ganglion cells degenerated during the first few days *in vitro*, leaving a small, thinly spread array of DRG neurons (g) that have matured and retained characteristic (myelinated) dorsal root (dr) connections to the cord (c). Note that the DRGs are of similar size, although only one (g₁) shows the characteristic "migration" away from the cord. Most cultures in control culture media showed even lower survival of DRG neurons. (C₁) Another cord–DRG explant after 1 month in the same culture medium, but NGF (1000 U/ml) was added at explantation. No additional NGF was provided during subsequent feedings. Note the remarkable enlargement of DRGs (g) relative to their initial size at explantation (A) and in contrast to the control culture (B). There are many hundreds of ganglion cells forming dense-packed clusters [see high-power view (C₂)]. (The major DRG volume increase was reached by the 2nd week *in vitro*.) The relatively dense appearance of dorsal cord (dc) is due partly to large numbers of myelinated axons that represent central branches of DRG neurons. The dense region in ventromedial cord (vc) is due primarily to a "necrotic core" that often develops in both treated and control explants. Scale bars: (A–C₁) 1 mm; (C₂) 20 μm. From Crain and Peterson (1974a).

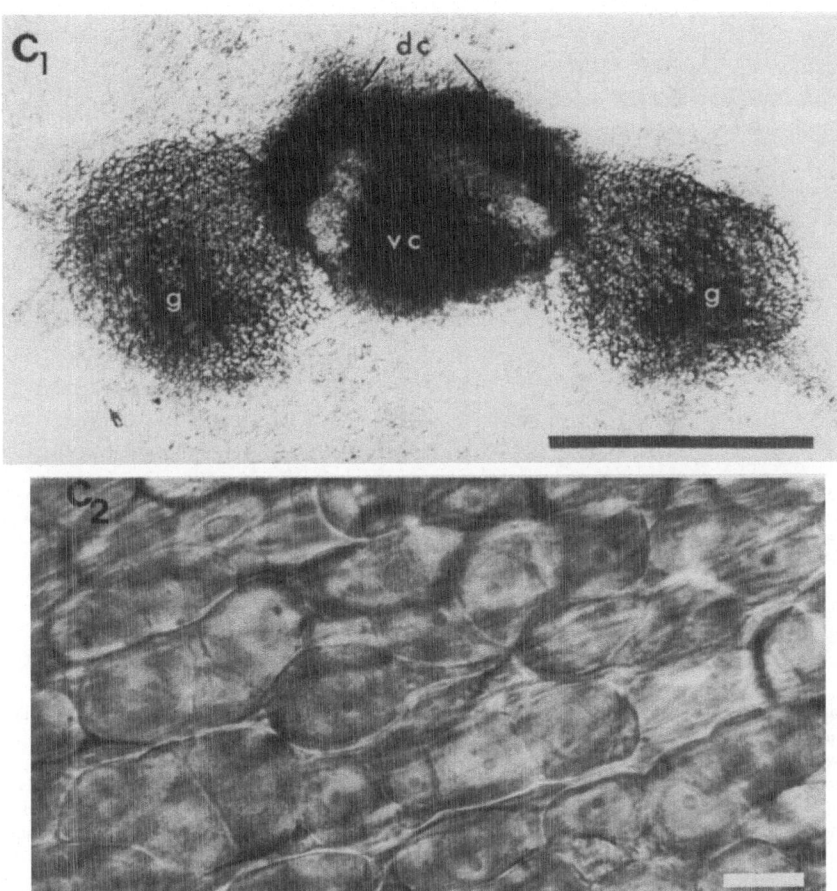

Figure 2. Continued.

be related to the disruption of the normal organization or function (or both) of MTs that probably occurs in the drug-treated neurons (see Section 2). Previous studies of embryonic DRG neurons exposed to colchicine and other plant alkaloids that inhibit MT assembly have shown that axon elongation is blocked at drug concentrations that produce marked depletion of MTs (e.g., Daniels, 1972, 1973; Yamada *et al.*, 1970, 1971). Our present data suggest that the growth and elongation of neurites may require not only the ability to assemble adequate numbers of MTs, but also the active regulation of MTs by MTOCs or other tubulin-associated systems (or both) involved in topographic distribution of specialized cell-surface components and maintenance of cellular polarity and asymmetry (see reviews by Brinkley *et al.*, 1981a,b; DeBrabander *et al.*, 1981a–c; Nicolson, 1979). Similar disruption of the organization and regulation of MTs

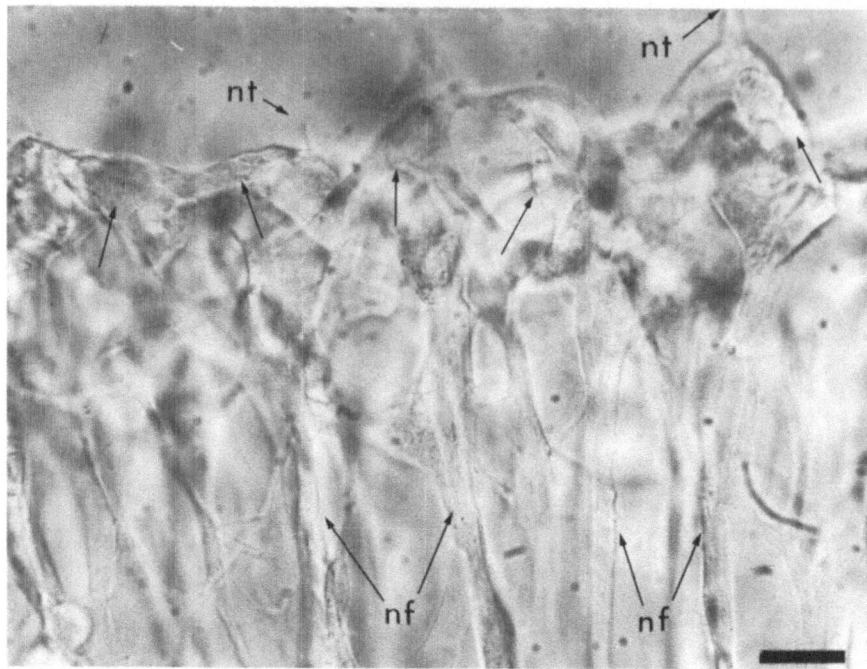

Figure 3. Blockade of elongation of neurites growing out from fetal mouse cord–DRG explant during exposure to taxol (1 μM + 500 U/ml NGF) for 2 days, following an initial 4-day period of growth in normal culture medium (+NGF). Photomicrograph of living culture (at 6 days *in vitro*) shows retracted and swollen supporting cells, e.g., Schwann cells (→), at the periphery of the neuritic outgrowth zone (about 1 mm from the edge of the explant). Neurites, in densely packed fascicles (nf), have ceased characteristic elongation. Note the few neurites (nt) projecting for a short distance beyond the outer rim of supporting cells. In the absence of taxol, large numbers of neurites in these fascicles would have splayed out and projected for hundreds of microns beyond the periphery of the growth zone shown here. Neurite elongation resumed within 1–2 days after transfer to normal medium supplemented with NGF [as in Figs. 4A and 5–7 (see the text)]. Scale bar: 20 μm.

Figure 4. Degeneration of DRG and cord neurons in a fetal mouse cord–DRG explant from which NGF was withdrawn after exposure to taxol + NGF. (A) Explant cultured for 5 weeks in medium with NGF (300 U/ml) and exposed to 1 μM taxol for 24 hr during days 4–5 *in vitro*. Neuronal development in DRG (G) and cord (C) tissues was not altered, although growth of Schwann cells was severely depressed. Dorsal (dc) and ventral (vc) regions of the cord contain abundant neurons. Note the profuse outgrowth of neurites (n) without Schwann cells (see also Figs. 5A, 6, and 7). (B$_1$, B$_2$) In contrast, extensive degeneration occurred in cord (C) and DRGs (G) in an explant in which NGF was withdrawn from the culture medium for 3 weeks following the exposure to 1 μM taxol during the 4th–8th days *in vitro*. NGF was present in the medium for 4 days prior to and during the exposure to taxol. The cord and DRG tissues have been reduced to thin layers, and neuritic outgrowth is sparse. The number of neurons per DRG has been reduced from more than 1000 [see (A)] to fewer than 100. (B$_2$) The arrow indicates a surviving DRG neuron (cf. Fig. 2C$_2$). The damage would have been even more severe had taxol been added at the time of explantation. Scale bars: (A, B$_1$) 500 μm; (B$_2$) 20 μm. From Peterson and Crain (1982a).

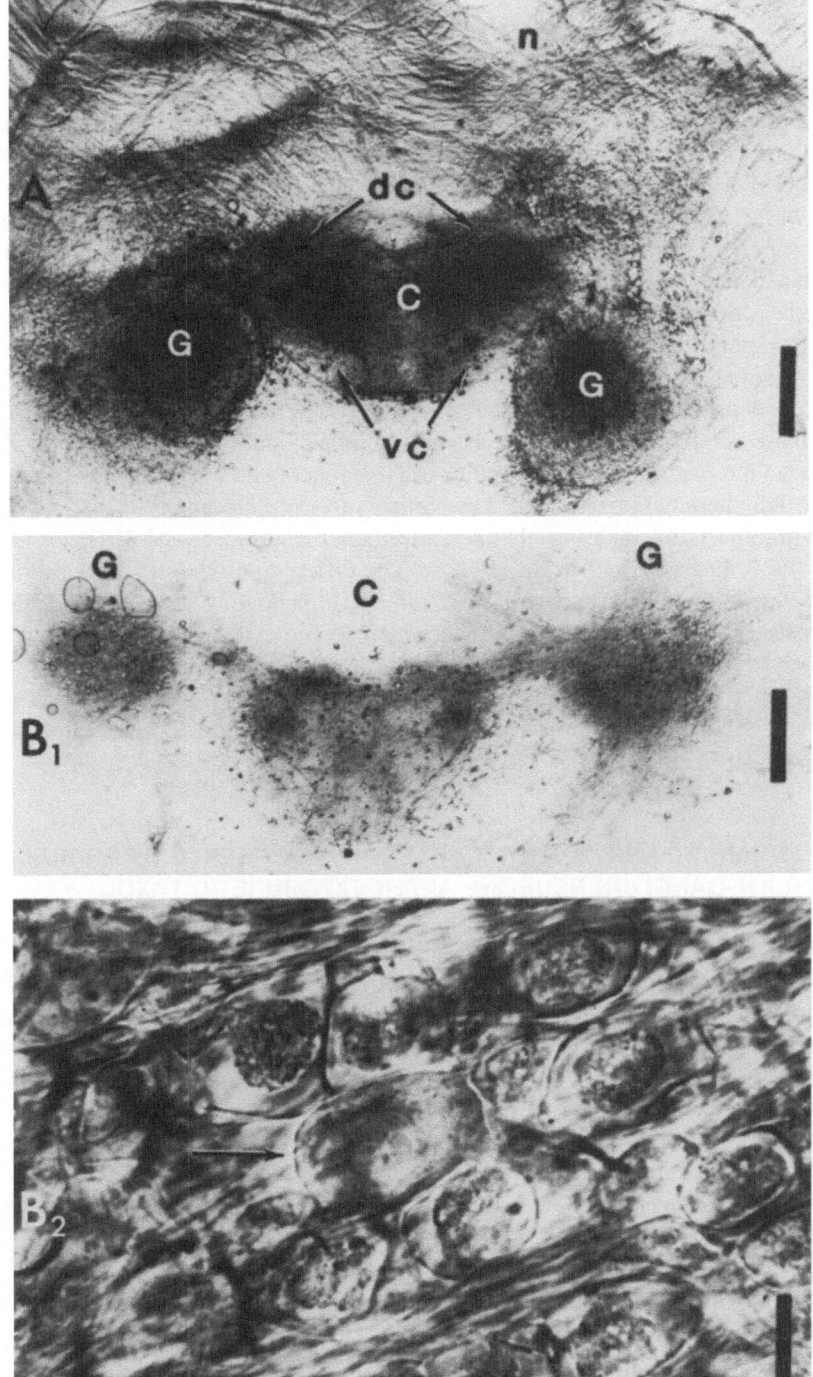

in fibroblasts exposed to taxol may account for the loss of ability of these cells to actively migrate even though they were still able to produce mobile surface projections, e.g., lamellipodia and filopodia (Schiff and Horwitz, 1980). These studies of the effects of taxol on fibroblasts, together with analyses of taxol effects on MT assembly in cell-free systems (Schiff *et al.*, 1979), led Schiff and Horwitz (1980) to suggest that "migrating cells must be able both to polymerize and depolymerize their cytoplasmic MTs in order to differentiate between their front and back ends." Further experiments are required, however, to demonstrate whether the neurite-growth-blocking effects of taxol are in fact mediated by similar specific interference with MTs or other tubulin-associated systems.

Schwann-cell and satellite-cell proliferation and development were also sharply depressed during exposure of fetal DRGs to taxol (Peterson *et al.*, 1980), but these surviving supporting cells showed little evidence of recovery following drug withdrawal, and many satellite cells degenerated. The DRG neuritic outgrowth in these cultures remained remarkably free of Schwann cells (and myelin) and fibroblasts for many weeks thereafter (see Figs. 4A, 5, and 7) (Peterson *et al.*, 1980; Peterson and Crain, 1982a). The differential recovery of the growth of neurons vs. supporting cells in fetal DRGs following a 4- to 6-day taxol exposure correlates well with electron-microscopic evidence of the gradual approach to more normal ultrastructure in these DRG neurons—in contrast to the persistence of an unusual abundance of MTs, as well as other cytoplasmic and nuclear abnormalities, in satellite and Schwann cells even 7 weeks after drug withdrawal (Masurovsky *et al.*, 1981b).

4. ENHANCED NERVE-GROWTH-FACTOR DEPENDENCE OF DORSAL-ROOT-GANGLION NEURONS AFTER EXPOSURE TO TAXOL

Most 13-day fetal mouse DRG neurons explanted into nutrient media containing high NGF levels (>300 U/ml) will survive and mature for months in culture even when exogenous NGF is withdrawn after the first 4 days *in vitro* (see Fig. 2C) (Crain and Peterson, 1974a). These data are consistent with a wide variety of evidence that chick, rodent, and human embryonic DRG neurons

--→

Figure 5.   Survival of dorsal—but not ventral—cord tissue in a cord–DRG explant in which NGF was maintained in the medium for 3 weeks after exposure to 1 μM taxol during the 4th–8th days *in vitro*. (A) Photomicrograph of living explant shortly after 4-day exposure to taxol. No significant loss of cord tissue has occurred at this stage, but ventral cord shows unusually marked granular appearance. (B) Photomicrograph of same explant, 5 days later. Note that the ventral region of the cord (vc) has been reduced to a thin layer, whereas the DRG (G) and dorsal cord (dc) tissues have retained characteristic multilayered neuronal populations [in contrast to the degeneration of the DRGs and entire cord in the absence of NGF supplementation (see Fig. 4B$_1$, B$_2$)]. Scale bar: 500 μm.

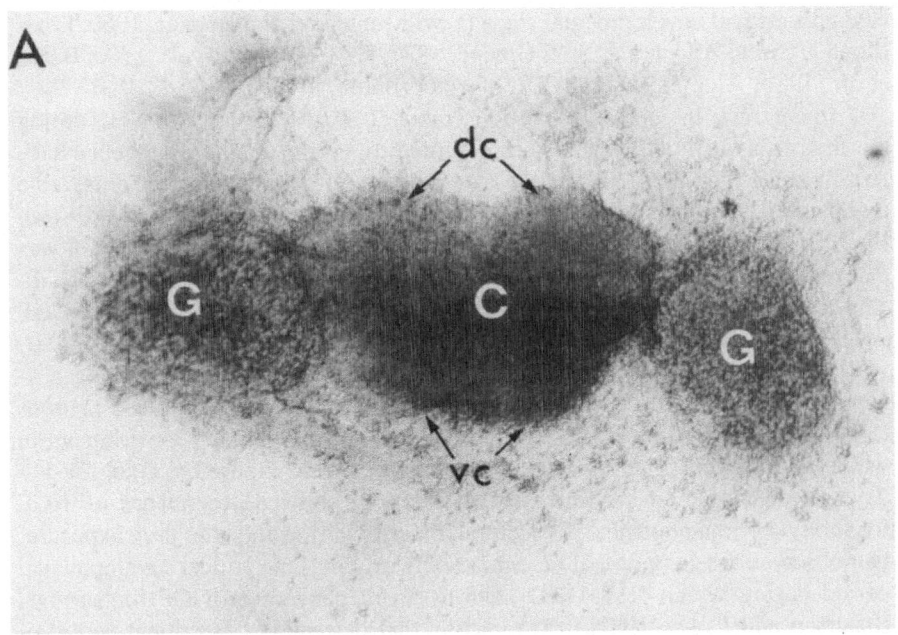

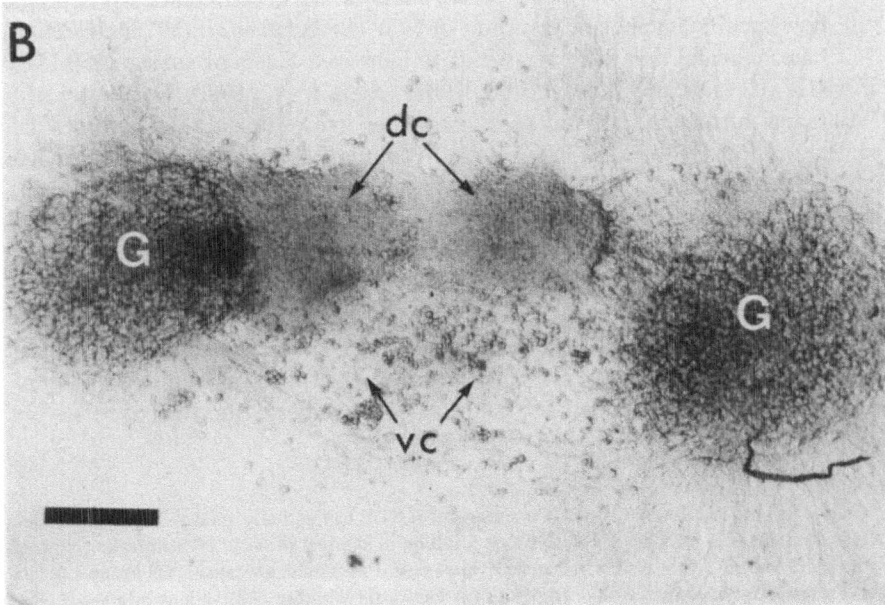

become less dependent on exogenous NGF for their survival as they mature beyond a critical developmental stage (Levi-Montalcini and Booker, 1960; Levi-Montalcini and Angeletti, 1968; Greene, 1977; E.M. Johnson *et al.*, 1980; Barde *et al.*, 1980; Aloe *et al.*, 1981; Rohrer and Barde, 1982).

In contrast, in cultures exposed to taxol (1–2 μM) along with NGF during the first 4 days in culture and then returned to control medium without NGF, less then 5% of the DRG neurons survived ($\approx$60 per ganglion). Comparable numbers of DRG neurons survive in control cultures grown without NGF (see Fig. 2B) (Crain and Peterson, 1974a). Even when the introduction of taxol was delayed until 4–8 days after explantation (to permit an initial period of NGF-stimulated DRG growth), similar widespread cytotoxic effects occurred if NGF was omitted after drug withdrawal (Fig. $4B_1$, B_2) (Peterson and Crain, 1982a).

If, however, NGF was maintained in the culture medium following a 4-day exposure to taxol, there was a greater than 90% survival of DRG neurons (Figs. 5–7) (see also Section 5). Furthermore, in preliminary experiments in which 3- to 5-week-old, NGF-enhanced explants were exposed to taxol for 4–6 days, a substantial fraction of the DRG neurons showed dependence on NGF for survival or maintenance of structural integrity during and after drug exposure. Our observations suggest that taxol markedly prolongs the critical developmental period during which fetal DRG neurons are dependent on NGF for survival (Peterson and Crain, 1982a). The factors that determine this critical period of NGF dependency during development are poorly understood. Adult DRG neurons still show specific uptake and transport of NGF (Stoeckel *et al.*, 1975). Injecting NGF into neonatal and adult rats leads to increased levels of substance P (SP) in DRGs (Kessler and Black, 1980, 1981; Otten *et al.*, 1980; Goedert *et al.*, 1981), and introduction of anti-NGF antibodies into postnatal rats decreases the content of SP in DRGs (Otten *et al.*, 1980). On the other hand, injecting postnatal rats with antibodies to NGF no longer depletes the DRG neuronal population, in sharp contrast to the degeneration of sympathetic-ganglion neurons that occurs even after antibody injections at more mature stages (Levi-Montalcini and Angeletti, 1966). Taxol exposure therefore appears to alter DRGs so that at least some of the more mature sensory neurons in culture become dependent for their survival on exogenous NGF,* resembling the NGF dependency of postnatal sympathetic-ganglion cells. These *in vitro* studies of taxol-altered DRG neurons suggest that in certain types of neurological disorders, survival of relatively

---

* Otten *et al.* (1983) have recently demonstrated that NGF can partially antagonize the neurotoxic effect of capsaicin on 2-day-old rat DRGs, resulting in survival of about 80% instead of 56% of the DRG neurons. This evidence that survival of postnatal capsaicin-treated DRG neurons *in situ* is enhanced by injection of NGF (0.5 μg/g) is consonant with our observations of marked NGF dependence in fetal mouse DRG neurons exposed to taxol after several weeks of maturation in culture (Peterson and Crain, 1982a).

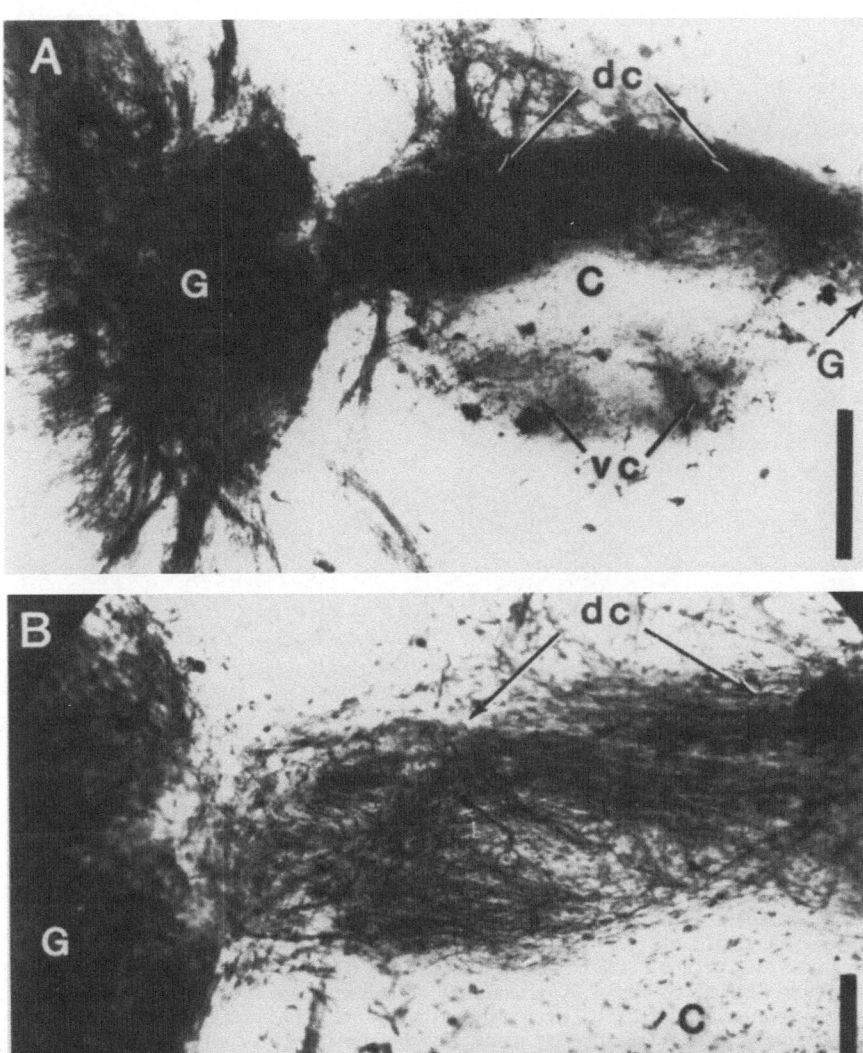

Figure 6. Cord–DRG explant exposed to taxol + NGF as in Fig. 5, showing abundant selective projection of DRG neurites throughout the dorsal region (dc) of cord (C). Horseradish peroxidase (HRP) was injected into the DRGs (G) by microiontophoresis after maintenance of the culture (with NGF) for 7 weeks following exposure to 1 μM taxol during the 4th–8th days *in vitro* (photomicrograph of whole-mount preparation, counterstained with toluidine blue.) Note that the HRP-labeled DRG neurites have arborized profusely throughout the dorsal cord and that few DRG neurites have invaded the ventral region (vc), as in control explants (see the text). (B) In higher-power view of dorsal cord, the monolayer of residual ventral cord tissue is out of focus. Abundant nerve cells were visible throughout dorsal cord tissue, whereas very few surviving neurons were seen in ventral tissue. Scale bars: (A) 500 μm; (B) 200 μm. From Peterson and Crain (1982a).

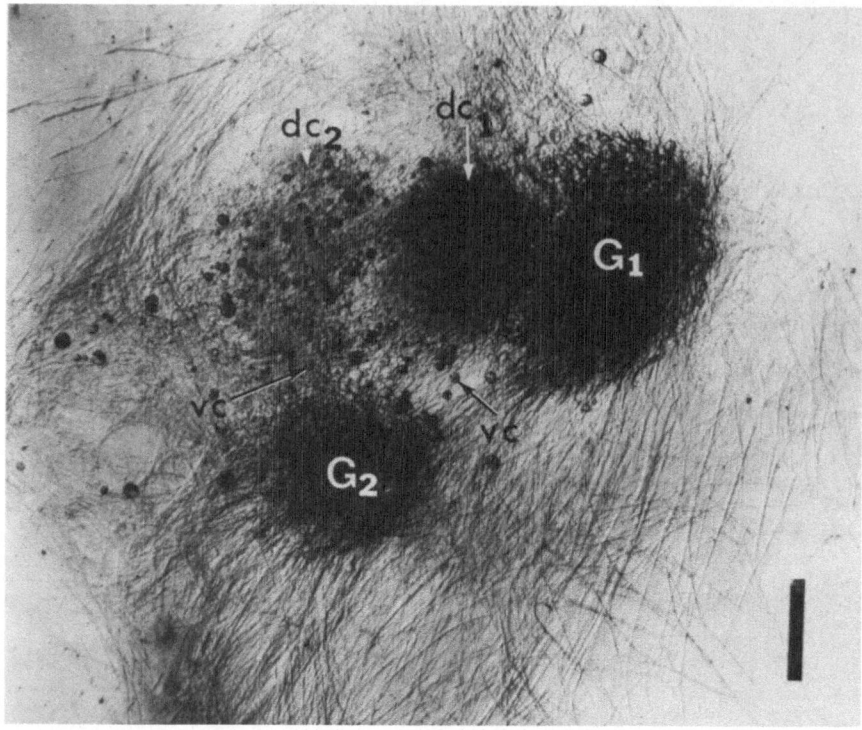

Figure 7. Selective survival of a portion of dorsal cord tissue (dc$_1$) innervated by attached DRG (G$_1$)—in contrast with severe degeneration of the contralateral region of dorsal cord (dc$_2$) lacking an attached DRG—in an explant in which NGF was maintained in the culture medium for 11 days after exposure to 1 μM taxol during the first 4 days *in vitro*. [The ventral region of the cord (vc) has been reduced to a thin layer (as in Figs. 5 and 6).] Degeneration of the deafferented side of the dorsal cord (dc$_2$) occurred even though an isolated DRG (G$_2$) was located nearby (this detached ganglion was actually located at the ventral edge of the cord tissue at the time of explantation). Neurites from the detached DRG were unable to invade the cord tissue (see the text), whereas neurites from attached DRGs arborized profusely within dorsal cord (as in Fig. 6). Scale bar: 500 μm.

mature DRG cells, as well as sympathetic-ganglion cells, may be enhanced by introduction of high levels of exogenous NGF.

Our demonstration that NGF attenuates the neurotoxic effects of taxol on fetal DRG explants in culture is of interest in relation to evidence that NGF prevents the cytotoxic effects of vinblastine on neonatal rodent sympathetic ganglia (SGs) *in situ* (Menesini-Chen *et al.*, 1977; Calissano *et al.*, 1976; E.M. Johnson, 1978). However, attempts to demonstrate this effect of NGF on SGs exposed to vinblastine in culture were not successful (Calissano *et al.*, 1976).

Although taxol promotes the assembly of MTs (Parness and Horwitz, 1981; Schiff *et al.*, 1979), whereas vinblastine inhibits such assembly (Dustin, 1978), the toxic effects of both drugs on these neurons may be due to similar alterations in the normal organization of MTs or of other tubulin-associated systems— alterations that are counteracted by NGF.

5. ENHANCED DORSAL-ROOT-GANGLION DEPENDENCE OF FETAL DORSAL-CORD NEURONS AFTER EXPOSURE TO TAXOL

CNS neurons in deafferented spinal-cord explants from rodent fetuses develop and are maintained quite normally for months in control culture media (Peterson *et al.*, 1965; Peterson and Crain, 1982b; Crain and Peterson, 1964, 1967, 1974a,b, 1982; Crain, 1976), and dorsal-horn neurons remain receptive to innervation by cocultured DRG neurons introduced after 2–3 weeks *in vitro* (Crain and Peterson, 1982). In contrast, during and after exposure to taxol (1–2 μM), most fetal dorsal-cord neurons degenerated unless they were innervated by neuritic projections from NGF-enhanced DRGs (Peterson and Crain, 1982a). Even when introduction of taxol was delayed until 4 days after explantation of fetal cord-DRG explants, so as to permit an initial period of NGF-stimulated DRG neuritic growth and arborization within dorsal-cord tissue, similar widespread neurotoxic effects occurred if NGF was omitted after drug withdrawal (Fig. 4B$_1$, B$_2$). On the other hand, when the recovery medium was supplemented with NGF after 4 days of treatment with taxol + NGF, there was not only excellent survival of DRG neurons (as noted in Section 4), but also remarkably good survival of dorsal-cord tissue, whereas ventral cord was generally reduced to a very thin layer with few surviving neurons (Figs. 5–7) (Peterson and Crain, 1982a). In the latter experiments, microiontophoretic injection of horseradish peroxidase (HRP) into the DRGs of the taxol-exposed cultures showed that DRG neurites ramified and arborized profusely throughout the dorsal target regions of the attached cord, and only occasional DRG neurites invaded the ventral cord (Fig. 6)—similar to the selective DRG projections that occur regularly in control cord–DRG explants (Smalheiser *et al.*, 1982; Crain, 1982). Not only did dorsal-cord tissue in 4-day taxol-treated cord–DRG explants show maintenance of a large neuronal population after return to high-NGF media, but also characteristic dorsal-horn network responses could be evoked with DRG stimuli (Crain and Peterson, 1974a; Crain, 1976) in bioelectrical tests made at 3 weeks *in vitro*.

In cord cross-sections explanted with only *one* attached DRG and exposed to taxol + NGF, a significant portion of the dorsal cord on the side lacking an attached DRG showed severe degeneration—even when an isolated DRG was located close to the edge of the cord explant [Fig. 7(G$_2$)]. Neurites from nearby detached DRGs were unable to project into the cord tissue, since meningeal barriers had not been removed (Crain and Peterson, 1982; Peterson and Crain, 1982b). In a few cases in which portions of dorsal cord survived on the side lacking an attached DRG, HRP labeling showed that DRG neurites had projected across the midline in these explants and arborized in the contralateral cord. Furthermore, in cultures of completely deafferented cord cross-sections, both dorsal and ventral cord degenerated completely after a 4-day taxol exposure, even though high concentrations of NGF were present during and after drug exposure (at levels that permitted survival and recovery of isolated DRGs). These data rule out the possibility of a direct effect of NGF on the survival of dorsal-cord neurons (Peterson and Crain, 1982a).

Our analyses indicate that exposure of fetal cord–DRG explants to taxol greatly enhances the dependence of DRG neurons on NGF and the dependence of dorsal-cord neurons on factors provided by DRG neurite projections; in the absence of sustained increases in these extrinsic influences, taxol exposure results in severe cytotoxic effects on DRG and dorsal-cord neurons. Perhaps in drug-free cultures of fetal mouse explants, DRG and dorsal cord neurons are able to utilize small amounts of essential trophic factors present in our regular serum–embryo extract nutrient medium, or provided by glial or other supporting cells, and thus to maintain their integrity in the absence of NGF supplementation, afferent inputs, or target tissues (see Section 6). Some target tissues in the CNS have been shown to degenerate after deafferentation during critical periods of embryonic development *in vivo* (Hughes, 1968; J.I. Johnson *et al.,* 1972; Jacobson, 1978), but this, to the best of our knowledge, is the first demonstration of *in vitro* conditions under which neurons in CNS explants become dependent on their afferent inputs for survival (Peterson and Crain, 1981, 1982a).

6. CONCLUDING REMARKS

These studies provide a valuable new tissue-culture model to analyze some of the trophic factors that regulate survival of peripheral and central neurons during normal development and in certain types of neurological disorders. Furthermore, since the primary effects of taxol appear to be on the tubulin–MT system, and since numerous abnormal bundles of MTs form in aberrant locations in neurons and other cells exposed to this drug, it will be of interest to determine whether the altered requirements for survival of DRG and cord neurons in organotypic explants are related to the disorganization of MTs or of other tubulin-associated systems in these cells (see Section 2)—or both—or, alternatively, whether they are related to adverse effects of taxol on Schwann cells and glia (Masurovsky *et al.,* 1981b) that may interfere with their trophic support of the neuronal complex (e.g., Thoenen and Barde, 1980; Varon *et al.,* 1981). In preliminary tests with other mitotic inhibitors, e.g., cytosine arabinoside and fluorodeoxyuridine, in cord–DRG explants, we have not observed marked increases in NGF dependence of DRG neurons or DRG dependence of dorsal-cord neurons (Peterson and Crain, 1982a and unpublished results). Comparative tests with neurotoxic plant alkaloids that have modes of action on tubulin different from those of taxol (e.g., agents that interfere with MT assembly) may help to determine whether the alterations in survival requirements of taxol-treated DRG and cord neurons are due to either (1) general impairment of either or both MT and other tubulin-associated systems of these cells, e.g., interfering with axonal transport,* or (2) a more specific effect of taxol, e.g., on an MT-associated system that mediates nerve-cell responses to hormones or other trophic factors.

DRG and cord cultures may facilitate testing the hypothesis that taxol reduces the effectiveness of MT- or tubulin-associated systems that mediate the biological responses of neurons to hormonal or other trophic agents. Substantial evidence has been obtained in recent years suggesting that some types of transmembrane signaling may be mediated by local alterations in cytoskeletal elements linked to the cell surface (e.g., Edelman, 1976, 1982; Nicolson, 1979; McClain and Edelman, 1980; Crossin and Carney, 1981a,b).† Taxol has in fact been shown to interfere selectively with the characteristic stimulation of thymidine incorporation into fibroblasts by epidermal growth factor (EGF) and thrombin (Crossin and Carney, 1981a,b). On the basis of these and related studies with colchicine and other drugs that disrupt MTs by different mechanisms, Crossin and Carney (1981a,b) suggest that taxol may inhibit the growth-factor initiation of DNA synthesis by preventing the localized depolymerization of MTs that appear to be a required step in the normal biological action of EGF, thrombin, and other hormones on these cells. Introduction of much higher levels of specific trophic factors into taxol-exposed cord–DRG cultures, then, may provide sufficiently high concentration gradients to overcome the presumptive drug-induced impairment of the ability of neurons to respond to these survival factors.

Finally, it should be noted that ventral-cord tissues, as well as dorsal-cord tissues, appear to develop relatively normally in isolated explants of fetal spinal cord or cord with attached DRGs (Crain and Peterson, 1964, 1967, 1974a,b,

---

* Studies of drug-treated and axotomized adrenergic neurons *in situ*, in which injury was counteracted by exogenous NGF, suggested that these experimental manipulations had interfered with uptake of NGF or retrograde transport, of NGF from DRG terminals to their perikarya (e.g., Hendry, 1975; Purves and Nja, 1976; West and Bunge, 1976; Nja and Purves, 1978; see also the reviews by Thoenen and Barde, 1980; Harper and Thoenen, 1980, 1981; Levi-Montalcini, 1982).

† These hormonal effects of cytoskeletal systems may involve intracellular "second" messengers, as suggested by Thoenen and Barde (1980) in regard to possible enhancing effects of NGF on assembly of MTs in nerve cells (see also Heumann *et al.*, 1981; cf. Calissano and Cozzari, 1974; Levi *et al.*, 1975; Monaco *et al.*, 1977; Marchisio *et al.*, 1980).

1982; Peterson and Crain, 1970, 1972, 1979, 1982b; Peterson *et al.*, 1965; Crain, 1976), and motoneurons in these cord explants are capable of innervating skeletal muscle after delayed presentation of muscle tissue (Peterson and Crain, 1972, 1979; Crain, 1976). Perhaps motoneurons, as well as dorsal-cord neurons, in isolated cord explants are able to utilize small amounts of essential trophic factors present in the serum–embryo extract culture medium, or provided by glial or other supporting cells, and thus to maintain their integrity in the absence of target tissues. Just as exposure of fetal mouse cord–DRG explants to taxol appears to alter dorsal-cord neurons so that they become markedly dependent for their survival on factors provided by NGF-enhanced DRG inputs, perhaps taxol-treated ventral-cord neurons may also become much more dependent for their survival on trophic factors provided by muscle, other target cells, or CNS inputs. This hypothesis can be tested by determining the degree to which the neurotoxic effects of taxol on ventral-cord neurons may be counteracted by introduction of extracts or exudates derived from muscle (see Section 1) and other target tissues, just as NGF rescues drug-treated DRG neurons (see Section 4) and DRG factors rescue drug-treated dorsal-cord neurons (see Section 5).

Studies of the enhanced dependence of fetal mouse neurons on neurono-trophic survival factors after taxol exposure in organotypic cultures have stimulated us to consider the possibility that certain types of neurological degenerative disorders might involve defective regulation of a putative endogenous factor with activity similar to that of taxol (see Section 2) in specific groups of neurons. The "unifying hypothesis" of Appel (1981) ". . . (that) the cause of ALS, Parkinsonism and Alzheimer disease" is due to lack of neuronotrophic survival factors that are retrogradely transported to certain types of presynaptic neurons from their target tissues therefore provides a useful perspective for understanding the molecular defects that may underlie these neurological degenerative disorders of the CNS—as well as the degeneration of sensory and autonomic ganglion neurons observed in FD (see Section 1). The hypothesis would be strengthened, however, by the following modifications: (1) emphasis on the possible importance of orthogradely, as well as retrogradely, transported neuronotrophic survival factors [comparable to the trophic influences of motoneurons on the maintenance of mature skeletal-muscle fibers (e.g., Carlson, 1972; Gutmann, 1976; Peterson and Crain, 1972, 1979; Crain and Peterson, 1974b) and DRG inputs on embryonic dorsal-cord (Peterson and Crain, 1982a) and dorsal-column nuclei (J. I. Johnson *et al.*, 1972) and (2) recognition that some of these and related degenerative disorders may be due to a defect in the ability of certain types of neurons to respond to or utilize the neuronotrophic factors, rather than to a deficit in the supply of specific trophic factors.

ACKNOWLEDGMENTS. These studies were supported by research grants to S.M.C. from the National Institute of Neurological and Communicative Disorders and

Stroke (NINCDS) (NS-14990), the National Science Foundation (BNS-821-847), and the Dysautonomia Foundation, and to E.R.P. from the NINCDS (NS-08770). Taxol was provided by Dr. S.B. Horwitz and NGF by Collaborative Research, Inc. Tissue-culture facilities were kindly provided by Dr. M.B. Bornstein. Thanks are due to B. Crain for skillful technical assistance in the HRP labeling and electrophysiological experiments.

REFERENCES

Aloe, L., Cozzari, C., Calissano, P., and Levi-Montalcini, R., 1981, Somatic and behavioural postnatal effects of fetal injections of nerve growth factor antibodies in the rat, *Nature (London)* **291:**413–415.

Appel, S.H., 1981, A unifying hypothesis for the cause of amyotrophic lateral sclerosis, Parkinsonism, and Alzheimer disease, *Ann. Neurol.* **10**(6):499–505.

Barde, Y.-A., Edgar, D., and Thoenen, H., 1980, Sensory neurons in culture: Changing requirements for survival factors during embryonic development, *Proc. Natl. Acad. Sci. U.S.A.* **77:**1199–1203.

Bennett, M.R., Lai, K., and Nurcombe, V., 1980, Identification of embryonic motoneurons *in vitro:* Their survival is dependent on skeletal muscle, *Brain Res.* **190:**537–542.

Brinkley, B.R., Cox, S.M., Pepper, D.A., Wible, L., Brenner, S.L., and Pardue, R.L., 1981a, Tubulin assembly sites and the organization of cytoplasmic microtubules in cultured mammalian cells, *J. Cell Biol.* **90**(3):554–562.

Brinkley, B.R., Cox, S.M., and Fistel, S.H., 1981b, Organizing centers for cell processes, in: *Cytoskeletons and the Architecture of Nervous Systems, Neurosci. Res. Program Bull.* **19:**108–124.

Burke, B.E., and DeLorenzo, R.J., 1982, Ca^{2+}- and calmodulin-dependent phosphorylation of endogenous synaptic vesicle tubulin by a vesicle-bound calmodulin kinase system, *J. Neurochem.* **38:**1205–1218.

Calissano, P., and Cozzari, C., 1974, Interaction of nerve growth factor with the mouse-brain neurotubule protein(s), *Proc. Natl. Acad. Sci. U.S.A.* **71:**2131–2135.

Calissano, P., Monaco, G., Levi, A., Menesini Chen, G.M., Chen, J.S., and Levi-Montalcini, R., 1976, New developments in the study of NGF–tubulin interaction, in: *Contractile Systems in Non-Muscle Tissues* (S.V. Perry, A. Margreth, and R.S. Adelstein, eds.), Elsevier/North-Holland, Amsterdam. pp. 201–211.

Carlson, B.M., 1972, *The Regeneration of Minced Muscles*, S. Karger, Basel.

Corvaja, N., Calissano, P., and DiLuzio, A., 1982, Effect of taxol and NGF on pheochromocytoma cells (clone PC12), *Caryologie* **35:**121–122.

Corvaja, N., DiLuzio, A., Biocca, S., Cattaneo, A., and Calissano, P., 1983, Morphological and ultrastructural changes in PC12 pheochromocytoma cells induced by a combined treatment with NGF and taxol, *Exp. Cell Res.* (in press).

Crain, S.M., 1976, *Neurophysiologic Studies in Tissue Culture*, Raven Press, New York.

Crain, S.M., 1982, Role of CNS target cues in formation of specific afferent synaptic connections in organotypic cultures, in: *Neuroscience Approached through Cell Culture*, Vol. II (S.R. Pfeiffer, ed.), CRC Press, Florida, pp. 1–32.

Crain, S.M., and Peterson, E.R., 1964, Complex bioelectric activity in organized tissue cultures of spinal cord (human, rat and chick), *J. Cell. Comp. Physiol.* **64:**1–15.

Crain, S.M., and Peterson, E.R., 1967, Onset and development of functional interneuronal connections in explants of rat spinal cord-ganglia during maturation in culture, *Brain Res.* **6:**750–762.

Crain, S.M., and Peterson, E.R., 1974a, Enhanced afferent synaptic functions in fetal mouse spinal cord–sensory ganglion explants following NGF-induced ganglion hypertrophy, *Brain Res.* **79**:145–152.

Crain, S.M., and Peterson, E.R., 1974b, Development of neural connections in cultures, in: *Symposium: The Trophic Function of the Neuron* (D.B. Drachman and A.S. Smith eds.), *Ann. N. Y. Acad. Sci.* **228**:6–34.

Crain, S.M., and Peterson, E.R., 1982, Selective innervation of target regions within fetal mouse spinal cord and medulla explants by isolated dorsal root ganglia in organotypic co-cultures, *Dev. Brain Res.* **2**:341–362.

Crain, S.M., Peterson, E.R., Leibman, M., and Schulman, H., 1980, Dependence on nerve growth factor of early human fetal dorsal root ganglion neurons in organotypic cultures, *Exp. Neurol.* **67**:205–214.

Crossin, K.L., and Carney, D.H., 1981a, Evidence that microtubule depolymerization early in the cell cycle is sufficient to initiate DNA synthesis, *Cell* **23**:61–71.

Crossin, K.L., and Carney, D.H., 1981b, Microtubule stabilization by taxol inhibits initiation of DNA synthesis by thrombin and by epidermal growth factor, *Cell* **27**:341–350.

Daniels, M.P., 1972, Colchicine inhibition of nerve fiber formation *in vitro, J. Cell Biol.* **53**:164–176.

Daniels, M.P., 1973, Fine structural changes in neurons and nerve fibers associated with colchicine inhibition of nerve fiber formation *in vitro, J. Cell Biol.* **58**:463–470.

DeBrabander, M., Geuens, G., Nuydens, R., Willebrords, R., and DeMey, J., 1981a, Taxol induces the assembly of free microtubules in living cells and blocks the organizing capacity of the centrosomes and kinetochores, *Proc. Natl. Acad. Sci. U.S.A.* **78**(9):5608–5612.

DeBrabander, M., Geuens, G., DeMey, J., and Joniau, M., 1981b, Nucleated assembly of mitotic microtubules in living PTK2 cells after release from nocodazole treatment, *Cell Motil.* **1**:469–483.

DeBrabander, M., Geuens, G., Nuydens, R., Willebrords, R., and DeMey, J., 1981c, Microtubule assembly in living cells after release from nocodazole block: The effects of metabolic inhibitors, taxol and pH, *Cell Biol. Int. Rep.* **5**:913–920.

Dustin, P., 1978, *Microtubules,* Springer-Verlag, Berlin.

Edelman, G.M., 1976, Surface modulation in cell recognition and cell growth, *Science* **192**:218–226.

Edelman, G.M., 1982, Cell surface modulation, in: *Signal Transduction across Cellular Membranes, Neurosci. Res. Program Bull.* **20**:376–382.

Giller, E.L., Schrier, B.K., Shainberg, G., Fisk, H.R., and Nelson, P.G., 1973, Increased choline acetyltransferase activity in combined cultures of spinal cord and muscle cells from the mouse, *Science* **182**:588–589.

Giller, E.L., Neale, J.G., Bullock, P.N., Schrier, B.K., and Nelson, P.G., 1977, Choline acetyltransferase activity of spinal cord cell cultures increased by co-culture with muscle and by muscle-conditioned medium, *J. Cell Biol.* **74**:16–29.

Goedert, M., Stoeckel, K., and Otten, U., 1981, Biological importance of the retrograde axonal transport of nerve growth factor in sensory neurons, *Proc. Natl. Acad. Sci. U.S.A.* **78**:5895–5898.

Gozes, I., and Littauer, U.Z., 1979, The α-subunit of tubulin is preferentially associated with presynaptic membrane, *FEBS Lett.* **99**:86–90.

Greene, L.A., 1977, Quantitative *in vitro* studies on the nerve growth factor (NGF) requirement of neurons. II. Sensory neurons, *Dev. Biol.* **58**:106–113.

Gutmann, E., 1976, Neurotrophic relations, *Annu. Rev. Physiol.* **38**:177–216.

Hamburger, V., 1975, Cell death in the development of the lateral motor column of the chick embryo, *J. Comp. Neurol.* **160**:535–546.

Harper, G.P., and Thoenen, H., 1980, Nerve growth factor: Biological significance, measurement, and distribution, *J. Neurochem.* **34**:5–16.

Harper, G.P., and Thoenen, H., 1981, Target cells, biological effects, and mechanism of action of nerve growth factor and its antibodies, *Annu. Rev. Pharmacol. Toxicol.* **21**:205–229.

Hemmendinger, L.M., Garber, B.B., Hoffman, P.C., and Heller, A., 1981, Target-specific process formation by embryonic mesencephalic dopamine neurons *in vitro, Proc. Natl. Acad. Sci. U.S.A.* **78**:1264–1268.

Hendry, I.A., 1975, The effects of axotomy on the development of the rat superior cervical ganglion, *Brain Res.* **90**:235–244.

Heumann, R., Schwab, M., and Thoenen, H., 1981, A second messenger required for nerve growth factor biological activity?, *Nature (London)* **292**:838–840.

Hollyday, M., and Hamburger, V., 1976, Reduction of the naturally occurring motoneuron loss by enlargement of the periphery, *J. Comp. Neurol.* **170**:311–320.

Hughes, A.F.W., 1968, *Aspects of Neural Ontogeny,* Academic Press, New York.

Jacobs, M., 1982, Multiple tubulins in a single neurone: The key to microtubule function?, *Trends in Neuroscience* **5**:64–65.

Jacobson, M., 1978, *Developmental Neurobiology,* Plenum Press, New York.

Johnson, E.M., 1978, Destruction of the sympathetic nervous system in neonatal rats and hamsters by vinblastine: Prevention by concomitant administration of nerve growth factor, *Brain Res.* **141**:105–118.

Johnson, E.M., Gorin, P.D., Brandeis, L.D., and Pearson, J., 1980, Dorsal root ganglion neurons are destroyed by exposure *in utero* to maternal antibody to nerve growth factor, *Science* **210**:916–918.

Johnson, J.I., Hamilton, T.C., and Hsung, J.-C., 1972, Gracile nucleus absent in adult opossums after leg removal in infancy, *Brain Res.* **38**:421–424.

Kessler, J.A., and Black, I.B., 1980, Nerve growth factor stimulates the development of substance P in sensory ganglia, *Proc. Natl. Acad. Sci. U.S.A.* **77**:649–652.

Kessler, J.A., and Black, I.B., 1981, Nerve growth factor stimulates development of substance P in the embryonic spinal cord, *Brain Res.* **208**:135–145.

Lamb, A.H., 1981, Target dependency of developing motoneurons in *Xenopus laevis, J. Comp. Neurol.* **203**(2):157–171.

Levi, A.M., Cimino, M., Mercanti, D., Chen, J.S., and Calissano, P., 1975, Interaction of nerve growth factor with tubulin: Studies on binding and induced polymerization, *Biochim. Biophys. Acta* **399**:50–60.

Levi-Montalcini, R., 1982, Developmental neurobiology and the natural history of nerve growth factor, *Annu. Rev. Neurosci.* **5**:341–362.

Levi-Montalcini, R., and Angeletti, P.U., 1963, Essential role of the nerve growth factor in the survival and maintenance of dissociated sensory and sympathetic embryonic nerve cells *in vitro, Dev. Biol.* **7**:653–659.

Levi-Montalcini, R., and Angeletti, P.U., 1966, Immunosympathectomy, *Pharmacol. Rev.* **18**:619–628.

Levi-Montalcini, R., and Angeletti, P.U., 1968, Biological aspects of the nerve growth factor, in: *Ciba Foundation Symposium: Growth of the Nervous System* (G.E.W. Wolstenholme and M. O'Connor, eds.), Churchill, London, pp. 126–147.

Levi-Montalcini, R., and Booker, B., 1960, Excessive growth of the sympathetic ganglia evoked by a protein isolated from mouse salivary glands, *Proc. Natl. Acad. Sci. U.S.A.* **46**: 373–384.

Marchisio, P.C., Naldini, L., and Calissano, P., 1980, Intracellular distribution of nerve growth factor in rat pheochromocytoma PC 12 cells: Evidence for a perinuclear and intranuclear location, *Proc. Natl. Acad. Sci. U.S.A.* **77**:1656–1660.

Masurovsky, E.B., Peterson, E.R., Crain, S.M., and Horwitz, S.B., 1981a, Microtubule arrays in taxol-treated mouse dorsal root ganglion–spinal cord cultures, *Brain Res.* **217**:392–398.

Masurovsky, E.B., Peterson, E.R., Crain, S.M., and Horwitz, S.B., 1981b, Morphological alterations in satellite and Schwann cells after exposure of fetal mouse dorsal root ganglia–spinal cord cultures to taxol, *IRCS Med. Sci.* **9**:968–969.

Masurovsky, E.B., Peterson, E.R., Crain, S.M., and Horwitz, S.B., 1982, Taxol-induced microtubule formations in fibroblasts of fetal mouse dorsal root ganglion–spinal cord cultures, *Biol. Cell* **46**:213–216.

Masurovsky, E.B., Peterson, E.R., Crain, S.M., and Horwitz, S.B., 1983, Morphologic alterations in dorsal root ganglion neurons and supporting cells of organotypic mouse spinal cord–ganglion cultures exposed to taxol, *Neuroscience* **10**:491–509.

McCaffery, C.A., Bennett, M.R., and Dreher, B., 1982, The survival of neonatal rat retinal ganglion cells *in vitro* is enhanced in the presence of appropriate parts of the brain, *Exp. Brain Res.* **48**:377–386.

McClain, D.A., and Edelman, G.M., 1980, Density-dependent stimulation and inhibition of cell growth by agents that disrupt microtubules, *Proc. Natl. Acad. Sci. U.S.A.* **77**:2748–2752.

Menesini Chen, M.G., Chen, J.S., Calissano, P., and Levi-Montalcini, R., 1977, Nerve growth factor prevents vinblastine destructive effects on sympathetic ganglia in newborn mice, *Proc. Natl. Acad. Sci. U.S.A.* **74**:5559–5563.

Monaco, G.P., Calissano, P., and Mercanti, D., 1977, Effect of nerve growth factor on *in vitro* preformed microtubules, evidence for a protective effect against vinblastine, *Brain Res.* **129**:265–274.

Nicolson, G.L., 1979, Topographic display of cell surface components and their role in transmembrane signaling, *Curr. Top. Dev. Biol.* **13**:305–338.

Nja, A., and Purves, D., 1978, The effects of nerve growth factor and its antiserum on synapses in the superior cervical ganglion of the guinea pig, *J. Physiol.* **277**:53–75.

Nurcombe, V., and Bennett, M.R., 1981, Embryonic chick retinal ganglion cells identified "*in vitro*," *Exp. Brain Res.* **44**:249–258.

Oppenheim, R.W., Chu-Wang, I.-Wu., and Maderdrut, J.L., 1978, Cell death of motoneurons in the chick embryo spinal cord. III. The differentiation of motoneurons prior to their induced degeneration following limb bud removal, *J. Comp. Neurol.* **177**:87–112.

Otten, U., Goedert, M., Mayer, N., and Lembeck, F., 1980, Requirement of nerve growth factor for development of substance P-containing sensory neurones, *Nature (London)* **287**:158–159.

Otten, V., Lorez, H.P., and Businger, F., 1983, Nerve growth factor antagonizes the neurotoxic action of capsaicin on primary sensory neurones, *Nature (London)* **301**:515–517.

Parness, J., and Horwitz, S.B., 1981, Taxol binds to polymerized tubulin *in vitro*, *J. Cell Biol.* **91**:479–487.

Pearson, J., 1979, Familial dysautonomia (a brief review), *J. Autonom. Nerv. System* **1**:119–126.

Pearson, J., and Pytel, B., 1978, Quantitative studies of ciliary and sphenopalatine ganglia in familial dysautonomia, *J. Neurol. Sci.* **39**:123–130.

Pearson, J., Axelrod, F., and Dancis, J., 1974, Current concepts of dysautonomia: Neuropathological defects, *Ann. N. Y. Acad. Sci.* **228**:288–300.

Pearson, J., Pytel, B.A., Grover-Johnson, N., Axelrod, F., and Dancis, J., 1978, Quantitative studies of dorsal root ganglia and neuropathic observations on spinal cords in familial dysautonomia, *J. Neurol. Sci.* **35**(1):77–92.

Peterson, E.R., and Crain, S.M., 1970, Innervation in cultures of fetal rodent skeletal muscle by organotypic explants of spinal cord from different animals, *Z. Zellforsch.* **106**:1–21.

Peterson, E.R., and Crain, S.M., 1972, Regeneration and innervation in cultures of adult mammalian skeletal muscle coupled with fetal rodent spinal cord, *Exp. Neurol.* **36**:136–159.

Peterson, E.R., and Crain, S.M., 1979, Maturation of human muscle after innervation by fetal mouse spinal cord explants in longterm cultures, in: *Regeneration of Striated Muscle* (A. Mauro, ed.), Raven Press, New York, pp. 429–441.

Peterson, E.R., and Crain, S.M., 1981, NGF-dependent recovery of fetal mouse DRG and dorsal cord neurons after chronic taxol exposure in organotypic cultures, *Soc. Neurosci. Abstr.* **7**:551.

Peterson, E.R., and Crain, S.M., 1982a, Nerve growth factor attenuates neurotoxic effects of taxol on spinal cord–ganglion explants from fetal mice, *Science* **217**:377–379.

Peterson, E.R., and Crain, S.M., 1982b, Preferential growth of neurites from isolated fetal mouse dorsal root ganglia in relation to specific regions of co-cultured spinal cord explants, *Dev. Brain Res.* **2**:363–382.

Peterson, E.R., Crain, S.M., and Murray, M.R., 1965, Differentiation and prolonged maintenance of bioelectrically active spinal cord cultures (rat, chick and human), *Z. Zellforsch* **66**:130–154.

Peterson, E.R., Crain, S.M., Masurovsky, E.B., Horwitz, S.B., and Schiff, P.B., 1980, Neuronal recovery after sustained taxol-induced block of neuritic outgrowth in fetal mouse dorsal root ganglion–spinal cord cultures, *J. Cell Biol.* **87**:77a.

Prestige, M.C., and Wilson, M.A., 1974, A quantitative study of the growth and development of the ventral root in normal and experimental conditions, *J. Embryol Exp. Morphol.* **32**:819–833.

Prochiantz, A., DiPorzio, V., Kato, A., Berger, B., and Glowinski, J., 1979, *In vitro* maturation of mesencephalic dopaminergic neurons from mouse embryos is enhanced in presence of their striatal target cells, *Proc. Natl. Acad. Sci. U.S.A.* **76**:5387–5391.

Purves, D., and Nja, A., 1976, Effect of nerve growth factor on synaptic depression after axotomy, *Nature (London)* **260**:535–536.

Rohrer, H., and Barde, Y.-A., 1982, Presence and disappearance of nerve growth factor receptors on sensory neurons in culture, *Dev. Biol.* **89**:309–315.

Schiff, P.B., and Horwitz, S.B., 1980, Taxol stabilizes microtubules in mouse fibroblast cells, *Proc. Natl. Acad. Sci. U.S.A.* **77**:1561–1566.

Schiff, P.B., and Horwitz, S.B., 1981, Tubulin: A target for chemotherapeutic agents, in: *Molecular Actions and Target for Cancer Chemotherapeutic Agents* (Sartorelli, A.C., Lazo, J.F., and Bertino, J., eds.), Academic Press, New York, pp. 483–507.

Schiff, P.B., Fant, J., and Horwitz, S.B., 1979, Promotion of microtubule assembly *in vitro* by taxol, *Nature (London)* **277**:665–667.

Schwartz, J.P., and Breakefield, X., 1980, Altered nerve growth factor in fibroblasts from patients with familial dysautonomia, *Proc. Natl. Acad. Sci. U.S.A.* **77**:1154–1158.

Simone, L.D., Brenner, S.L., Wible, L.J., Turner, D.S., and Brinkley, B.R., 1981, Taxol-induced microtubule initiation and assembly in mammalian cells, *J. Cell Biol.* **91**:337a.

Smalheiser, N.R., Crain, S.M., and Bornstein, M.B., 1981, Development of ganglion cells and their axons in organized cultures of fetal mouse retinal explants, *Brain Res.* **204**:159–178.

Smalheiser, N.R., Peterson, E.R., and Crain, S.M., 1982, Specific neuritic pathways and arborizations formed by fetal mouse dorsal root ganglion cells within organized spinal cord explants in culture: A peroxidase-labeling study, *Dev. Brain Res.* **2**:383–395.

Smith, R.G., and Appel, S.H., 1981, Evidence for a skeletal muscle protein that enhances neuron survival, neurite extension, and acetylcholine (ACh) synthesis, *Soc. Neurosci. Abstr.* **7**:144.

Smith, R.G., and Appel, S.H., 1982, Comparison in spinal neurons of morphologic and cholinergic factor activities from muscle extract *in vitro*, *Soc. Neurosci. Abstr.* **8**:398.

Smith, R.G., and Appel, S.H., 1983, Extracts of skeletal muscle promote increased neurite outgrowth and cholinergic activity of rat spinal motor neurons, *Science* **219**:1079–1081.

Soifer, D., and Czosnek, H., 1980, The possible origin of neuronal plasma membrane tubulin, in: *Microtubules and Microtubule Inhibitors* (M. DeBrabander and J. DeMey, eds.), Elsevier/North-Holland, Amsterdam, pp. 429–447.

Stephens, R.E., 1983, Ciliary membrane reconstitution: Membrane tubulin–lipid interaction, *Biophys. J.* **41**:20a.

Stoeckel, K., Schwab, M.E., and Thoenen, H., 1975, Specificity of retrograde transport of nerve growth factor (NGF) in sensory neurons: A biochemical and morphological study, *Brain Res.* **89**:1–14.

Strocchi, P., Brown, B.A., Young, J.D., Bonventure, J.A., and Gilbert, J.M., 1981, The characterization of tubulin in CNS membrane fractions, *J. Neurochem.* **37**:1295–1307.

Thoenen, H., and Barde, Y.A., 1980, Physiology of nerve growth factor, *Physiol. Rev.* **60**:1284–1335.

Varon, S., and Bunge, R.P., 1978, Trophic mechanisms in the peripheral nervous system, *Annu. Rev. Neurosci.* **1**:327–361.

Varon, S., Skaper, S.D., and Manthorpe, M., 1981, Trophic activities for dorsal root and sympathetic neurons in media conditioned by Schwann and other peripheral cells, *Dev. Brain Res.* **1**:73–87.

Wani, M.C., Taylor, H.L., Wall, M.E., Coggon, P., and McPhail, A.T., 1971, Plant antitumor agents. VI. The isolation and structure of taxol, a novel antileukemic and antitumor agent from *Taxus brevifolia, J. Am. Chem. Soc.* **93**:2325–2327.

West, N., and Bunge, R., 1976, Prevention of the chromatolysis response in rat superior cervical ganglion neurons by nerve growth factor, *Soc. Neurosci. Abstr.* **2**:1038.

Yamada, K.M., Spooner, B.S., and Wessells, N.K., 1970, Axon growth: Roles of neurofilaments and microtubules, *Proc. Natl. Acad. Sci. U.S.A.* **66**:1206–1212.

Yamada, K.M., Spooner, B.S., and Wessells, N.K., 1971, Ultrastructure and function of growth cones and axons of cultured nerve cells, *J. Cell Biol.* **49**:614–635.

Zisapel, N., Levi, M., and Gozes, I., 1980, Tubulin: An integral protein of mammalian synaptic vesicles, *J. Neurochem.* **34**:26–32.

The Interaction of Nerve Growth Factor with Its Specific Receptors

Arne Sutter, Markus Hosang, Ronald D. Vale, and
Eric M. Shooter

1. INTRODUCTION

Although it has long been known that the development of nerve cells is influenced by interactions with target cells and surrounding nonneuronal cells, it is only recently that the problem has been explored at the molecular level. It is becoming increasingly clear that at least some of these interactions are mediated by macromolecular substances, usually proteins, secreted by one or more of the cell types (Varon and Bunge, 1978; Gottlieb and Glaser, 1980). The classic example of such substances is nerve growth factor (NGF). Although originally described for its role in regulating the development of sympathetic and some sensory neurons (Levi-Montalcini and Angeletti, 1968), NGF is equally involved in target cell–neuron communication, which controls the growth of axons and the maintenance and metabolic state of the differentiated neuron (Thoenen and Barde, 1980). Different experimental protocols have been used to define these various roles of NGF. Thus, removal of circulating NGF with anti-NGF antibody completely inhibits the development of sympathetic neurons in newborn rats or mice (Levi-Montalcini and Angeletti, 1968) or partially inhibits sensory-neuron

ARNE SUTTER, MARKUS HOSANG, RONALD D. VALE, and ERIC M. SHOOTER ● Department of Neurobiology, Stanford University School of Medicine, Stanford, California 94305. Dr. Sutter's present address is: Freie Universität Berlin, Institüt für Pharmakologie, Universitätsklinikum Charlottenburg, D-1000, Berlin 33, Federal Republic of Germany.

development in the fetus (Gorin and Johnson, 1979). Growing sympathetic axons recover from transection only if sufficient NGF is available at the site of injury (Hendry, 1975). Injection of [^{125}I]-NGF into a target organ, e.g., the smooth muscle of the iris, demonstrates the specific uptake of the labeled NGF by sympathetic terminals and its retrograde flow to the neuronal cell body, while biochemical analyses of these cell bodies reveals the regulation of neurotransmitter synthesis by the transported NGF (Thoenen and Barde, 1980). In carrying out these functions, NGF not only interacts with the receptors on the plasma membrane of the responsive neurons, as in development, but also is internalized by the neuronal processes in developing and mature neurons (Thoenen and Barde, 1980). It is the nature of these receptor interactions, which lead to internalization of NGF and the generation of appropriate intracellular signals, that is discussed in this chapter in the context of trying to understand how NGF alters gene expression in the responsive neurons. These studies have been made possible by the availability of purified NGF (Varon *et al.*, 1967).

NGF is a small dimeric, basic protein (Server and Shooter, 1977), the amino acid sequence of which shows some homology to that of insulin (Angeletti and Bradshaw, 1971; Frazier *et al.*, 1972). It is synthesized in one sympathetic target organ, the mouse submaxillary gland, in relative abundance, and the last stage of this process results in the formation of a high-molecular-weight NGF complex (7 S NGF) containing the NGF protein, the processing enzyme, and another more acidic protein (Varon *et al.*, 1968). The complex is greatly stabilized by zinc ions (Pattison and Dunn, 1975; Bothwell and Shooter, 1978) and provides a means for the protection of NGF from proteolytic modification (Mobley *et al.*, 1976) as well as potentially for the transport of NGF in saliva (Burton *et al.*, 1978) or the circulation. Since NGF in the 7 S NGF is biologically inactive (Harris-Warrick *et al.*, 1980), dissociation of the complex must occur to release biologically active NGF.

2. NERVE-GROWTH-FACTOR RECEPTORS ON SENSORY AND SYMPATHETIC NEURONS

The interaction of NGF with sympathetic and sensory neurons is mediated by specific receptors. After the initial discovery of the specific cell-surface receptor for NGF (Bannerjee *et al.*, 1973; Herrup and Shooter, 1973; Frazier *et al.*, 1974), recent studies have attempted to characterize the binding of NGF in more detail to understand the events that occur during and after the NGF–receptor interaction. The binding of ^{125}I-labeled NGF ([^{125}I]-NGF) to sympathetic and sensory neuronal membranes is not homogeneous (Frazier *et al.*, 1974). On sensory neurons from embryonic chick, Sutter *et al.*, (1979a) detected two saturable binding components by steady state binding as shown in the Scatchard plot in Fig. 1. The higher-affinity component, Site I, with an apparent equilibrium dissociation constant of 2×10^{-11} M, differs in affinity by 100-fold from the lower-affinity component, Site II, the K_d of the latter being 2×10^{-9} M. Sensory neurons have about 15-fold more Site II than Site I receptors at 37°C. The same two classes of binding sites are observed at 2°C, although fewer Site I but not Site II receptors are observed at this low temperature, indicating that Site I receptors do not arise as the result of internalization. The cell surface localization of Site I receptors was also shown in a NGF dependent cytotoxicity assay (Zimmermann et al., 1978). The marked difference in the rate constants of dissociation is sufficient to account for the different affinities of Site I and Site II receptors. Association kinetics of both sites are rapid and probably diffusion controlled. Dissociation from Site II is rapid while from Site I is slow, and for this reason the receptors are described as rapid or slow receptors, respectively.

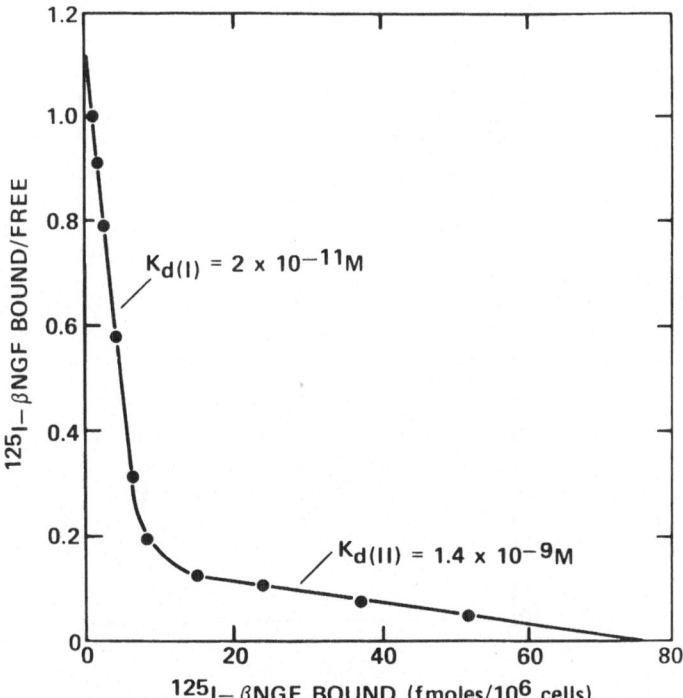

Figure 1. Scatchard analysis of steady state binding of [^{125}I]-NGF to 8-day-old chick embryo sensory-ganglion cells. Cells (0.27×10^6 in 100 μl) were incubated at 37°C for 45 min with various concentrations of [^{125}I]-NGF (3 pM to 3.7 nM). Triplicate determinations of binding were made at each point, nonspecific binding was subtracted, and data were transformed into a Scatchard plot. Reproduced from Sutter *et al.*, (1979b) with permission.

Thus, when sensory cells are allowed to reach steady state with [^{125}I]-NGF at concentrations below 10^{-11} M where binding is largely to Site I, monophasic, slow-dissociation kinetics are observed on addition of unlabeled NGF (Fig. 2). At higher preloading concentrations of [^{125}I]-NGF, biphasic dissociation curves are observed, the proportion of the rapidly released [^{125}I]-NGF, from Site II increasing with increasing preloading concentrations. Good agreement is found for the apparent equilibrium dissociation constants derived from kinetic and steady state data. Similar observations have been made with chick sympathetic neurons (Godfrey and Shooter, 1979). However, the proportion of rapidly dissociating binding is always somewhat less than predicted from the steady state data, and this discrepancy increases if the initial binding of [^{125}I]-NGF is continued beyond the time required to reach maximal binding values (Sutter *et al.*, 1979a). Such behavior is indicative—as is the marked curvature in the slowly

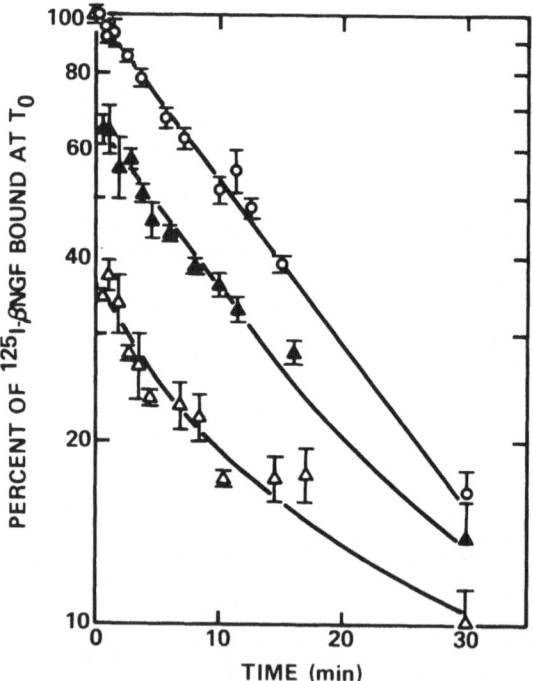

Figure 2. Dissociation of [^{125}I]-NGF after preequilibration with different [^{125}I]-NGF concentrations at 37°C. Cells (3 × 10^6/ml) were preincubated for 30 min with 0.6 × 10^{-11} M (○), 1.6 × 10^{-10} M (△), and 1.36 × 10^{-9} M (▲) [^{125}I]-NGF. The dissociation of [^{125}I]-NGF was induced by the addition of 3.8 × 10^{-7} M unlabeled NGF. The specific binding at t_0 was measured in sextuplets and after different times of dissociation in triplicates of 100 μl each. The data are corrected for nonspecific binding. Reproduced from Sutter *et al.*, (1979a) with permission.

dissociating binding at long time intervals (Fig. 2)—of an additional binding component appearing slowly with time (see below). Oxidation of the tryptophan residues in NGF leads to loss of biological activity (Frazier et al., 1973). The biological activity of the derivative in which the most reactive Trp-21 residue is oxidized is about 5% of that of native NGF (Cohen et al., 1980). Its binding to both Site I and Site II is reduced by the same extent, suggesting that the binding domains in the two sites are similar.

Does this heterogeneity of binding reflect the presence of differnt classes of receptors, or is it the result of negatively cooperative interactions within a single class of receptors? Since biphasic dissociation curves are observed in the presence of excess unlabeled NGF, dissociation is clearly a function of receptor occupancy prior to addition of unlabeled NGF. This shows that the low-affinity Site II is not generated by a process of negative cooperativity after binding of NGF to the higher-affinity Site I receptor. The rate of dissociation of ^{125}I-NGF from Site I is enhanced in the presence of unlabeled NGF, but the accelerated dissociation is slow compared to that of Site II. Furthermore, dissociation is also enhanced when unlabeled NGF is added at concentrations which allow receptor occupancy to decrease in the dissociation step rather than to increase (Sutter et al., 1979a). Enhanced dissociation rates from site I receptors under chase conditions are indicative of some cooperativity of site I binding. The enhancement of the dissociation rate from Site I is also observed in membrane preparations (Riopelle et al., 1980). Here, the degree of enhancement can be lessened by vigorous shaking, suggesting that an "unstirred layer," which might trap released [^{125}I]-NGF in the absence but not the presence of unlabeled NGF, may account for part of the phenomenon. Models of hormone receptor binding which could explain the binding behaviour of site I receptors without evoking diffusion barriers or negative cooperativity are discussed in the section on receptor conversion.

Longer-term binding studies reveal some of the processes that affect the NGF receptor. Tait et al. (1981) also observed that after longer preincubation times dissociation of slowly dissociating label follows first order kinetics only in the first 30–40 minutes. Beyond that time the dissociation rate constant decreases and dissociation can not be accelerated any longer by the addition of unlabeled NGF. Negative cooperativity can not explain this complex binding behaviour. A process termed "sequestriation," whereby bound [^{125}I]-NGF is not released after a long incubation with unlabeled NGF, has been described on both sympathetic and sensory neurons (Olender and Stach, 1980; Olender et al., 1981). Significant amounts of [^{125}I]-NGF can be sequestered in a process that occurs at both high and low temperature but that is blocked by inhibitors of metabolic energy. It is unlikely to represent internalization, since it occurs at low temperature. It has been argued that the process is limited to Site I receptors and that it may reflect spontaneous covalent linkage between NGF and its receptor.

A different approach reveals heterogeneity in the Site II binding (A. Sutter, unpublished data). As noted earlier, dissociation from Site II receptors in the presence of unlabeled NGF is rapid at both 37 and 0°C. However, when sensory cells that have reached steady state at 37°C with respect to [^{125}I]-NGF binding are incubated at low temperature with an excess of unlabeled NGF, they display a residual binding component with slow dissociation kinetics ($t_{\frac{1}{2}}$ at 2°C of >180 min), yet with an affinity comparable to that of Site II (Fig. 3). This "chase-stable" [^{125}I]-NGF binding component, unlike binding to Sites I and II, does not appear at 0°C or when cellular ATP is depleted by energy poisons and therefore represents a secondary event like internalization of [^{125}I]-NGF initially bound to Site II. At saturating concentrations, it can account for up to 50% of the Site II binding. The rate of appearance of this extra binding component is slow, as expected from its low affinity and slow rate of dissociation, and this again distinguishes it from Site I and II binding, which are rapid. It seems likely that the upward curvature of the semilog dissociation plot of the slowly dissociating label at longer time intervals is due to the generation of this third binding component.

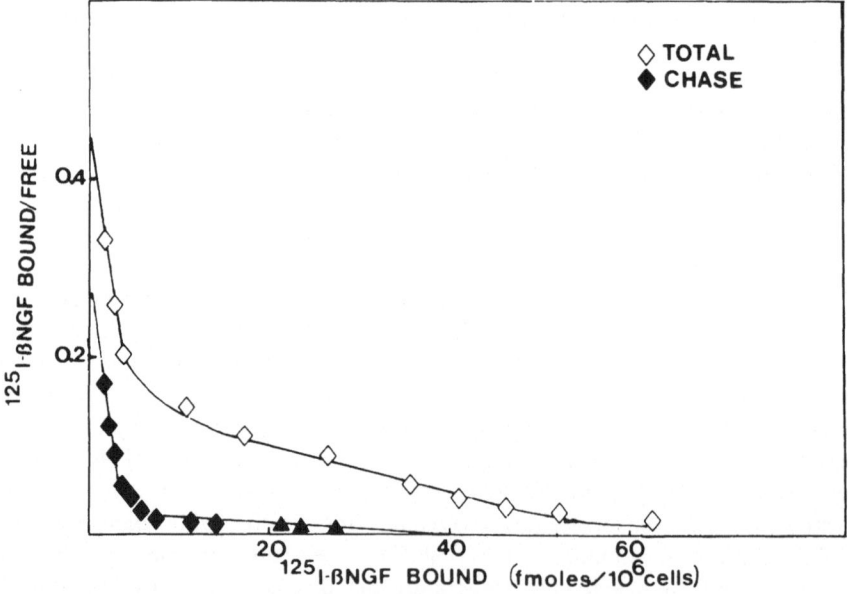

Figure 3. Heterogeneity of the low-affinity binding component shown in a "steady-state-chase" experiment. Sensory cells were incubated at 37°C with 5×10^{-12} to 2×10^{-9} M [^{125}I]-NGF. After 45 min, total specific binding ($\Diamond$) as well as chase-stable specific binding ($\blacklozenge$) (binding resistant to a 10-min incubation at 0° with 5×10^{-7} M unlabeled NGF) were determined and transformed into Scatchard plots.

It is worth emphasizing at this point that since the process of NGF-induced neurite outgrowth occurs at NGF concentrations below 10^{-11} M, it is reasonable to assume that it is linked to occupancy of Site I rather than Site II receptors (Sutter *et al.*, 1979b). In support of this idea, it may be noted that Site I receptors appear only on neurons in the chick dorsal-root ganglia (the glial cells have only the Site II receptor, Zimmermann *et al.*, 1983) and that the Site I high-affinity binding appears only on the sensory neurons at embryonic day 6, the day at which they first become responsive to NGF (Zimmermann *et al.*, 1978).

3. NERVE-GROWTH-FACTOR RECEPTORS ON PC12 PHEOCHROMOCYTOMA CELLS

This clonal, NGF-responsive cell line possesses many of the adrenergic properties characteristic of the pheochromocytoma and adrenal medullary chromaffin cells (Greene and Tischler, 1976), but when exposed to NGF for periods of several days acquires a number of the properties of sympathetic neurons including neurite outgrowth. The PC12 cells possess specific NGF receptors, but their equilibrium binding properties cannot be as readily interpreted as those of chick sensory or sympathetic neurons. Part of the reason is that NGF receptors on PC12 cells are down-regulated in a manner similar to the process described for other hormone systems. Binding of $[^{125}I]$-NGF to PC12 cells reaches a maximum after 90 min and then declines markedly to plateau at a value of 25% of maximal binding by 10 hr (Layer and Shooter, 1983). The decrease in binding, due to loss of receptors from the cell membrane, is accompanied by lysosomal degradation of the $[^{125}I]$-NGF, and both processes are inhibited at 0°C. Moreover, internalization of the NGF–receptor complex occurs quite rapidly, so that after only 15 min exposure, a significant fraction of the bound $[^{125}I]$-NGF is found inside the cell (Bernd and Green, 1983). The complications due to internalization of $[^{125}I]$-NGF are minimized at 0°C, and under these conditions, one class of receptors is observed with an equilibrium dissociation constant of 2.9×10^{-10} M (Herrup and Thoenen, 1979). Such receptors correspond to the low-affinity Site II receptors on sensory and sympathetic neurons. Competition binding assays at 37°C and with low $[^{125}I]$-NGF concentrations (15 pM) also show a displacement curve with a single point of inflection at approximately 10^{-9} M, in sharp contrast to the behavior of sensory neurons, where the inflection point falls at 2×10^{-11} M as anticipated for the Site I receptors (Cohen *et al.*, 1980). The Scatchard plot of $[^{125}I]$-NGF binding to PC12 cells does, however, show slight curvature at both 0 and 37°C indicative of a small number of high-affinity receptors, a point that is made clearer by examination of the dissociation kinetics of $[^{125}I]$-NGF from PC12 receptors.

Like sensory and sympathetic neurons, PC12 cells have two NGF-receptor populations, one with fast and one with slow dissociation kinetics (Landreth and Shooter, 1980; Schechter and Bothwell, 1981). At 37°C in the presence of unlabeled NGF, $[^{125}I]$-NGF dissociates from the fast receptors with a half-time of 30 sec. As with sensory and sympathetic neurons, the proportion of binding to each receptor varies with the initial $[^{125}I]$-NGF concentration used in the preloading of the cells. At low temperature, the dissociation rates are slowed down to such an extent that while complete release of $[^{125}I]$-NGF from the fast receptors is observed in 30 min, very little, if any, dissociation occurs from slow receptors. Thus, by incubating cells previously loaded with $[^{125}I]$-NGF with an excess of unlabeled NGF for 30 min at 4°C, NGF bound to fast receptors is selectively released. Scatchard analysis of the two components of binding then reveals the equilibrium dissociation constants of the fast and slow receptors. As anticipated, fast receptors have an equilibrium dissociation constant of 10^{-9} M. Slow receptors show two affinities of binding, one with an equilibrium dissociation constant of 5×10^{-11} M and the other of 10^{-9} M, reminiscent of the components observed with sensory neurons after relatively long intervals of binding. The lower-affinity component of the slow PC12-receptor binding is much reduced in the presence of inhibitors of metabolic energy, suggesting that it reflects internalized $[^{125}I]$-NGF. The population of PC12 receptors defined as slowly dissociating receptors by chase with unlabeled NGF at 4°C therefore comprises two components, one being the cell-surface slow receptor and the other internalized NGF. Evidence that at least part of the slow-receptor binding is from cell-surface receptors comes from the characteristics of the slow dissociation process (Landreth and Shooter, 1980) and from cross-linking studies (see Section 6). It has been suggested that slow and fast receptors on PC12 cells have the same affinities because of the canceling effects of their association and dissociation rate constants (Schechter and Bothwell, 1981).

Again, it should be emphasized that while the dose–response curve for NGF-induced neurite outgrowth from PC12 cells is shifted to higher NGF concentrations compared to the curve for sensory neurons, optimal neurite induction still occurs at concentrations less than 10^{-10} M, consistent with significant occupancy of the slow (high-affinity) rather than the fast (low-affinity) receptors (Cohen et al., 1980; Gunning et al., 1981).

4. THE QUESTION OF RECEPTOR CONVERSION

As noted earlier, the results obtained with the binding of $[^{125}I]$-NGF to sensory, sympathetic, and PC12 cells cannot be explained on the basis of two independent populations of receptors or by negatively cooperative phenomena. Landreth and Shooter (1980) have suggested that on PC12 cells, the slow receptor may be generated by the binding of NGF to the fast receptor. This hypothesis was generated by the finding of a lag phase in the appearance of slow receptors during the binding of very low concentrations of [^{125}I]-NGF (14 pM) and by a severalfold increase in the amount of slow binding at the expense of fast binding when PC12 cells loaded with [^{125}I]-NGF for a brief period were further incubated in the absence of [^{125}I]-NGF. Since at least 90% of the [^{125}I]-NGF bound to the slow receptors is released by dissociation in the presence of unlabeled NGF, it is unlikely that the continued generation of slow receptors is due to internalization. However, it is possible that some dissociation of receptor-bound [^{125}I]-NGF does occur under these conditions with rebinding preferentially to the slow receptors. [^{125}I]-NGF bound to slow receptors on PC12 cells is resistant to mild digestion with trypsin at 0°C, whereas that bound to fast receptors is not. Schechter and Bothwell (1981) observed degradation of unoccupied fast but not slow receptors by trypsin prior to addition of [^{125}I]-NGF, suggesting that both receptor populations exist on naïve PC12 cells. If this is true, then receptor conversion may not be required to generate slow receptors.

The binding of a receptor to an effector molecule in the cell membrane is one example of a sequential event after ligand binding that could alter the affinity of the ligand–receptor complex. This type of model, particularly where the affinity of the ligand–receptor complex increases after interaction with the effector, can account for the curvilinear Scatchard plot and enhanced dissociation kinetics in insulin binding. In a search for such an effector molecule in the NGF receptor system, studies of both lectin-binding (see Section 5) and chemical cross-linking (see Section 6) have been carried out.

5. EFFECTS OF LECTIN AND ANTI-NERVE GROWTH FACTOR ANTIBODY ON NERVE GROWTH FACTOR BINDING

Among a range of lectins studied, only wheat germ agglutinin (WGA) affected NGF binding. When added after [^{125}I]-NGF, WGA has no effect on total binding, but substantially increases the ratio of [^{125}I]-NGF bound to slow compared to fast receptors (Vale and Shooter, 1982). The WGA-induced receptor conversion occurs rapidly at both 37 and 4°C (Fig. 4) and confers trypsin resistance on the occupied slow receptors. This together with similar association and dissociation kinetics makes the WGA-induced slow receptor, at least, comparable to the naturally ocurring slow receptor. The receptor conversion also confers on the WGA-induced slow receptor the property of being insoluble in Triton X-100 (Fig. 4). This effect was first observed with the slow receptors on PC12 cells (Schechter and Bothwell, 1981) and was interpreted as indicating a cytoskeletal attachment for this receptor. Anti-NGF antibody, when added after the binding of [^{125}I]-NGF, also significantly increases the proportion of slow receptors and of Triton-X-100-insoluble binding, although the latter is less complete than with

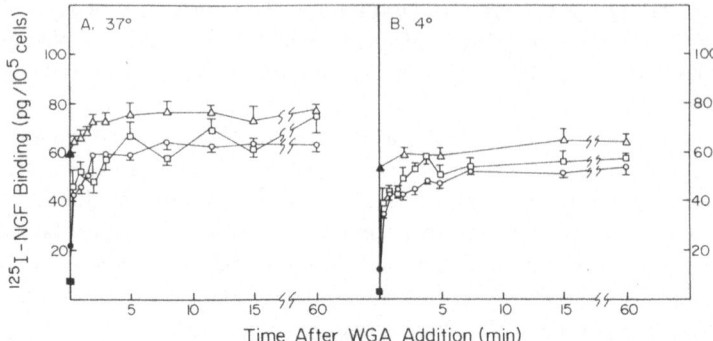

Figure 4. Kinetics of the effects of WGA on NGF binding. PC12 cells were incubated with either 150 pM [^{125}I]-NGF at 37°C (A) or 190 pM [^{125}I]-NGF at 4°C (B). At 30 min after [^{125}I]-NGF addition, WGA (50 μg/ml) was added and total (△), slowly dissociating (○), and Triton-X-100-insoluble (□) binding were determined as a function of time after WGA addition. The 0-min time points (▲, ●, ■) represents these binding parameters before WGA addition. Values are the means of triplicate determinations. (I) Standard deviations.

WGA (R. Vale and E.M. Shooter, unpublished data). Since both effects of the antibody are abolished if the F_{ab} fraction is used, it would appear that cross-linking of the occupied fast NGF receptor by NGF antibody or by WGA is important in changing receptor affinities or dissociation kinetics. Clustering of NGF receptors after addition of a fluorescent-labeled NGF has been observed directly (Levi *et al.*, 1980). The increase in Triton X-100 insolubility of the [^{125}I]-NGF–receptor complex aftr cross-linking raises the interesting possibility that the effector molecule may be a cytoskeleton-associated protein.

6. CHARACTERIZATION OF THE MOLECULAR SIZES OF THE NERVE-GROWTH-FACTOR RECEPTOR

An alternative approach to the question of receptor conversion is to explore the molecular characteristics of the two receptor types using the photoaffinity cross-linking method (Massagué *et al.*, 1981). The solubilized NGF receptor from sympathetic neurons behaves as a single species in sedimentation analyses with a molecular weight of 135,000 (Costrini and Bradshaw, 1979). When [^{125}I]-NGF is covalently cross-liked to sympathetic neuronal membranes with hydroxysuccinimidyl-4-azidobenzoate (HSAB), two species are observed with molecular weights, for the [^{125}I]-NGF–receptor complexes, of 143,000 and 112,000, respectively (Massagué *et al.*, 1981). On the basis of the appearance of four common [^{125}I]-labeled peptides in both these species, it was suggested that the smaller receptor is derived from the latter by limited proteolysis.

Cross-linking of [^{125}I]-NGF to PC12 cells with HSAB also reveals two different-size [^{125}I]-NGF–receptor complexes, with molecular weights of 158,000 and 100,000 (Fig. 5). Binding and cross-linking of [^{125}I]-NGF to both these species is abolished in the presence of excess unlabled NGF [Fig. 5 (4)]. When the [^{125}I]-NGF-loaded PC12 cells are incubated in excess unlabeled NGF at 0°C prior to cross-linking, only one molecular-weight species remains, the 158,000 species [Fig. 5 (2 and 3)]. This experiment identifies the 158,000 species as the slow receptor (chase-stable at 0°C) and the smaller 100,000 species as the fast receptor. In line with this identification, the same cross-linking method shows that binding of [^{125}I]-NGF to the molecular-weight-158,000 species is trypsin-resistant at 0°C, whereas binding to the 100,000 species is not. Furthermore,

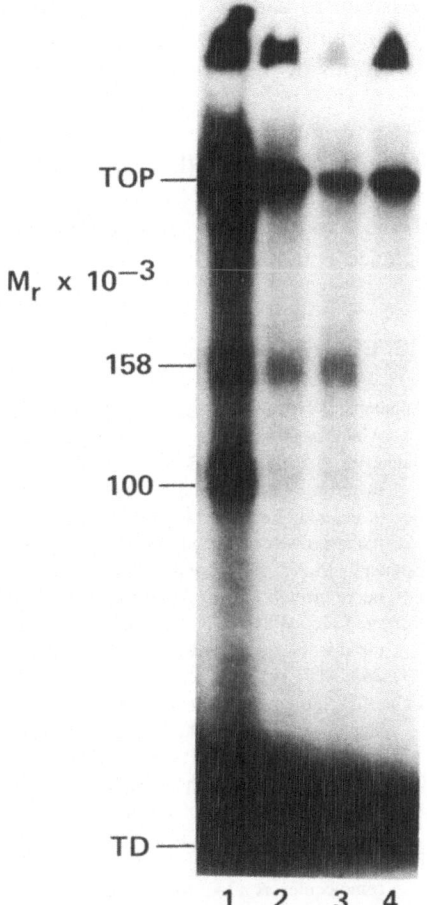

Figure 5. Autoradiograph demonstration that high-molecular-weight NGF receptors are slow and low-molecular-weight NGF receptors are fast. PC12 cells (2.0 × 10^6 cells/ml) were incubated at 23°C for 60 min in 10 mM HEPES/Krebs–Ringer medium, pH 7.35 containing 1 mg/ml bovine serum albumin in the presence of 0.8 nM [^{125}I]-NGF. At the end of the incubation, cells were cooled to 0.5°C, and cross-linking with 50 μM HSAB was performed either immediately (1) or following incubation with 0.8 μM unlabeled NGF at 0.5°C for 5 min (2) or 30 min (3) or at 37°C for 30 min (4). Cross-linking was effected by adding to the cells HSAB, 5 mM, freshly dissolved in dimethylsulfoxide, at a 1 : 100 dilution, incubating for 3 min on ice in the dark, and then photolysing for 7 min in a quartz cuvette in the cold with a 200 W Hg Orien high-pressure arc lamp equipped with a 330 nm cutoff filter. Photolysis was stopped and cells were washed by dilution into excess 10 mM Tris–saline, pH 7.0, and centrifugation at 1000g for 5 min. Samples (50 μg) of the resulting pellets were solubilized by boiling for 5 min in the presence of 1% sodium dodecyl sulfate and 50 mM dithiothreitol. They were then subjected to electrophoresis on a 6% polyacrylamide slab gel. The autoradiograph shown was obtained after a 10-day exposure of the resulting fixed, dried gel. Protein standards used in a parallel run were: myosin (M$_r$ 200,000), β-galactosidase (M$_r$ 116,000), phosphorylase b (M$_r$ 94,000), bovine serum albumin (M$_r$ 68,000), ovalbumin (M$_r$ 45,000), and carbonic anhydrase (M$_r$ 31,000). The M$_r$'s of the cross-linked [^{125}I]-NGF–receptor complexes are indicated on the left.

[^{125}I]-NGF bound to the 158,000- but not the molecular-weight-100,000 species is displaced by very low concentrations of unlabeled NGF, showing that at least a significant fraction of the higher-molecular-weight slow receptor is also a high-affinity receptor. These data can be interpreted in a number of different ways. For example, the high-affinity molecular-weight-158,000 species may be the only biologically important receptor, while the lower-affinity, lower-molecular-weight 100,000 species is a trivial proteolytic product of the larger species. Alternatively, the latter may also have specific biological functions, and this seems likely, since the process of retrograde transport of [^{125}I]-NGF involves both high- and low-affinity components (Dumas *et al.*, 1979). In terms of the receptor-conversion hypothesis, the data would suggest that the high-affinity molecular-weight-158,000 receptor does indeed contain additional protein components compared to the smaller low-affinity receptor. Although none of these possibilities can be precluded at the moment, the combination of kinetic and equilibrium binding studies with the further biochemical characterization of the two forms of the receptor should allow each of them to be explored in detail.

ACKNOWLEDGMENTS. This work was supported by grants from the National Institutes of Health (NINCDS, NS 04270) and the American Cancer Society (BC 325B). R.D.V. is a trainee in the Medical Scientist Training Program (GM 07365), and M.H. is supported in part by a fellowship from the Swiss National Science Foundation.

REFERENCES

Angeletti, R.H., and Bradshaw, R.A., 1971, Nerve growth factor from mouse submaxillary gland: Amino acid sequence, *Proc. Natl. Acad. Sci. U.S.A.* **68**:2417–2420.

Bannerjee, S.P., Snyder, S.H., Cuatrecasas, P., and Greene, L.A., 1973, Binding of nerve growth factor in sympathetic ganglia, *Proc. Natl. Acad. Sci. U.S.A.* **70**:2519–2523.

Bernd, P., and Greene, L.A., 1983, Electron microscopic radioautographic localization of iodinated nerve growth factor within and on PC12 cells, *J. Neurosci.* **3**:631–643.

Bothwell, M.A., and Shooter, E.M., 1978, Thermodynamics of the interaction of subunits of 7S nerve growth factor, *J. Biol. Chem.* **253**:8458–8464.

Burton, L.E., Wilson, W.H., and Shooter, E.M., 1978, Nerve growth factor in mouse saliva— rapid isolation procedures for and characterization of 7S nerve growth factor, *J. Biol. Chem.* **253**:7807–7812.

Cohen, P., Sutter, A., Landreth, G.E., Zimmermann, A., and Shooter, E.M., 1980, Oxidation of tryptophan-21 alters the biological activity and receptor binding characteristics of mouse nerve growth factor, *J. Biol. Chem.* **255**:2949–2954.

Costrini, N.V., and Bradshaw, R.A., 1979, Binding characteristics and apparent molecular size of detergent-solubilized nerve growth factor receptor of sympathetic ganglia, *Proc. Natl. Acad. Sci. U.S.A.* **76**:3242–3245.

Dumas, M., Schwab, M.E., and Thoenen, H., 1979, Retrograde axonal transport of specific macromolecules as a tool for characterizing nerve terminal membranes, *J. Neurobiol.* **10**:179–197.

Frazier, W.A., Angeletti, R.H., and Bradshaw, R.A., 1972, Nerve growth factor and insulin, *Science* **176**:482–488.

Frazier, W.A., Hogue-Angeletti, R.A., Sherman, R., and Bradshaw, R.A., 1973, Topography of mouse 2.5S nerve growth factor: Reactivity of tyrosine and trytophan, *Biochemistry* **12**:3281–3293.

Frazier, W.A., Boyd, L.F., and Bradshaw, R.A., 1974, Properties of the specific binding of ^{125}I nerve growth factor to responsive peripheral neurons, *J. Biol. Chem.* **249**:5513–5519.

Godfrey, E., and Shooter, E.M., 1979, Nerve growth factor receptors on chick sympathetic ganglion cells, *Soc. Neurosci. Abstr.* **5**:67.

Gorin, P., and Johnson, E.M., 1979, Experimental autoimmune model of nerve growth factor deprivation: Effects on developing peripheral sympathetic and sensory neurons, *Proc. Natl. Acad. Sci. U.S.A.* **76**:5382–5386.

Gottlieb, D. I., and Glaser L., 1980, Cellular recognition during neural development, *Annu. Rev. Neurosci.* **3**:303–318.

Greene, L.A., and Tischler, A.S., 1976, Establishment of a noradrenergic clonal line of rat adrenal pheochromocytoma cells which respond to nerve growth factor, *Proc. Natl. Acad. Sci. U.S.A.* **73**:2424–2428.

Gunning, R.W., Landreth, G.E., Layer, P., Ignatius, M., and Shooter, E.M., 1981, Nerve growth factor-induced differentiation of PC12 cells: Evaluation of changes in RNA and DNA metabolism, *J. Neurosci.* **1**:368–379.

Harris-Warrick, R.M., Bothwell, M.A., and Shooter, E.M., 1980, Subunit interactions inhibit the binding of β nerve growth factor to receptors on embryonic chick sensory neurons, *J. Biol. Chem.* **255**:11,284–11,289.

Hendry, I.A., 1975, The response of adrenergic neurons to axotomy and nerve growth factor, *Brain Res.* **94**:87–97.

Herrup, K., and Shooter, E.M., 1973, Properties of the NGF receptor of avian dorsal root ganglia, *Proc. Natl. Acad. Sci. U.S.A.* **70**:3384–3388.

Herrup, K., and Thoenen, H., 1979, Properties of the nerve growth factor receptor of a clonal line of rat pheochromocytoma (PC12) cells, *Exp. Cell Res.* **121**:71–78.

Landreth, G.E., and Shooter, E.M., 1980, Nerve growth factor receptors on PC12 cells: Ligand induced conversion from low to high affinity states, *Proc. Natl. Acad. Sci. U.S.A.* **77**:4751–4755.

Layer, P., and Shooter, E.M., 1983, Binding and degradation of nerve growth factor by PC12 pheochromocytoma cells, *J. Biol. Chem.* **258**:3012–3018.

Levi, A., Schechter, Y., Neufeld, E.J., and Schlessinger, J., 1980, Mobility, clustering and transport of nerve growth factor in embryonic sensory cells and in a sympathetic neuronal cell line, *Proc. Natl. Acad. Sci. U.S.A.* **77**:3469–3473.

Levi-Montalcini, R., and Angeletti, P.U., 1968, Nerve growth factor, *Physiol. Rev.* **48**:534–569.

Massague, J., Guillette, B.J., Czech, M.P., Morgan, C.J., and Bradshaw, R.A., 1981, Identification of a nerve growth factor receptor protein in sympathetic ganglia membranes by affinity labeling, *J. Biol. Chem.* **256**:9419–9424.

Mobley, W.C., Schenker, A., and Shooter, E.M., 1976, Characterization and isolation of proteolytically modified nerve growth factor, *Biochemistry* **15**:5543–5551.

Olender, E.A., and Stach, R.W., 1980, Sequestration of ^{125}I-labeled β nerve growth factor by sympathetic neurons, *J. Biol. Chem.* **225**:9338–9343.

Olender, E.A., Wagner, B.J., and Stach, R.W., 1981, Sequestration of I-125-labeled beta-nerve growth-factor by embryonic sensory neurons, *J. Neurochem.* **37**:436–442.

Pattison, S.E., and Dunn, M.F., 1975, On the relationship of zinc ion to the structure and function of the 7S nerve growth factor protein, *Biochemistry* **14**:2573–2581.

Riopelle, R.J., Klearman, M., and Sutter, A., 1980, Nerve growth factor receptors—analysis of the interaction of beta-NGF with membranes of chick embryo dorsal root ganglia, *Brain Res.* **199**:63–77.

Schechter, A.L., and Bothwell, M.A., 1981, Nerve growth factor receptors on PC12 cells: Evidence for two receptor classes with differing cytoskeletal associations, *Cell* **24:**867–874.

Server, A.C., and Shooter, E.M., 1977, Nerve growth factor, *Adv. Protein Chem.* **31:**339–409.

Sutter, A., Riopelle, R.J., Harris-Warrick, R.M., and Shooter, E.M., 1979a, Nerve growth factor receptors: Characterization of two distinct classes of binding sites on chick embryo sensory ganglia cells, *J. Biol. Chem.* **254:**5972–5982.

Sutter, A., Riopelle, R.J., Harris-Warrick, R.M., and Shooter, E.M., 1979b, The heterogeneity of nerve growth factor receptors, in: *Transmembrane Signalling* (M. Bitensky, R.J. Collier, D.F. Steiner, and C.F. Fox, eds.), Alan R. Liss, New York, pp. 659–667.

Tait, J.F., Weinman, S.A., and Bradshaw, R.A., 1981, Dissociation kinetics of I-125-nerve growth-factor from cell-surface receptors—acceleration by unlabeled ligand and its relationship to negative cooperativity, *J. Biol. Chem.* **256:**11,086–11,092.

Thoenen, H., and Barde, Y.A., 1980, Physiology of nerve growth factor, *Physiol. Rev.* **60:**1284–1335.

Vale, R.D., and Shooter, E.M., 1982, Alteration of binding properties and cytoskeletal attachment of nerve growth factor receptors in PC12 cells by wheat germ agglutinin, *J. Cell Biol.* **94:**710–717.

Varon, S.S., and Bunge, R.P., 1978, Trophic mechanisms in the peripheral nervous system, *Annu. Rev. Neurosci.* **1:**327–361.

Varon, S., Nomura, J., and Shooter, E.M., 1967, Subunit structure of a high molecular weight form of the nerve growth factor from mouse submaxillary gland, *Proc. Natl. Acad. Sci. U.S.A.* **57:**1782–1789.

Varon, S., Nomura, J., and Shooter, E.M., 1968, Reversible dissociation of the mouse nerve growth factor protein into different subunits, *Biochemistry* **7:**1296–1303.

Zimmermann, A., and Sutter, A., 1983, β-Nerve growth factor (β-NGF) receptors on glial cells. Cell-cell interaction between neurones and Schwann cells in cultures of chick sensory ganglia, *EMBO J.* **2:**879–885.

Zimmermann, A., Sutter, A., Samuelson, J., and Shooter, E.M., 1978, A serological assay for the detection of cell surface receptors for nerve growth factor, *J. Supramol. Str.* **9:**351–361.

IV

New Neuronal Growth Factors

Session Chairman: Hans Thoenen

Multiple Sites for the Regulation of Neurite Outgrowth

Frank Collins

1. INTRODUCTION

It is my impression that the work of Wessells's students on the culturing of dissociated parasympathetic neurons (Helfand *et al.*, 1976) provided the inspiration and methods for much of the recent work on new nerve growth factors. This was certainly true in the case of my own research presented herein. Their paper described a method for culturing dissociated parasympathetic neurons from the ciliary ganglion of chicken embryos using culture medium conditioned by prior exposure to a monolayer of embryonic heart cells and a substrate coated with polyornithine. These methods allowed one to study the development in culture of single, dissociated neurons that were of interest because they were (1) parasympathetic, (2) a pure population of motor neurons, and (3) neurons the developmental history of which in the embryo had already been described in detail.

The ability to compare the behavior of these neurons in culture with their known developmental history in the embryo has provided useful insights, as discussed below. Since these neurons are parasympathetic and supposedly unresponsive to nerve growth factor (NGF), these results also focused attention on the possibility that the culture conditions were providing a new nerve growth factor, analogous to NGF, but specific for parasympathetic neurons.

FRANK COLLINS ● Department of Anatomy, University of Utah School of Medicine, Salt Lake City, Utah 84132.

More significantly, the methods used in their study provided a means to approach the characterization and isolation of such new growth factors. Previous work had suggested that nonneuronal cells cocultured with neurons could enhance neural development in ways not attributable to the release of NGF (Burnham *et al.*, 1972). The introduction of conditioned medium (CM) provided a convenient means to obtain the products of such nonneuronal cells in sufficient quantity to attempt the characterization of the active cell products.

When I came to Wessells's laboratory as a postdoctoral fellow, I set out to characterize the manner in which heart-conditioned medium (HCM) promoted the development of ciliary-ganglion neurons in culture. This work has yielded unexpected results that I believe provide insights into some of the points at which neural development is susceptible to external control. I am taking this opportunity to summarize the work that my skilled undergraduate assistants, Jim Garrett, Mark Lee, and Andrea Dawson, and I have done on this subject and to relate our observations to the larger context of neuroembryogenesis.

2. RESULTS

2.1. Substrate-Conditioning Factors

2.1.1. Requirement for Polyornithine and Conditioned Medium

Dissociated ciliary-ganglion neurons do not attach to a plastic tissue-culture substrate. Their lack of attachment provides a convenient method for separating the neurons in a ganglion from the non-neuronal cells that adhere well to the plastic (Collins, 1980). To extend neurites, however, the nerve cells must be provided with a substrate to which they can adhere. This requirement is met by pretreating the substrate with the positively charged synthetic polypeptide polyornithine or polylysine, either of which promotes rapid and firm adhesion of the ciliary-ganglion nerve-cell bodies (Helfand *et al.*, 1976).

Even in the presence of sufficient polyornithine to promote firm attachment, the nerve cells do not extend axons. The nerve-cell bodies remain rounded and refractile, actively extending filopodia around their periphery, but not forming growth cones, and finally disintegrating 18–24 hr after being placed in culture (Collins, 1978a). This absence of outgrowth was observed using standard culture media, such as Ham's F12 or Dulbecco's minimal essential medium (MEM), supplemented with 10% fetal calf or horse serum. However, when the neurons were exposed to these same media "conditioned" by prior exposure to a confluent monolayer of embryonic heart cells, extensive axonal networks were produced and survival of the nerve cells in culture was prolonged (Helfand *et al.*, 1976).

2.1.2. Rapid Effect of Conditioned Medium on Outgrowth

Despite their active surface motility, dissociated ciliary-ganglion neurons, cultured on a polyornithine-coated substrate in standard culture media, fail to extend axons during the 18 to 24-hr period in which they remain intact. However, within 15–20 min after the addition of HCM, such neurons show the first clear signs of altered surface motility, which are followed within 60 min by the formation of elongating axons by over 80% of the nerve-cell bodies (Collins, 1978a). The ability to induce rapid and synchronous axon outgrowth in a high percentage of single dissociated neurons makes this system very useful for studying the early events of axon initiation. A detailed description of these processes, visualized by time-lapse phase microscopy, has been published (Collins, 1978a).

The unexpected rapidity of the stimulation of outgrowth by HCM suggested that this effect could not be due simply to prolonged neuronal survival (Collins, 1978a). This was later confirmed by studies that led to the eventual separation of the agent that induces outgrowth from the agent in CM that promotes survival (Collins, 1978b).

2.1.3. Action of Conditioned Medium via the Substrate

Material in CM that becomes bound to the polyornithine-coated substrate is sufficient to support axon outgrowth in the absence of the other components of CM (Collins, 1978b). Moreover, CM from which the substrate-binding components are removed loses its ability to stimulate outgrowth (Collins, 1978b). These results indicate that this material is necessary and sufficient for the induction of outgrowth. Various other experiments, described in detail elsewhere (Collins, 1978b), indicate that this material *must* attach to the substrate before it can exert its effects on the nerve cells.

These observations led us to refer to this material as a substrate-conditioning factor (SCF) and to propose that the material may be related to components of the cell surface or extracellular matrix that support the extension of axons through embryonic tissues during development (Collins, 1978b). It seemed to us that polyornithine was acting to anchor SCF to the substrate, a necessary precondition for the action of SCF, but that this anchoring function could probably also be served by normal components of the cell surface or extracellular matrix. These predictions caused us to test the ability to induce axon outgrowth of heart-cell products thought to be derived directly from the cell surface–extracellular matrix, namely, the microexudate deposited by cells in culture onto their substrate (Poste *et al.*, 1973; Culp and Buniel, 1976).

Embryonic heart-cell microexudate, in the absence of polyornithine, induced neurite outgrowth from dissociated neurons in a manner very similar to that of

SCF bound to polyornithine (Collins, 1980). This suggests that embryonic heart cells in culture deposit onto their substrate and release into their medium surface components that can form an excellent substrate for axon extension.

Since these results were published, the widespread occurrence of polyornithine-bound inducers of axon outgrowth has been confirmed in other culture systems using CMs or tissue extracts from a range of cell types and dissociated neurons from various vertebrate and invertebrate sources (Wong et al., 1981; Adler et al., 1981; Miki et al., 1981; Lander et al., 1982). The ability of cell-derived microexudates to promote axon outgrowth has also been observed with additional types of neurons (Hawrot, 1980; Fujii et al., 1982). It appears, then, that many cell types in culture, including established cell lines, release into their medium or deposit onto their substrate materials that induce outgrowth in a manner analogous to that of embryonic heart-cell SCF and microexudate.

Our characterization of the chemical properties of SCF indicate that it is a discrete component of CM—a large macromolecule or aggregate with a sufficiently high negative charge to account for its binding to polyornithine. Other negatively charged macromolecules, such as RNA, DNA, mucin, chondroitin, and heparan sulfates, and hyaluronic acid, have none of the axonal-growth-inducing activity of SCF, indicating that the action of SCF is not simply due to a physical modification of the charge on the substrate (Collins, unpublished observations). Recently, evidence has been presented linking an SCF-like activity to cell-surface-associated heparan sulfate proteoglycan (Lander et al., 1982). It will be of interest to determine whether these results also apply to the CMs from other sources.

2.1.4. Substrate-Conditioning-Factor Guidance of Elongating Axons by the Mechanism of Haptotaxis

If just a narrow pathway of the substrate, instead of the entire substrate, is coated with CM, elongating axons remain confined within this pathway (Collins and Garrett, 1980). This observation was the first clear evidence that the material deposited onto the polyornithine substrate could influence axon elongation and not just the initiation of outgrowth. In these experiments, which used dissociated ciliary-ganglion neurons, occasional axons were seen to cross outside the border of the SCF-coated pathway. Once they had done so, such axons were capable of continued elongation on those regions of the substrate that had not been pretreated with CM (Collins and Garrett, 1980). On the assumption that such axons were not atypical, the results suggested that confinement of the great majority of the axons within the pathway was not due to an absolute inability to elongate outside the pathway, but rather to a preference to remain in contact with the material deposited onto the substrate from CM. Other experiments indicated that confinement of outgrowth was not due to a physical barrier to

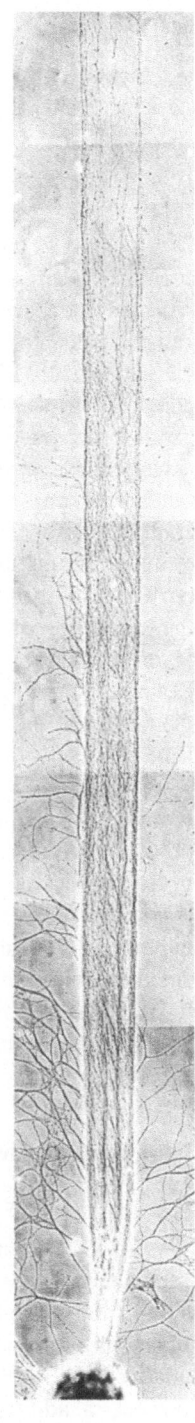

Figure 1. Axonal guidance by a pathway of SCF on the substrate. A pathway of SCF was prepared on the substrate by the method of Collins and Garrett (1980). An intact ciliary ganglion from an 8-day chick embryo was placed on the pathway at the bottom. This photograph was taken after 3 days in culture.

axons at the borders between the SCF-treated and untreated regions of the substrate, but rather to response of the axons to the differences in chemical composition between different regions of the substrate (Collins and Garrett, 1980). This mechanism of axonal guidance is referred to as *haptotaxis* (Katz and Lasek, 1980).

We have since made additional, related observations that fully support the conclusions suggested in our earlier study (Collins and Lee, in prep.). We have repeated the earlier observations on dissociated neurons using whole ciliary ganglia placed at one end of an SCF-coated pathway that is narrower than the ganglion itself (Fig. 1). Axonal outgrowth from that region of the ganglion in contact with the SCF pathway remains confined within the pathway until the axons become so numerous near the ganglion that newly emerging axons can no longer make direct contact with the SCF. Only then do axons begin to cross the border and leave the SCF pathway (Fig. 1, bottom). Whole ganglia, in contrast to dissociated neurons, are able to extend numerous axons directly onto a polyornithine substrate even without pretreatment with CM. This observation provides additional evidence that confinement of elongation to the pathway is the expression of a preference for axons to remain in contact with the SCF rather than an absolute necessity for them to do so in order to continue elongating. Confinement of outgrowth to the pathway is a result of growth cones, on reaching the border of the SCF pathway, turning so as to remain in contact with the SCF (Fig. 2).

We have also created pathways using our most purified preparations of SCF. Pathways made using these preparations (which contain less than 1% of the protein originally present in serum-free HCM) are able to guide elongating axons in the manner described above. This suggests that the same component of CM may be responsible both for the induction of outgrowth and for guidance by haptotaxis (Collins and Lee, in prep.). An important additional point is that the small amount of material that forms the pathway using such purified preparations is still detected by the growth cones even in the presence of the much greater amounts of material deposited onto all parts of the substrate from the culture medium itself.

2.2. Nerve Growth Factors

2.2.1. A Soluble Factor That Increases the Rate of Axon Elongation

An SCF-coated substrate is sufficient to induce axon outgrowth in the absence of any other components of CM. However, the length of axons formed is significantly less and growth cones appear less well spread when nerve cells are plated on such a conditioned substrate alone rather than in complete CM (Collins and Dawson, 1982). We have found that a second component of HCM must be present in addition to SCF to restore the more extensive axonal lengths and

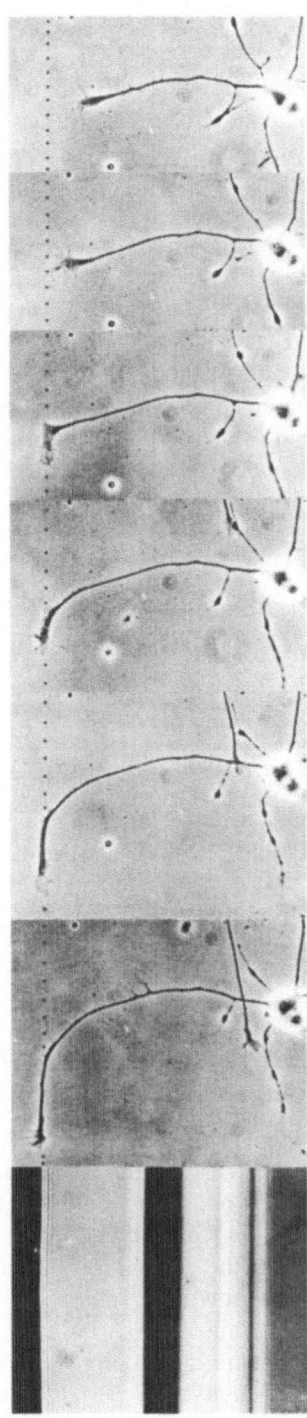

Figure 2. Turning response of a growth cone on reaching the border of an SCF pathway. A pathway of SCF was prepared by the method of Collins and Garrett (1980). The location of the border of the SCF pathway is indicated in the bottom panel by the sharp transition from white (region of SCF treatment) to black (untreated region). The remaining panels from top to bottom illustrate the behavior of a single growth cone as it approaches, reaches, and finally turns at the border so as to remain in contact with the SCF.

spread growth cones characteristic of growth in whole CM. When this second component is added to nerve cells that have initiated outgrowth on an SCF-coated substrate, it increases the rate of axonal elongation approximately 3-fold within 30–60 min of its addition (Collins and Dawson, 1982). This elongation-promoting agent does not bind to the polyornithine substrate and for this reason can be readily separated from SCF, which does bind (Collins and Dawson, 1982). The activity of the elongation-promoting agent is trypsin-sensitive and is also affected by other treatments that characteristically damage proteins (Collins and Dawson, 1982).

Thus, there are two apparently quite distinct components of CM that affect neurite outgrowth. The first component, which we have referred to as SCF, must bind to the polyornithine substrate in order to act. Bound SCF is required for the induction of outgrowth, is sufficient by itself to give substantial outgrowth, and is capable in the appropriate culture conditions of guiding elongating axons by haptotaxis. The second component, an elongation-promoting agent, does not bind to the substrate and is incapable of inducing outgrowth in the absence of SCF, but rapidly increases the rate of axonal elongation and also the degree of spreading of axonal growth cones.

2.2.2. Relationship of These Activities to Neuronal Survival

HCM also prolongs the survival of ciliary-ganglion neurons in culture (Helfand *et al.*, 1976). The survival-promoting activity is readily separated from SCF, and the latter apparently lacks any ability to prolong neuronal survival (Collins, 1978b). On the other hand, the methods used to prepare fractions of CM that contain the elongation-promoting activity have so far failed to separate it from the survival factor (Collins and Dawson, 1982). However, no thorough-going attempt was made in this study to separate the two activities, and so the question of their possible relationship is still open.

2.2.3. An Effect of Nerve Growth Factor on Parasympathetic-Axon Growth

It is often difficult to evaluate the significance for animals of work done in culture. This seems especially true of the elongation-promoting activity discussed above. Despite the rapid and substantial effect of this activity on the rate of elongation, some would tend to dismiss the finding as a result of simply improving the culture conditions. This criticism implies that the elongation-promoting agent affects the general level of cell activity at a point far removed from the process of outgrowth.

While there is no evidence at present bearing on the site of action of the elongation-promoting activity in CM, there is other evidence that suggests that this activity may be relevant to neural development. This evidence consists of the observation that NGF has the same effect. This does not simply mean that

NGF increases the rate of sympathetic- or sensory-axon elongation in culture, analogous to the effect of the elongation-promoting agent on ciliary-ganglion neurons, although this is the case (Levi-Montalcini and Angeletti, 1963; Greene 1977a,b; Collins and Dawson, 1983). Of more significance for this discussion is our recent observation that *NGF also increases the rate of axonal elongation and the extent of growth-cone spreading of parasympathetic ciliary-ganglion neurons* in a manner quite similar to the activity of HCM (Collins and Dawson, 1983).

Addition of NGF to dissociated parasympathetic ciliary-ganglion neurons results, within 60 min of its addition, in at least 2-fold increase in average neurite length and an accompanying enlargement and spreading of neuronal growth cones. These effects occur over a concentration range of NGF from 0.1 to 10 ng/ml and are blocked by affinity-purified antibody to NGF. Epidermal growth factor, fibroblast growth factor, and angiotensin do not have these effects, although insulin at high concentrations is able to induce a response similar to that of NGF. Dissociated sympathetic-chain neurons also respond to NGF with increased neurite lengths, and in addition, NGF considerably extends the survival time of these neurons in culture. However, the effect of NGF on ciliary-ganglion neurons is limited to neurite outgrowth, and NGF does not promote the survival of these parasympathetic neurons.

Max *et al.* (1978) reported that ciliary-ganglion neurons *in vivo* were able to take up NGF at their synaptic terminals and retrogradely transport NGF back to the ganglionic-cell body. Thus, ciliary-ganglion neurons are capable of actively taking up NGF *in vivo* and also of responding to NGF *in vitro* with increased rates of axonal elongation. Effects such as these are thought to require high-affinity receptors for NGF (Stoeckel *et al.*, 1974; Paravicini *et al.*, 1975).

These results suggest either that NGF may be involved with normal development of ciliary-ganglion neurons or alternatively that NGF is cross-reacting with receptor sites normally activated by a related parasympathetic nerve growth factor, such as that found in HCM (Collins and Dawson, 1983). The latter notion is supported by the additional observation that for parasympathetic neurons, NGF is less effective in promoting elongation than the heart-cell elongation-promoting agent, while the reverse is true for sympathetic neurons (Collins and Dawson, 1983). Since insulin at high concentrations can mimic the effect of both NGF and the heart-cell elongation-promoting agent, it is possible that there is a class of insulin-like nerve growth factors, of which NGF is one, that are normally specific for different types of neurons, but not absolutely so.

2.3. Intrinsic Factors

The previous work has demonstrated two ways in which agents external to the nerve cell can regulate axonal outgrowth: by altering the composition of the substrate and by increasing the ability of the nerve cell to maintain actively elongating axons. The activity of these external agents depends, of course, on the ability of the nerve cell to respond to them. This ability was taken for granted in the previous studies, an assumption that was appropriate so long as neurons of the same embryonic age were consistently used. However, it became apparent that when neurons from older embryos were tested, they had undergone a developmental change that dramatically altered their response to CM (Collins and Lee, 1982).

From embryonic stages 35 to 40, there was a marked decline in the ability of dissociated ciliary-ganglion neurons to extend axons in culture. By stage 40, fewer than 20% of the nerve cells extended axons compared to more than 80% of stage 35 nerve cells. Also, the axons of the older neurons were shorter and elongated much more slowly (Collins and Lee, 1982).

The inability of neurons from stage 40 or older ganglia to extend axons readily was observed using heart-cell microexudate or SCF-coated polyornithine as substrate and various culture media, including HCM. This loss of ability was also observed using intact stage 40 ganglia as well as dissociated neurons. This developmental loss is reversible, since stage 40 ganglia, if removed from the embryo and cultured for several days, can subsequently be dissociated into neurons as capable as stage 35 or younger neurons of extending axons rapidly.

There is a close temporal correlation between the period in which these nerve cells form functional synapses in the embryo (stages 35–40) and the period in which they lose the ability to extend axons readily when placed in culture. These and related observations suggested that the formation of functional peripheral synapses by the ciliary-ganglion neurons led to the loss of ability to extend axons. When these connections were broken by removal from the embryo, the ability to extend axons was eventually restored (Collins and Lee, 1982).

It is not clear whether the developmental change represents a loss of the ability of the nerve cells to extend axons under any circumstances or whether they have simply lost the ability to respond to our particular culture conditions, such as SCF and microexudate. The latter alternative implies that there exists some as yet unknown agent the presence of which becomes necessary (between stages 35 and 40) for outgrowth. Such a possibility will be difficult to rule out, making the two alternatives indistinguishable until either the unknown factor is discovered or the biochemical nature of the developmental change is defined.

3. SUMMARY AND DISCUSSION

Taking the broadest view, these investigations implicate three different sites at which the process of axon outgrowth can be regulated.

3.1. Substrate-Conditioning Factors

Cells in culture release into the medium and deposit onto the substrate components of the cell surface or extracellular matrix that can act as an excellent substrate for axon outgrowth. When present as a pathway on the substrate, these components can determine, by the mechanism of haptotaxis, the course taken by elongating axons.

There is strong evidence that elongating axons in insect embryos choose to associate specifically with particular cells in a way that determines the course taken by these axons in reaching their targets (Raper *et al.*, 1983a,b). Other evidence, both in the embryo and in culture, supports the notion that the course taken by embryonic neruons can be determined by their preferential association with specific cell types or substrate pathways (Attardi and Sperry, 1963; Katz and Lasek, 1980; Collins and Garrett, 1980; Bonhoeffer and Huf, 1982).

These considerations imply a close similarity between axonal guidance and other forms of morphogenetic cell recognition, which involve an interaction between specific cell-surface components (Moscona, 1974). Specific recognition occurs not only between axons and embryonic nonneuronal cells, but also between axon and axon, resulting in selective fasciculation. It is important to remember that the specificity of these interactions is probably not absolute, but relative. Axons may prefer to grow along a given substrate pathway or nerve bundle, but are capable of taking another if the preferred route is not available. This lack of absolute specificity would also apply to the *in vitro* manifestation of the axonal-growth-promoting or -guiding activity of such surface components.

One would expect a range of different compounds to be involved in these specific and complex cell interactions, some perhaps belonging to a class of specific variants of a particular type of molecule (Edelman, 1983). Further study of the SCFs from a variety of cell types is likely to reveal much about the chemistry and mode of action of some of the surface components responsible for supporting and directing axon outgrowth during development.

3.2. Nerve Growth Factors

The second type of agent in CM affecting outgrowth increases the ability of the nerve cell to maintain rapidly elongating axons. This agent acts in solution, not as part of the substrate, and cannot by itself substitute for an inadequate substrate. The activity of this agent is closely mimicked by that of NGF, even, surprisingly enough, for parasympathetic ciliary-ganglion neurons. The detailed results are consistent with there being a class of related, insulinlike agents, of which NGF is one, that normally affect specific types of neurons, but are capable of cross-reacting with other types *in vitro*.

These soluble growth factors are likely to be released by the target and serve to enhance elongation and probably also play a role in target location by the mechanisms of chemotaxis or oriented outgrowth, as demonstrated for NGF *in vitro* and *in vivo* (Levi-Montalcini *et al.*, 1978; Gundersen and Barrett, 1979). In contrast, materials resembling the SCFs are probably produced and act only locally, supporting and directing outgrowth along the pathway to the target and acting within the target by promoting surface recognition of appropriate synaptic partners.

3.3. Intrinsic Factors

At a specified time in their development, neurons can undergo a striking change in their response to environmental agents. We have described one example of such a change in which ciliary-ganglion neurons lose their ability to extend axons rapidly in culture just after they form functional synapses *in vivo*. Other examples of such developmental changes are the sudden dependence of neurons on target-derived trophic factors, which seems to account for the timing of neuronal death that occurs during normal development (Berg, 1982), and the loss of responsiveness of sensory neurons to NGF, which is accompanied by the loss of surface receptors for that agent (Rohrer and Barde, 1982).

In the case of ciliary-ganglion neurons, the developmental loss of ability to extend axons readily seems to be triggered by the formation of functional synapses, and the change is reversed when these peripheral connections are broken. The change may reflect the shutdown of synthetic machinery no longer required for axonal growth.

As this example and the others mentioned above clearly indicate, neuronal development is not a function of external trophic factors alone. Rather, changes within the neuron itself regulate the response to such environmental agents. It is this shifting interaction between the cell and its environment that guides the course of neural development.

REFERENCES

Adler, R., Manthorpe, M., Skaper, S.D., and Varon, S., 1981, Polyornithine-attached neurite-promoting factors (PNPF's): Culture sources and responsive neurons, *Brain Res.* **206:**129–144.

Attardi, D.G., and Sperry, R.W., 1963, Preferential selection of central pathways by regenerating optic fibers, *Exp. Neurol.* **7:**46–64.

Berg, D.K., 1982, Cell death in neuronal development, in: *Neuronal Development* (N.C. Spitzer, ed.), Plenum Press, New York, pp. 297–331.

Bonhoeffer, F., and Huf, J., 1982, *In vitro* experiments on axon guidance demonstrating an anterior–posterior gradient on the tectum, *Eur. Mol. Biol. Organization J.* **1:**427–431.

Burnham, P., Raiborn, C., and Varon, S., 1972, Replacement of nerve-growth factor by ganglionic non-neuronal cells for the survival *in vitro* of dissociated ganglionic neurons, *Proc. Natl. Acad. Sci. U.S.A.* **69:**3556–3560.

Collins, F., 1978a, Axon initiation by ciliary neurons in culture, *Dev. Biol.* **65:**50–57.

Collins, F., 1978b, Induction of neurite outgrowth by a conditioned medium factor bound to the culture substratum, *Proc. Natl. Acad. Sci. U.S.A.* **75**:5210–5213.

Collins, F., 1980, Neurite outgrowth induced by the substrate associated material from nonneuronal cells, *Dev. Biol.* **79**:247–252.

Collins, F., and Dawson, A., 1982, Conditioned medium increases the rate of neurite elongation: Separation of this activity from the substratum-bound inducer of neurite outgrowth, *J. Neurosci.* **2**:1005–1010.

Collins, F., and Dawson, A., 1983, An effect of nerve growth factor on parasympathetic neurite outgrowth, *Proc. Natl. Acad. Sci. U.S.A.* **80**:2091–2094.

Collins, F., and Garrett, J.E., Jr., 1980, Elongating nerve fibers are guided by a pathway of material released from embryonic nonneuronal cells, *Proc. Natl. Acad. Sci. U.S.A.* **77**:6626–6628.

Collins, F., and Lee, M.R., 1982, A reversible developmental change in the ability of ciliary ganglion neurons to extend neurites in culture, *J. Neurosci.* **2**:424–430.

Culp, L.A., and Buniel, J.F., 1976, Substrate-attached serum and cell proteins in adhesion of mouse fibroblasts, *J. Cell Physiol.* **88**:89–106.

Edelman, G.M., 1983, Cell adhesion molecules, *Science* **219**:450–457.

Fujii, D.K., Massoglia, S.L., Savion, N., and Gospodarowicz, D., 1982, Neurite outgrowth and protein synthesis by PC12 cells as a function of substratum and nerve growth factor, *J. Neurosci.* **2**:1157–1175.

Greene, L.A., 1977a, Quantitative *in vitro* studies on the nerve growth factor requirement of neurons. I. Sympathetic neurons, *Dev. Biol.* **58**:96–105.

Greene, L.A., 1977b, Quantitative *in vitro* studies on the nerve growth factor requirement of neurons. II. Sensory neurons, *Dev. Biol.* **58**:106–113.

Gundersen, R.W., and Barrett, J.N., 1979, Neuronal chemotaxis: Chick dorsal-root axons turn toward high concentrations of nerve growth factor, *Science* **206**:1079–1080.

Hawrot, E., 1980, Cultured sympathetic neurons: Effects of cell-derived and synthetic substrata on survival and development, *Dev. Biol.* **74**:136–151.

Helfand, S.L., Smith, G.A., and Wessells, N.K., 1976, Survival and development in culture of dissociated neurons from ciliary ganglia, *Dev. Biol.* **50**:541–547.

Katz, M.J., and Lasek, R.J., 1980, Guidance cue patterns and cell migration in multicellular organisms, *Cell Motil.* **1**:141–157.

Lander, A.D., Fujii, D.K., Gospodarowicz, D., and Reichardt, L.F., 1982, Characterization of a factor that promotes neurite outgrowth: Evidence linking activity to heparan sulfate proteoglycan, *J. Cell Biol.* **94**:574–585.

Levi-Montalcini, R., and Angeletti, P.U., 1963, Essential role of nerve growth factor in the survival and maintainence of dissociated sensory and sympathetic embryonic nerve cells *in vitro*, *Dev. Biol.* **7**:653–659.

Levi-Montalcini, R., Menesini Chen, M.G., and Chen, J.S., 1978, Neurotropic effects of the nerve growth factor in chick embryos and in neonatal rodents, *Zoon* **6**:201–212.

Max, S.R., Schwab, M., Dumas, M., and Thoenen, H., 1978, Retrograde axonal transport of nerve growth factor in the ciliary ganglion of the chick and the rat, *Brain Res.* **159**:411–415.

Miki, N., Hayashi, Y., and Higashida, H., 1981, Characterization of chick gizzard extract that promotes neurite outgrowth in cultured ciliary neurons, *J. Neurochem.* **37**:627–633.

Moscona, A.A., 1974, Surface specification of embryonic cells: Lectin receptors, cell recognition, and specific cell ligands, in: *The Cell Surface in Development* (A.A. Moscona, ed.), John Wiley, New York, pp. 67–99.

Paravicini, U., Stoeckel, K., and Thoenen, H., 1975, Biological importance of retrograde axonal transport of NGF in adrenergic neurons, *Brain Res.* **84**:279–291.

Poste, G., Greenham, L.W., Mallucci, L., Reeve, P., and Alexander, D.J.A., 1973, The study of cellular "microexudates" by ellipsometry and their relationship to the cell coat, *Exp. Cell Res.* **78**:303–313.

Raper, J.A., Bastini, M., and Goodman, C.S., 1983a, Pathfinding by neuronal growth cones in grasshopper embryos. I. Divergent choices made by the growth cones of sibling neurons, *J. Neurosci.* **3:**20–30.

Raper, J.A., Bastiani, M., and Goodman, C.S., 1983b, Pathfinding by neuronal growth cones in grasshopper embryos. II. Selective fasciculation onto specific axonal pathways, *J. Neurosci.* **3:**31–41.

Rohrer, H., and Barde, Y.-A., 1982, Presence and disappearance of nerve growth factor receptors on sensory neurons in culture, *Dev. Biol.* **89:**309–315.

Stoeckel, K., Paravicini, U., and Thoenen, H., 1974, Specificity of the retrograde axonal transport of nerve growth factor, *Brain Res.* **76:**413–421.

Wong, R.G., Hadley, R.D., Kater, S.B., and Hauser, G.C., 1981, Neurite outgrowth in molluscan organ and cell cultures: The role of conditioning factors, *J. Neurosci.* **1:**1008–1021.

Chapter 15

Nerve Growth Factors in Chick and Rat Tissues

Ted Ebendal, Lars Olson, Åke Seiger, and Makonnen Belew

1. INTRODUCTION

Coculture experiments have suggested the existence of tissue-released chemical factors necessary for the initiation and support of nerve-fiber growth (Ebendal and Jacobson, 1977a; Ebendal, 1979, 1981). Such factors may be important in establishing and maintaining the innervated state of various tissues *in vivo*, and their identity, distribution, and level are of obvious interest for gaining insight into mechanisms of growth and development of neurons.

Nerve growth factor (NGF) (Levi-Montalcini and Angeletti, 1968; Thoenen and Barde, 1980) is so far the best-characterized substance that stimulates neuron growth and axonal extension. We demonstrated by bioassay that little if any NGF is present in the normal adult rat iris, but that NGF appears after sensory denervation *in situ* (Ebendal *et al.*, 1980). Rat tissues were also found to contain a nerve-growth-stimulating factor with properties different from those of NGF (Ebendal *et al.*, 1980; Richardson and Ebendal, 1982). Moreover, tissues from the chick embryo are fairly rich in a nerve-growth-promoting substance different in the bioassay from NGF (Ebendal and Jacobson, 1977b; Ebendal and Belew, 1980; Ebendal *et al.*, 1979, 1982b).

To supplement the information obtained by the bioassay, we have evolved

TED EBENDAL ● Department of Zoology, Uppsala University, S-751 22 Uppsala, Sweden. LARS OLSON and ÅKE SEIGER ● Department of Histology, Karolinska Institute, S-104 01 Stockholm, Sweden. MAKONNEN BELEW ● Department of Biochemistry (Faculty of Science), Uppsala University, S-751 23 Uppsala, Sweden.

a new radioimmunoassay (RIA) for NGF, using affinity-purified antibodies. We now demonstrate that the bioassay and the RIA give fully comparable results regarding the presence of NGF in normal and sensory-denervated rat irides. In the chick, both methods yield negative results for crude embryo extract, but indicate the existence of a chick NGF in fractions obtained after partial purification.

2. MATERIALS AND METHODS

2.1. Bioassay

Sympathetic, spinal, and ciliary ganglia were dissected from the Day 9 chick embryo and embedded in a gel of native collagen fibrils (Ebendal, 1979; Ebendal et al., 1979). The cultures were incubated for 2 days under standard tissue-culture conditions and then examined in dark-field illumination through an inverted microscope.

2.2. Preparation of Affinity-Purified Antibodies to β-Nerve Growth Factor

NGF was prepared from the submandibular glands of male N.M.R.I. mice (Alab, Stockholm) according to the procedure of Mobley et al. (1976). The active material was eluted from the second carboxymethyl (CM)–Sepharose column (Fig. 1) by a linear gradient of NaCl to obtain immunogenically pure βNGF (Chapman et al., 1979). Peak tubes were used to immunize rabbits (multiple intradermal injections) by standard methods. The antiserum obtained was passed over an affinity column of βNGF coupled to CNBr-activated Sepharose 4B. The adsorbed antibodies were displaced from the column by 4.5 $MgCl_2$ in acetate buffer at pH 5 as described by Stoeckel et al. (1976).

2.3. Radioimmunoassay

Two types of RIA methods were tested. For the first one [a competitive binding assay (Wide, 1981)], βNGF was linked with 1–10-μm Sephadex particles by CNBr and then allowed to adsorb antibodies to βNGF from the antiserum. The samples were mixed with iodinated βNGF and allowed to compete for the binding sites on the solid-phase antibodies (see Wide, 1981). However, even with the use of Tween 20, we experienced the difficulties described by Suda et al. (1978) for the competition assay for βNGF and had to abandon this method.

The second RIA was based on the sandwich binding assay [the two-site assay of Suda et al. (1978)] (see Wide, 1981). Polystyrene tubes (10 × 60 mm)

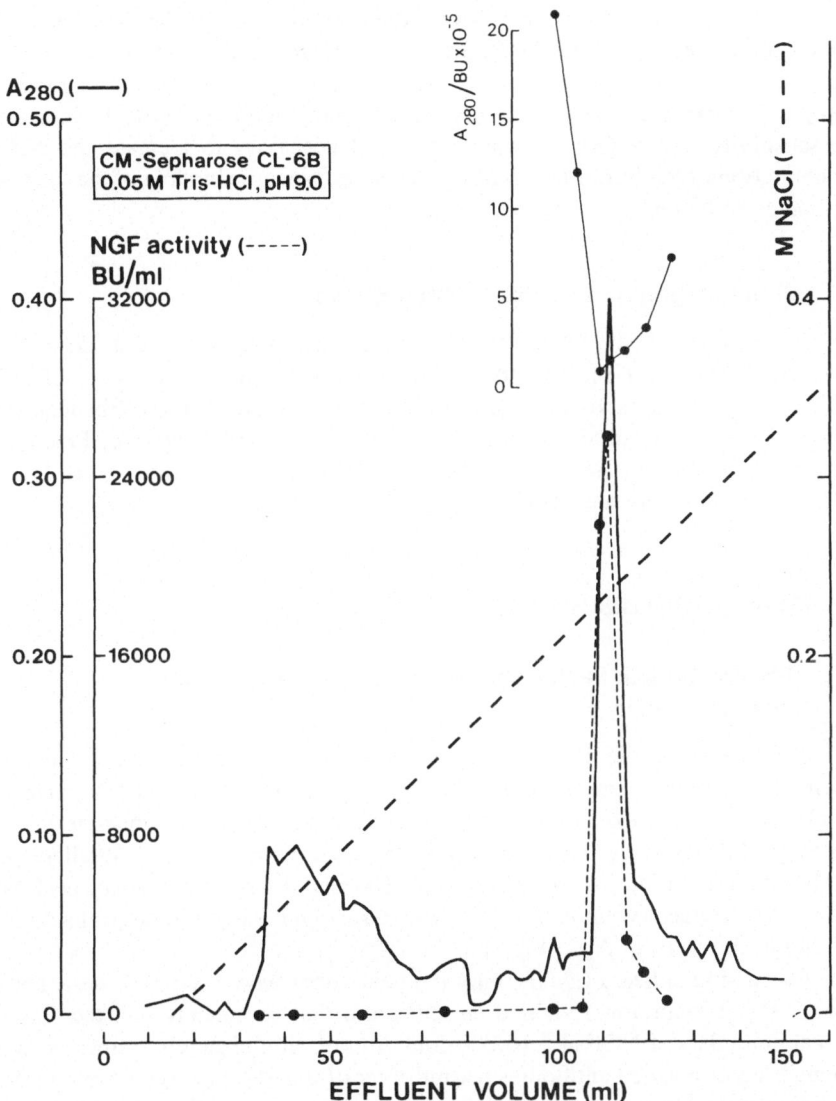

Figure 1. Elution of βNGF from the second CM–Sepharose column using the procedure of Mobley *et al.* (1976). Submandibular glands (10 g) were homogenized and taken through the first CM–Sepharose column, after which the pH was lowered to 4.0 and NaCl was added to 0.4 M. The supernatant was applied to a second 1.6 × 20 cm CM–Sepharose CL-6B column and the pH was raised to 8.0. The βNGF was eluted by a linear gradient of NaCl as described by Chapman *et al.* (1979). (BU) Biological units.

were coated with affinity-purified anti-βNGF immunoglobulin (1.2 μg in 200 μl) and sample was added for 4 days. After washing, antibodies labeled with ^{125}I by the chloramine-T method (Greenwood *et al.*, 1963) were added for an additional period of 3 days, followed by extensive washing with buffer for another 3 days. The extended periods of incubation were found necessary to increase the sensitivity of the RIA. It was also found necessary to increase the NaCl concentration of the buffer to 0.5 M to prevent nonspecific binding of the labeled antibodies following the tissue extracts.

2.4. Chromatography of Chick-Embryo Extract

Embryonic chicks (18-day) were homogenized and extracted as described by Ebendal *et al.* (1979). The crude extract was submitted to two steps of ion-exchange chromatography and gel filtration for partial purification (Ebendal and Belew, 1980). Relevant details are given in the Fig. 7 and 8 captions. Fractions were pooled, concentrated, and dialyzed against culture medium before being tested in the bioassay and RIA.

3. RESULTS AND DISCUSSION

3.1. β-Nerve Growth Factor and Anti-β-Nerve Growth Factor Immunoglobulin

The elution profile of βNGF from the second CM–Sepharose column was essentially similar to that described by Mobley *et al.* (1976) and Chapman *et al.* (1979), with the exception that the weakly reddish material (inactive in the bioassay) did not leave the column until the beginning of the salt gradient [at an effluent volume of 40–50 ml (Fig. 1)]. The βNGF eluted as a sharp peak (at 108–112 ml effluent volume in Fig. 1) with a steep increase in biological activity parallel to the peak in A_{280} (Fig. 1).

For raising antisera βNGF, only the peak tubes were used. The best rabbit (No. 6 at our laboratory) yielded an antiserum that completely blocked 1 biological unit (BU) of NGF at a 16,000-fold dilution (in the ganglion-fiber assay). About 1 mg of purified antibodies per milliliter of this serum was recovered after affinity chromatography on a βNGF–Sepharose column. The antibodies at a concentration of 200 ng/ml inhibited 20 ng/ml of NGF (Fig. 2). These values are in agreement with those reported by Stoeckel *et al.* (1976).

After storage at −20°C, a βNGF concentration of 6–20 ng/ml (determined from an absorbance of 1.6 at 280 nm for 1 mg/ml) gave the optimal fiber-halo response in spinal (Fig. 3) and sympathetic ganglia (Fig. 4). Complete inhibition of NGF activity by 200 ng/ml of anti-NGF immunoglobulin is shown in Fig. 5.

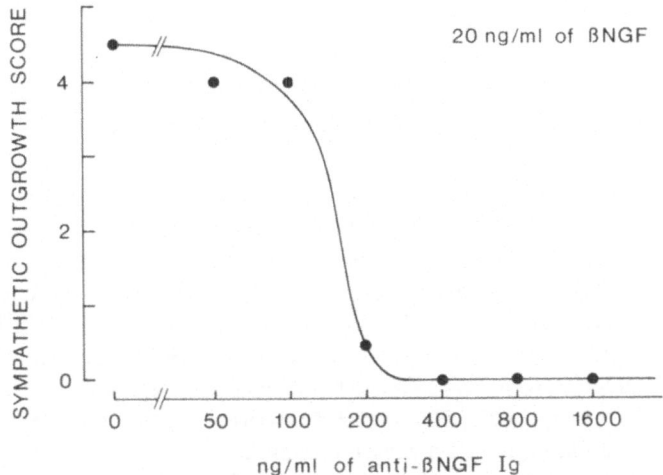

Figure 2. Inhibition of fiber outgrowth by affinity-purified antibodies to βNGF. Sympathetic ganglia were stimulated in a collagen gel by 20 ng/ml of mouse βNGF and in the presence of various amounts of the immunoglobulin. Fiber densities (from 0 to 5) were ranked on a blind basis 2 days after the cultures were initiated. The βNGF was purified as illustrated in Fig. 1, and the antibodies were purified from rabbit antiserum essentially as described by Stoeckel *et al.* (1976).

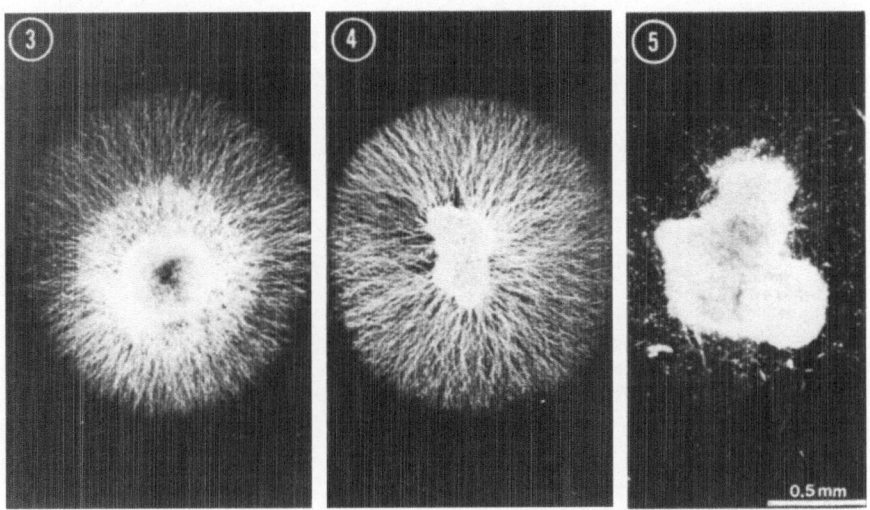

Figure 3. A spinal ganglion (from a 9-day-old chick embryo) responding to 20 ng/ml of mouse βNGF by forming a fiber halo in a collagen gel (2-day incubation; dark-field microscopy).

Figure 4. As in Fig. 3, but a sympathetic (lumbar paravertebral) ganglion.

Figure 5. As in Fig. 4, but with the addition of 200 ng/ml of affinity-purified antibodies (cf. Fig. 2). The magnification is the same as in Figs. 3 and 4.

3.2.   Radioimmunoassay of β-Nerve Growth Factor in the Rat Iris

The two-site RIA was sensitive to 5–15 pg βNGF in a sample volume of 100 μl (Fig. 6).

Irides from adult Sprague–Dawley rats were pooled in groups of three, carefully homogenized in 300 μl RIA buffer, and centrifuged. The supernatants were collected for RIA, and counts were converted to NGF levels by comparison with linearized data obtained from dilutions of mouse βNGF. No attempt was made to correct calculations for the possible difference in antibody affinity between mouse and rat NGF (cf. Harper and Thoenen, 1980). Known amounts of βNGF were added to some iris samples to estimate recoveries (Suda *et al.*, 1978; Harper and Thoenen, 1980). Nearly 100% recovery was found in all groups of iris extracts, validating the data presented herein.

As shown by the bioassay (Ebendal *et al.*, 1980), the normal iris contained little NGF. The RIA detected about 7 pg NGF per iris, a figure that should be interpreted with care, since it is near the lower limit of sensitivity of the method (Table I).

Three kinds of treatments have been shown (Ebendal *et al.*, 1980) to increase the level of NGF detectable by bioassay: culture of the iris, transplantation of

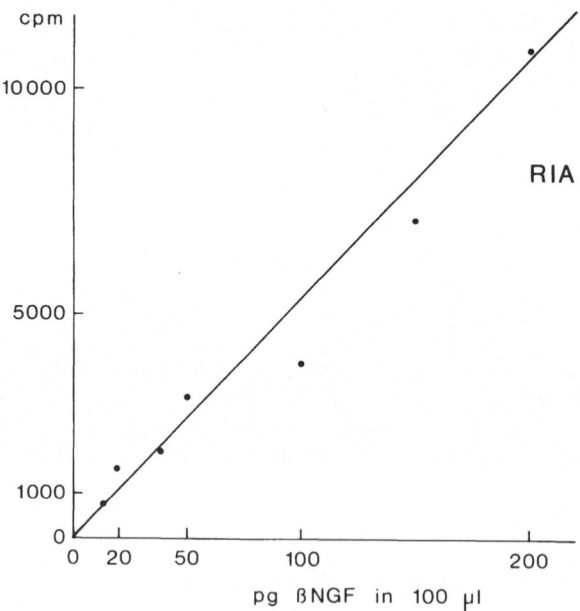

Figure 6.   Results obtained by the two-site RIA for serial dilutions of βNGF. Background binding of the [125I]-labeled anti-NGF antibodies is subtracted from the cpm values given.

Table I. Amount of Nerve Growth Factor in the Rat Iris as Determined by Radioimmunoassay

Irides	NGF per iris (pg)[a]	Level of significance
Normal iris	7 ± 0.5 (8)	$P \approx 0.05^b$
Transplanted for 2 days	110 ± 13 (4)	$P < 0.001^c$
Sensorially and sympathetically denervated for 10 days	118 ± 12 (4)	$P < 0.001^c$

[a] Means ± S.D. (number of irides).
[b] U test against background cpm.
[c] Compared with the level of NGF in normal irides.

the iris, and sensory denervation of the iris *in situ*. In this experiment, the two latter treatments were tested with the RIA (Table I). In irides transplanted to the anterior eye chamber, the RIA demonstrated a rise to 110 pg NGF per iris after 2 days, a highly significant increase (Table I). It is noteworthy that the transplants were edematous and when tested in a *competition* RIA gave low counts, indicating severalfold higher levels of NGF than in the two-site RIA. Such an overestimation is consistent with the findings of Suda *et al.* (1978), which indicate that NGF-binding components of serum can substantially affect the recovery of NGF in competition RIAs.

Perhaps the most interesting group consisted of sensory (plus sympathetically) denervated irides. Here, too, a more than 10-fold increase in the NGF content (to 118 pg/iris) was found by RIA (Table I). This confirms our earlier bioassay estimations of 80–160 pg/mg denervated iris [equivalent to one iris (Ebendal *et al.*, 1980)]. The actual mechanism whereby the level of NGF is regulated requires further study. Denervation may lead to increased synthesis, to decreased degradation, to retrieval from a storage pool, or simply to the accumulation of NGF in the end organ due to the block in retrograde axonal transport away from the area.

Further results from this study, including sympathetic and parasympathetic denervations, will be presented elsewhere (Ebendal *et al.*, 1983).

3.3. Bioassay and Radioimmunoassay for Nerve Growth Factor in Fractions of Chick-Embryo Extract

Partial purification of one or more nerve-growth-promoting substances in embryonic chick tissues was described earlier (Ebendal *et al.*, 1979; Ebendal and Belew, 1980). The activity of the embryonic extract, as well as the nerve-growth-stimulating activity seen around chick-tissue explants (Ebendal and Jacobson, 1977b; Ebendal, 1979), stimulates both NGF-sensitive ganglia and the ciliary ganglion. Moreover, activity is not inhibited by antiserum to mouse NGF. This suggests that NGF in its typical form is not involved in the growth stimulation. Two further observations prompted us to renew our search for NGF in chick-embryo extract: First, a living iris explant from the chicken, like that from the rat, produced NGF in culture (Ebendal *et al.*, 1982), which was not inhibited

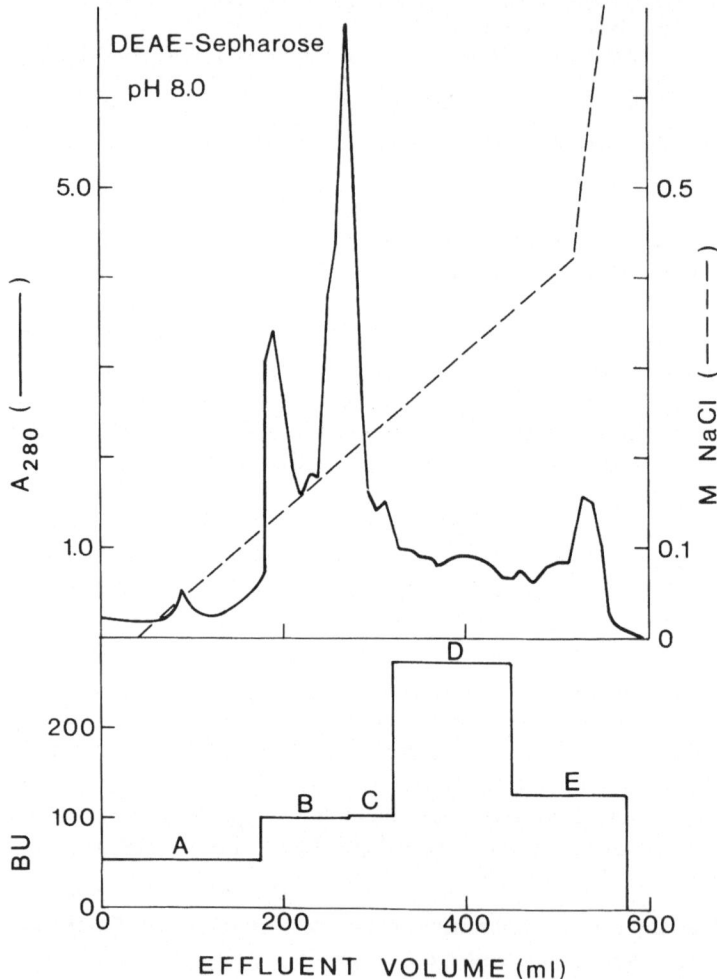

Figure 7. Elution pattern of chick-embryo extract adsorbed onto DEAE–Sepharose CL-6B. A sample of 350 ml extract (made from 35 g frozen 18-day-old chick embryos) rejected from CM–Sepharose at neutral pH was applied to the 2.4 × 17 cm column at a flow rate of 25 ml · hr^{-1}. After washing with buffer (0.02 M Tris-HCl, pH 8.0), a linear salt gradient was applied and fractions collected and pooled (A–E) for bioassay. The nerve-outgrowth-stimulating activity in BU (1 BU/ml stimulating a dense circular fiber halo) is given at bottom. The most active fraction (pool D) was taken for gel filtration as shown in Fig. 8.

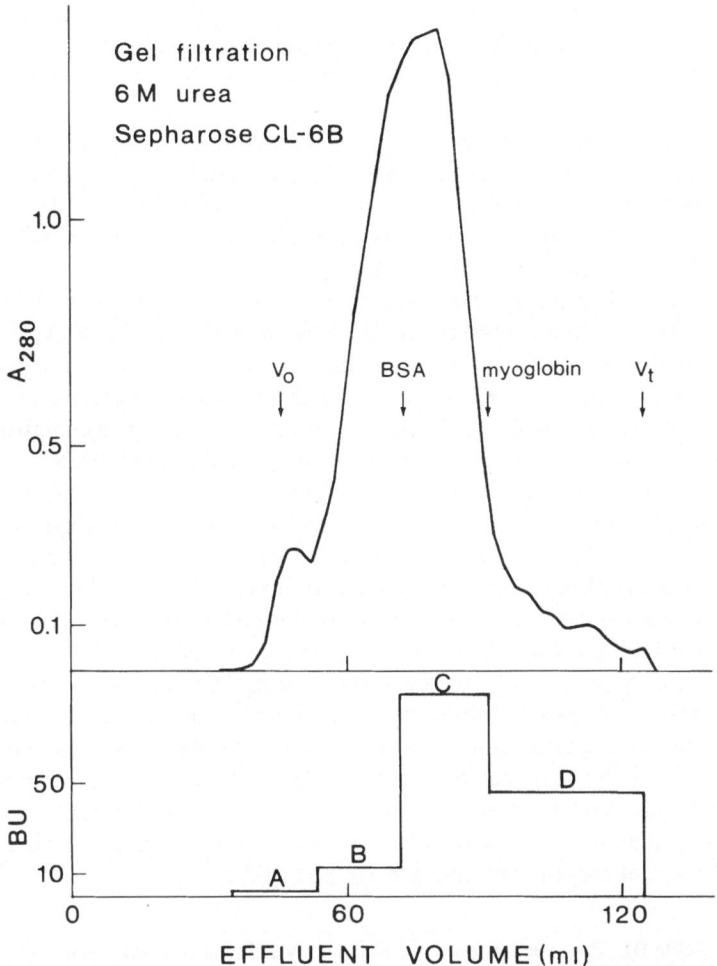

Figure 8. Gel filtration of pool D from the DEAE–Sephadex column in Fig. 7. Seventy percent of pool D was concentrated to 6 ml, solid urea was added to 6 M, and the sample was applied at a flow rate of 5 ml · h⁻¹ to 1.6 × 62 cm column of Sepharose CL-6B equilibrated with 0.02 M Tris-HCl (pH 8.0) containing 6 M urea. Elution was continued with the same buffer and the collected fractions pooled as indicated (A–D). Bovine serum albumin (BSA) and myoglobin were run under identical conditions to serve as markers. The bioassay results are given as BU at bottom. The results from the RIA of these pools are presented in Table II.

by antiserum to mouse NGF. Second, chick-heart fibroblasts maintained in a serum-free medium produced a factor similar to NGF (Norrgren *et al.*, 1983).

Extract prepared from the entire 18-day-old chick embryo (Ebendal *et al.*, 1982b) was chromatographed on CM–Sepharose at neutral pH, and the excluded material was pooled. Nearly all the nerve-growth-promoting activity was recovered in this fraction. It was active on ciliary as well as on sympathetic and spinal ganglia. Antiserum to NGF exerted no inhibitory effect, reproducing results obtained with the crude extract. Similarly, the RIA did not reveal NGF in this fraction.

The active material was applied to a diethylaminoethyl (DEAE)–Sepharose CL-6B column at pH 8.0, and all unadsorbed material was washed out. Most of the activity adsorbed to the column and was eluted with a linear salt gradient (Fig. 7). After this step, RIA analysis detected small amounts of NGF in the biologically most active pool (Table II).

The material was further studied by gel filtration in the presence of urea (Fig. 8). The activity (expressed in BU comparable to the NGF-evoked fiber response) eluted at a position corresponding to a molecular weight of less than 70,000 (Ebendal et al., 1979). Surprisingly, abundant activity eluted at positions corresponding to less than 20,000 daltons in this study [(Fig. 8) compare with myoglobin, having a molecular weight of 17,000]. Furthermore, the RIA detected NGF-like activity in this molecular-weight range (Table II). NGF may have been masked in the crude material, since the recovery was better than expected. However, it should be emphasized that the amounts detected by the RIA were small (less than 10 ng NGF, compared with about 50 BU by bioassay). This may be due to low affinity of chick NGF for antibodies to mouse NGF or to the fact that the biological activity was only partly attributable to NGF.

Preliminary results from the bioassay indicated that active material in pool C (Fig. 8) was also partly blocked by affinity-purified antibodies to βNGF. The ciliary-ganglion response to this fraction was not substantially inhibited by the antibodies. Possible explanations include (1) contamination by a chick NGF bound to acidic carrier substances and (2) minor antigenic similarities between mouse NGF and an acidic growth factor capable of stimulating the ciliary ganglion. We are at present examining these possibilities.

Table II. Radioimmunoassay for Nerve Growth Factor in Chick-Embryo Extract Fractionated by Gel Filtration in the Presence of 6 M Urea

Samples	NGF indicated by RIA (ng)	Apparent recovery (%)
Starting material		
Pool D from DEAE–Sepharose[a]	$3.5 + 0.25$[b]	(100)
Gel Filtration		
Pool A	0	0
Pool B	0	0
Pool C	0	0
Pool D	$8.6 + 1.8$[b]	246

[a] Amount corresponding to 70% of pool D in Fig. 7.
[b] Means ± S.D. (from two tubes).

ACKNOWLEDGMENTS. Dr. Leif Wide, Department of Clinical Chemistry, University Hospital, Uppsala, is thanked for his kind support and advice regarding RIA techniques. This work was aided by grants from the Swedish Science Research Council (B-BU 4024-102, S-FO 4024-101), the Swedish Medical Research Council (14X-0318, 14P-5867, 14X-0655, 25P-6326), "Magnus Bergvalls Stiftelse," "Karolinska Institutets fonder," and the "Expressen" Prenatal Research Foundation. Dr. Belew was supported by grants from the Fortia Research Fund and from the Foundation of Sven and Ebba-Christina Hagberg. We thank Bo Molin, Annika Kylberg, Stine Söderström, Ingrid Strömberg, and Lena Hultgren for technical assistance.

REFERENCES

Chapman, C.A., Banks, B.E.C., Carstairs, J.R., Pearce, F.L., and Vernon, C.A., 1979, The preparation of nerve growth factor from the prostate of the guinea-pig and isolation of immunogenically pure material from the mouse submandibular gland, *FEBS Lett.* **105**:341–344.

Ebendal, T., 1979, Stage-dependent stimulation of neurite outgrowth exerted by nerve growth factor and chick heart in cultured embryonic ganglia, *Dev. Biol.* **72**:276–290.

Ebendal, T., 1981, Control of neurite extension by embryonic heart explants, *J. Embryol. Exp. Morphol.* **61**:289–301.

Ebendal, T., and Belew, M., 1980, Chick heart factor controlling neurite extension, *Eur. J. Cell Biol.* **22**:409.

Ebendal, T., and Jacobson, C.-O., 1977a, Tissue explants affecting extension and orientation of axons in cultured chick embryo ganglia, *Exp. Cell Res.* **105**:379–387.

Ebendal, T., and Jacobson, C.-O., 1977b, Tests of possible role of NGF in neurite outgrowth stimulation exerted by glial cells and heart explants in culture, *Brain Res.* **131**:373–378.

Ebendal, T., Belew, M., Jacobson, C.-O., and Porath, J., 1979, Neurite outgrowth elicited by embryonic chick heart: Partial purification of the active factor, *Neurosci. Lett.* **14**:91–95.

Ebendal, T., Olson, L., Seiger, Å., and Hedlund, K.-O., 1980, Nerve growth factors in the rat iris, *Nature (London)* **286**:25–28.

Ebendal, T., Hedlund, K.-O., and Norrgren, G., 1982, Nerve growth factors in chick tissues, *J. Neurosci. Res.* **8**:153–164.

Ebendal, T., Norrgren, G., and Hedlund, K.-O., 1982b, Nerve growth-promoting activity in the chick embryo: Quantitative aspects, *Med. Biol.* **61**:65–92.

Ebendal, T., Olson, L., and Seiger, Å., 1983, The level of nerve growth factor as a function of innervation: A correlative radioimmunoassay and bioassay study of the rat iris, *Exp. Cell Res.* **148**:311–317.

Greenwood, F.C., Hunter, W.M., and Glover, J.S., 1963, The preparation of ^{131}I-labelled human growth hormone of high specific radioactivity, *Biochem. J.* **89**:114–123.

Harper, G.P., and Thoenen, H., 1980, The distribution of nerve growth factor in the male sex organs of mammals, *J. Neurochem.* **34**:893–903.

Levi-Montalcini, R., and Angeletti, P.U., 1968, The nerve growth factor, *Physiol. Rev.* **48**:534–569.

Mobley, W.C., Schenker, A., and Shooter, E.M., 1976, Characterization and isolation of proteolytically modified nerve growth factor, *Biochemistry* **15**:5543–5552.

Norrgren, G., Ebendal, T., Gebb, C., and Wikström, H., 1983, The use of Cytodex® 3 microcarriers and reduced-serum media for the production of nerve growth promoters from chicken hearts cells, *Dev. Biol. Stand.* **55**:(in press).

Richardson, P.M., and Ebendal, T., 1982, Nerve growth activities in rat peripheral nerve, *Brain Res.* **246:**57–64.

Stoeckel, K., Gagnon, C., Guroff, G., and Thoenen, H., 1976, Purification of nerve growth factor antibodies by affinity chromatography. *J. Neurochem.* **26:**1207–1211.

Suda, K., Barde, Y.-A., and Thoenen, H., 1978, Nerve growth factor in mouse and rat serum: Correlation between bioassay and radioimmunoassay determinations, *Proc. Natl. Acad. Sci. U.S.A.* **75:**4042–4046.

Thoenen, H., and Barde, Y.-A., 1980, Physiology of nerve growth factor, *Physiol. Rev.* **60:**1284–1335.

Wide, L., 1981, Use of particulate immunosorbents in radioimmunoassay, *Methods Enzymol.* **73:**203–224.

Chapter 16

Macromolecular Factors Involved in the Regulation of the Survival and Differentiation of Peripheral Sensory and Sympathetic Neurons

Hans Thoenen, Yves-Alain Barde, and David Edgar

1. INTRODUCTION

Neuronal cell death is a widespread physiological phenomenon that occurs during development of vertebrate peripheral and central nervous systems (Cowan, 1973; Jacobson, 1978). The topographically selective regulation of neuronal survival represents an important mechanism for the formation of the structure and function of the fully differentiated nervous system, and there is evidence that a major mechanism responsible involves the local production and release of trophic factors (see Barde *et al.*, 1983).

Although the final goal is to understand these events at the molecular level *in vivo,* appropriate tissue-culture techniques are invaluable tools for analyzing the changing requirements of neurons for trophic factors during development or as an assay system for purifying these factors. Purification is necessary in order to evaluate their physiological role, as has been demonstrated for the protein nerve growth factor (NGF) (see Thoenen and Barde, 1980; Harper and Thoenen, 1981). NGF is so far the only neurotrophic factor that has been characterized chemically and has been shown to have a biological function. It was isolated at a time (Cohen, 1960) when our general knowledge of developmental neuro-

---

HANS THOENEN, YVES-ALAIN BARDE, and DAVID EDGAR ● Department of Neurochemistry, Max-Planck-Institute for Psychiatry, D-8033 Martinsried, Federal Republic of Germany.

biology at the molecular level was essentially nonexistent, and purification was possible only because very rich sources of NGF were discovered (see Levi-Montalcini, 1966). However, it is now recognized that the large amounts of NGF present in the salivary gland of the male mouse or in the more recently discovered source of the bovine seminal plasma (see Harper and Thoenen, 1981; Harper et al., 1982) have no physiological importance for the regulation of the survival, differentiation, and maintenance of function of sympathetic and sensory neurons. Physiologically significant NGF is synthesized by and released from the peripheral target tissues of these neurons. However, the quantities present therein are so small (Korsching and Thoenen, 1983) that even with the most advanced contemporary technology, the isolation from these physiologically relevant sources would be extremely difficult.

In the following sections, we will discuss the necessary criteria that have to be applied to tissue-culture procedures in order that they may be used to establish the importance of the factors that operate during different periods of development. Moreover, we will report on the purification of a new neurotrophic macromolecule that supports the survival of specific populations of peripheral and probably also central neurons.

2. CULTURE SYSTEMS USED FOR THE CHARACTERIZATION AND PURIFICATION OF NEURONAL TROPHIC FACTORS

For many years, cultured explants of ganglia proved to be useful for experiments aimed at determining whether the outgrowth of neurites is preferentially either or both promoted by and directed toward explants of target organs (see Burnstock, 1974; Ebendal and Jacobson, 1977). However, for the purification and characterization of neurotrophic factors, neuronal explants have clear limitations. For instance, it cannot be determined whether added neurotrophic factors act directly on neurons or only indirectly via nonneuronal cells. Such limitations are obviated by cultures of dissociated neurons, provided sufficiently rigorous criteria are applied (see Barde et al., 1982b).

1. To avoid the selection of an arbitrary subpopulation of neurons, dissociation procedures have to be chosen that give a high yield of neurons (i.e., a representative population) from the sources investigated.
2. The nonneuronal cells must be eliminated from the cultures to eliminate the aforementioned possibility that the added trophic factors act via nonneuronal cells rather than directly on the neurons.
3. The concentrations of added trophic factors should always be "saturating"; i.e., supramaximal concentrations must be used to preclude ambiguities in the interpretation of results, which, for instance, could occur

if added tissue extracts or conditioned media (CMs) contain not only neurotrophic factors but also molecules that change the substrate of the culture dish. The latter phenomenon could lead to an augmented binding of the survival factor(s) to the substrate, resulting in a locally higher concentration in the immediate surrounding of the neurons and consequently mimicking a higher concentration of the survival factor in an extract or CM.

3. AGE-DEPENDENT CHANGES IN THE REQUIREMENT FOR DIFFERENT SURVIVAL FACTORS *IN VITRO*

In vivo, it is well established that the requirement of NGF for survival of both sympathetic and sensory neurons depends on the developmental stage (see Thoenen and Barde, 1980). This change in the age-dependent requirement is also reflected *in vitro* not only for NGF but also for other survival factors. For instance, the proportion of chick dorsal-root sensory neurons that survive with NGF alone at embryonic day 8 (E8) is relatively small (25–30%) (Barde *et al.,* 1980). It increases to a maximum level of about 45% between E10 and E12 and then declines again to virtual background levels at E16. In contrast, the time–course of sensitivity to rat-brain extract increases continuously from E8 to E16, reaching a maximal effect of 75–80% (Barde *et al.,* 1980). The combination of both NGF and brain factor results in the survival of 90–100% of the neurons, provided optimal substrate conditions (see Section 4) are chosen.

4. POTENTIATION OF THE EFFECT OF SURVIVAL FACTORS BY MACROMOLECULES BOUND TO THE CULTURE-DISH SUBSTRATE

In recent experiments, it has been shown that the NGF-mediated survival of chick sympathetic neurons can be increased markedly by protease-sensitive macromolecules that attach strongly to the polyornithine substrate of the culture dish (Edgar and Thoenen, 1982). These substrate-attached molecules have no survival-promoting activity of their own, but they enhance the age-dependent survival effect of NGF on chick sympathetic neurons to such an extent that virtually all neurons survive from E8 to E16. Interestingly, the extent of survival of sympathetic neurons mediated by both NGF and high potassium ion concentrations is potentiated to the same extent by the polyornithine-adsorbed macromolecules (Wakade *et al.,* 1983).

It is worth noting that the conditions of the production of this factor and the manner of interaction with the polyornithine substrate of the culture dish are identical to those of the neurite-outgrowth-promoting factor(s) (Collins, 1978;

Adler *et al.*, 1981). This factor(s) is produced not only by chick heart cells, but also by many other cell lines (Adler *et al.*, 1981). Particularly large quantities are produced by the schwannoma (RN22) cell line. Both the neurite-promoting and NGF and the high-potassium-potentiating activity can be completely removed from the CM by adsorption to polyornithine substrate (Adler *et al.*, 1981; Wakade *et al.*, 1983). It remains to be established whether the molecules that are responsible for the potentiation of the survival effect of NGF and high potassium, on one hand, and the neurite-outgrowth-promoting activity, on the other, are identical and, if this should be the case, whether the same active domain of the molecule is responsible for the two different actions.

In preliminary experiments, it has also been shown that the potentiating effect of the substrate-bound molecule(s) is not confined to the potentiation of the effect of NGF and high potassium on sympathetic neurons, but can also be shown for sensory neurons. In the latter case, the enhancement of neuronal survival seems to be particularly noticeable when the substrate-bound molecules act in conjunction with the newly isolated brain survival factor (Barde *et al.*, 1982a). The modulatory action of substrate-bound macromolecules, so far demonstrated only under tissue-culture conditions, could be of great importance for the elucidation of the physiological roles of new survival molecules.

5. APPROACHES TO THE PURIFICATION AND CHARACTERIZATION OF NEW NEUROTROPHIC FACTORS

The history of the research on NGF has demonstrated how essential it is to purify molecules present in CMs or tissue extracts to establish their physiological role (see Thoenen and Barde, 1980; Harper and Thoenen, 1981). The administration of antibodies to a molecule with potential neurotrophic activity is the most definitive way to establish its physiological role, i.e., to determine which neurons depend at which time of their development on which factor(s) for survival or maintenance of function or for both.

The detection of the particularly rich sources of NGF will most probably remain the exception and cannot be expected *eo ipso* for other neurotropic factors. The difficulties that arise in attempts to purify trophic factors from their physiological sources become apparent if one compares the quantities of NGF present in male mouse salivary glands with the quantities present in densely innervated sympathetic target organs. The male mouse submandibular gland contains about 1 mg NGF/g wet weight [there are large variations and strain differences (for a review, see Thoenen and Barde, 1980)]. The NGF content of the densely innervated sympathetic target organs such as the rat iris or the heart atrium amounts to 2 and 1 ng/g wet weight, respectively (Korsching and Thoenen, 1983). Thus, the concentration present in the salivary gland differs from that in the physiological target cells of sympathetic neurons by a factor of about 1 million. Moreover, the bioassay, which allows the detection of NGF in tissue homogenates and in standard solutions of purified NGF, is not sensitive enough to detect NGF in tissue extracts of intact, densely innervated tissues that contain the highest levels of NGF (Harper *et al.*, 1980). Thus, NGF would not be among the numerous factors shown to be present in CMs and tissue extracts (see Barde *et al.*, 1983) by the biological effects. The purification factor predicted for NGF from physiological sources did indeed have to be accomplished for the purification of a neurotrophic factor present in pig brain (Barde *et al.*, 1982a). The original observations dealing with the neurotrophic effect of CMs of glioma cells (Barde *et al.*, 1978) revealed that the quantities of the neurotrophic factor present would make the purification of this molecule(s) (assuming a molecular weight and a specific biological activity similar to NGF) extremely difficult. In subsequent experiments, however, it was demonstrated that mammalian brain extract contained an activity that resembled that of CMs (Barde *et al.*, 1980, 1982a). Although the specific activity of the initial homogenate was very low, the availability of virtually unlimited amounts of pig brain made the purification of the brain factor possible (Barde *et al.*, 1982a). It has a molecular weight of 12,300 and is very basic (pI ≥ 10.2). It is apparently homogeneous, as judged by electrophoresis in the presence of sodium dodecyl sulfate and a purification factor of about 1.4 million.

As already mentioned, the production of antibodies is an important step in the elucidation of the physiological role of a neurotrophic factor. The monoclonal-antibody technique (Köhler and Milstein, 1976) in principle no longer requires the availability of pure antigen as a prerequisite for the production of specific antibodies: monoclonal antibodies can be produced against impure mixtures of antigens. Furthermore, the quantity of antigen necessary for immunization can be reduced by the recently developed methods for *in vitro* immunization (Luben *et al.*, 1982). However, many reported successes using these techniques may be those fortunate cases wherein the molecules in question were excellent antigens. Macromolecules that are in reasonably pure form but are available only in relatively small quantities (nanomole range) and are not particularly good immunogens cannot be expected to permit the production of antibodies or allow extensive biological characterization. In these cases, alternative approaches that became possible by reason of recent methodological achievements in protein chemistry and molecular genetics have to be used (see Schmitt *et al.*, 1982), for example, micromethods for the determination of amino acid sequence, efficient synthesis of oligopeptides and oligonucleotides, and availability of vectors that allow the production of a molecule in prokaryotes or eukaryotes after insertion of the appropriate genetic information into the vectors. For the majority of these approaches, the availability of a small quantity of purified material is necessary. The (micro)determination of the amino acid sequence (Chang, 1981; Hood *et al.,* 1981) and consequently the selection of an appropriate, unique peptide sequence is necessary for the synthesis of large quantities of a (hydrophilic) antigenic peptide. The peptide, coupled to larger carrier molecules, can be used for the immunization of an animal or the activation of B cells in culture to produce mono- or polyclonal specific antibodies (for a review, see Walter and Doolittle, 1983). The availability of the amino acid sequence can further be used to synthesize oligonucleotides suitable as primers (Noyes *et al.,* 1979) for reverse transcriptase or the selection of complementary DNA (cDNA) clones from a cDNA or genomic library (Stetler *et al.,* 1982). The cloned cDNA of the molecule of interest can be inserted into appropriate vectors and the molecules can be synthesized in eukaryotic or prokaryotic systems as already demonstrated for insulin, growth hormones, and interferon (see Lomedico, 1982; Devos *et al.,* 1982; Gluzman, 1982). The molecules so produced would then allow extensive biological analysis including binding kinetics *in vitro,* "pharmacological" studies *in vitro* and *in vivo,* and the determination of the site of production of the neurotrophic factor.

6. CONCLUDING REMARKS

A large number of experimental observations support the concept that macromolecular factors are involved in the regulation of the survival and differentiation of neurons. Appropriate tissue-culture procedures are important tools for the characterization and purification of such factors: the scientific history of NGF impressively demonstrates the necessity for purifying molecules to establish their physiological role. The purification of NGF, and in consequence the production of specific antibodies against it, was possible only because NGF is present in very large quantities in organs in which it plays no physiological role as a neurotrophic factor. At physiologically relevant sites, the quantities of NGF are minute; indeed, only very recently has a supersensitive enzyme immunoassay made it possible to quantitate the NGF levels in densely innervated sympathetic effector organs and sympathetic ganglia. It is now clear that even with the most sophisticated purification procedures available, it would be extremely difficult to isolate NGF from these physiologically relevant sites of production; the necessary extent of purification would be even greater than that for a new neurotrophic factor isolated from mammalian brain, which, to achieve apparent homogeneity, had to be purified 1.4-million-fold. The quantities of this molecule that can be isolated from its natural sources are not sufficient for extensive biological characterization or for the evaluation of its possible therapeutic potential. To obtain sufficient quantities of this molecule for such purposes, it will be necessary to produce it by methods of gene technology that have recently become available.

REFERENCES

Adler, R., Manthorpe, M., Skaper, S.D., and Varon, S., 1981, Polyornithine neurite-promoting factors, *Brain Res.* **206**:129–144.

Barde, Y.-A., Lindsay, R.M., Monard, D., and Thoenen, H., 1978, New growth factor released by glioma cells supporting survival and growth of sensory neurones, *Nature (London)* **274**:818.

Barde, Y.-A., Edgar, D., and Thoenen, H., 1980, Sensory neurons in culture: Changing requirements for survival factors during embryonic development, *Proc. Natl. Acad. Sci. U.S.A.* **77**:1199–1203.

Barde, Y.-A., Edgar, D., and Thoenen, H., 1982a, Purification of a new neurotrophic factor from mammalian brain, *EMBO J.* **1**:549–553.

Barde, Y.-A., Edgar, D., and Thoenen, H., 1982b, Molecules involved in the regulation of neuron survival during development, in: *Neuroscience Approached through Cell Culture*, Vol. 1 (S. Pfeiffer, ed.), CRC Press, Boca Raton, Florida, pp. 69–86.

Barde, Y.-A., Edgar, D., and Thoenen, H., 1983, New neurotrophic factors, *Annu. Rev. Physiol.* **45**:601–612.

Burnstock, G., 1974, Degeneration and orientation of growth of autonomic nerves in relation to smooth muscle in joint tissue cultures and anterior eye chamber transplants, in: *Dynamics of Degeneration and Growth in Neurons* (K. Fuxe, L. Olson, and Y. Zollerman, eds.), Pergamon Press, Oxford, pp. 509–520.

Chang, J.-Y., 1981, N-terminal sequence analysis of polypeptide at the picomole level, *Biochem. J.* **199**:557–564.

Cohen, S., 1960, Purification of a nerve-growth promoting protein from the mouse salivary gland and its neurocytotoxic antiserum, *Proc. Natl. Acad. Sci. U.S.A.* **46**:302–311.

Collins, F., 1978, Induction of neurite outgrowth by a conditioned-medium factor bound to the culture substratum, *Proc. Natl. Acad. Sci. U.S.A.* **75**:5210–5213.

Cowan, W.M., 1973, Neuronal death as a regulative mechanism in the control of cell number in the nervous system, in: *Development and Aging in the Nervous System* (M. Rockstein, ed.), Academic Press, New York, pp. 19–41.

Devos, R., Cheroutre, H., Taya, Y., Degrave, W., Van Heuverswyn, H., and Fiers, W., 1982, Molecular cloning of human immune interferon cDNA and its expression in eukaryotic cells, *Nucleic Acids Res.* **10**:2487–2501.

Ebendal, T., and Jacobson, C.O., 1977, Tissue explants affecting extension and orientation of axons in cultured chick embryo ganglia, *Exp. Cell Res.* **105**:379–387.

Edgar, D., and Thoenen, H., 1982, Modulation of NGF-induced survival of chick sympathetic neurons by contact with a conditioned medium factor bound to the culture substrate, *Dev. Brain Res.* **5**:89–92.

Gluzman, Y., 1982, *Eukaryotic Viral Vectors*, Cold Spring Harbor Laboratory, Cold Spring Harbor, New York.

Harper, G.P., and Thoenen, H., 1981, Target cells, biological effects and mechanism of action of nerve growth factor and its antibodies, *Annu. Rev. Pharmacol. Toxicol.* **21**:205–229.

Harper, G.P., Pearce, F.L., and Vernon, C.A., 1980, The production and storage of nerve growth factor *in vivo* by tissues of the mouse, rat, guinea pig, hamster and gerbil, *Dev. Biol.* **34**:893–903.

Harper, G.P., Glanville, R.W., and Thoenen, H., 1982, The purification of nerve growth factor from bovine seminal plasma, *J. Biol. Chem.* **257**:8541–8548.

Hood, L., Hunkapiller, M., and Dreyer, W.J., 1981, Microchemical instrumentation: ICN–UCLA Symposium on Cellular Recognition, *J. Supramol. Struct. Cell Biochem.* **17**:27–36.

Jacobson, M. (ed.), 1978, *Developmental Neurobiology*, 2nd ed., Plenum Press, New York.

Köhler, G., and Milstein, C., 1976, Derivation of specific antibody-producing tissue culture and tumor lines by cell fusion, *Eur. J. Immunol.* **6**:511–519.

Korsching, S., and Thoenen, H., 1983, Levels of nerve growth factor in sympathetic ganglia and corresponding target organs of the rat: Correlation with the density of sympathetic innervation, *Proc. Natl. Acad. Sci. U.S.A.* **80:**3513–3516.

Levi-Montalcini, R., 1966, The nerve growth factor: Its mode of action on sensory and sympathetic nerve cells, *Harvey Lect.* **60:**217–259.

Lomedico, P.T., 1982, Use of recombinant DNA technology to program eukaryotic cells to synthesize rat proinsulin: A rapid expression assay for cloned genes, *Proc. Natl. Acad. Sci. U.S.A.* **79:**5798–5802.

Luben, R.A., Brazeau, P., Böhlen, P., and Guillemin, R., 1982, Monoclonal antibodies to hypothalamic growth hormone-releasing factor with picomoles of antigen, *Science* **218:**887–889.

Noyes, B.E., Mevarech, N., Stein, R., and Agarwal, K.L., 1979, Detection and partial sequence analysis of gastrin mRNA by using an oligodeoxynucleotide probe, *Proc. Natl. Acad. Sci. U.S.A.* **76:**1770–1774.

Schmitt, F.O., Bird, S.J., and Bloom, F.E. (eds.), 1982, *Molecular Genetic Neuroscience*, Neurosciences Research Program, Raven Press, New York.

Stetler, D., Das, H., Nunberg, J.H., Saiki, R., Sheng-Dong, R., Mullis, K.B., Weissman, S.H., and Erlich, H.A., 1982, Isolation of a cDNA clone for the human HLA-DR antigen chain by using a synthetic oligonucleotide as hybridization probe, *Proc. Natl. Acad. Sci. U.S.A.* **79:**5966–5970.

Thoenen, H., and Barde, Y.-A., 1980, Physiology of nerve growth factor, *Physiol. Rev.* **60:**1284–1335.

Wakade, A.R., Edgar, D., and Thoenen, H., 1983, Both nerve growth factor and high K^+ concentrations support the survival of chick embryo sympathetic neurons: Evidence for a common mechanism of action, *Exp. Cell Res.* **144:**377–384.

Walter, G., and Doolittle, R.F., 1983, Antibodies against synthetic peptides, *Genetic Engineering* (R. Williamson, ed.), **5,** Academic Press, New York, (in press).

Chapter 17

Trophic and Neurite-Promoting Factors for Cholinergic Neurons

Silvio Varon and Marston Manthorpe

1. INTRODUCTION

1.1. Neuronotrophic Factors

The term *neuronotrophic factor* (NTF) designates a biological macromolecule that is required for the maintenance and general growth capabilities of selected nerve cells (Varon, 1977; Varon and Adler, 1980). The prototype for such factors has been nerve growth factor (NGF), a protein discovered some 30 years ago that acts both *in vivo* and *in vitro* on sensory and sympathetic peripheral neurons (e.g., Levi-Montalcini and Angeletti, 1968; Varon, 1975a; Greene and Shooter, 1980; Thoenen and Barde, 1980; Varon and Skaper, 1983a,b). Virtually all the NGF studies have relied on the use of tissue-culture techniques for the quantitative measure and monitoring of its biological activities.

The existence of NGF has long suggested the possible occurrence of a whole family of trophic factors for various neurons. Two major sources have been postulated for such factors (Varon and Bunge, 1978; Varon and Somjen, 1979; Varon and Adler, 1980, 1981; Varon and Manthorpe, 1982): (1) the postsynaptic innervation territory of a neuron (peripheral tissues for peripheral and certain motor neurons, other nerve cells for intrinsic central neurons) and (2) the glial cells associated with the neuronal soma or its processes (Schwann cells for

SILVIO VARON and MARSTON MANTHORPE ● Department of Biology, School of Medicine, University of California, San Diego, La Jolla, California 92093.

peripheral neurons, astrocytes and oligodendrocytes for central ones). With the use of neuronal survival in monolayer cultures to monitor trophic sources, several new NTFs have indeed been recognized in the extracts of such tissues and in media preexposed to corresponding cell cultures ["conditioned" media (CMs)] (reviewed in Varon and Adler, 1981; Varon and Manthorpe, 1982). Practically all the trophic activities investigated are directed to ganglionic cells of the peripheral nervous system (PNS) or to spinal-cord motor neurons that, despite the location of their bodies in the central nervous system (CNS), extend their axons outside the CNS. All these new NTFs also appear to be proteins, or protein-associated molecules. Little progress, in contrast, has been reported thus far with trophic factors directed to intrinsic CNS neurons, namely, neurons that have both axons and somata confined to the CNS.

Few neuroscientists now question the existence of NTFs and their potential importance in the early development of the nervous system. Similar factors may continue to play crucial roles in the maintenance and hence possibly the *repair* of adult nerve cells as well (e.g., Varon, 1975b, 1977; Varon and Lundborg, 1983; Varon and Manthorpe, 1983; Varon et al., 1983d). Recent studies have demonstrated the presence and temporal accumulation *in vivo* of NTFs within the fluid that collects in wounds of adult rat PNS (Lundborg et al., 1982b; Longo et al., 1983a,b) and, more recently, CNS (Manthorpe et al., 1983b; Nieto-Sampedro et al., 1983). It is also reasonable to speculate that various neuronal diseases may reflect abnormalities in the supply or competence of NTFs or in the ability of the neuron to respond properly to them (Varon, 1975b; Varon et al., 1982). Both concepts are increasingly being considered by investigators concerned with neural-regeneration research (e.g., Aguayo et al., 1982), neuronal diseases (e.g., Appel, 1981) (see also Chapter 20), and toxin-induced neuropathies (e.g., Peterson and Crain, 1982). Moreover, "neuronotoxic" factors, defined by their ability to eliminate neuronal survival *in vitro* even in the presence of an adequate trophic supply, have now been observed in several source materials for trophic agents (Varon et al., 1981; Longo et al., 1983a; Manthorpe et al., 1982c, 1983b; Nieto-Sampedro et al., 1983), but their molecular nature, biological effects, and antithetical relation to trophic factors have yet to be investigated systematically.

1.2. Neurite-Promoting Factors

Explant cultures of neural tissue have often been used, particularly in earlier research, as test systems for exogenously supplied agents (e.g., Levi-Montalcini and Angeletti, 1968; Varon et al., 1972; Ebendal et al., 1978, 1979; Coughlin et al., 1978; McLennen and Hendry, 1980). The most immediately observable response of a neural explant is the outgrowth of neurites, which does document the presence of viable neurons within the explant but does not directly distinguish between exogenous agents that address neuronal survival (trophic factors) and those that might stimulate neurite outgrowth from independently maintained neurons [neurite-promoting factors (NPFs)]. NGF, although initially discovered for its effects on neuritic growth (hence its very name), is clearly an agent that can stimulate *both* neuronal survival *and* neuritic growth; i.e., it acts both as an NTF and as an NPF. The latter action requires the application of NGF directly to the distal end of a growing neurite (Campenot, 1977, 1982a,b) and may be seen with neurons that no longer require NGF for their survival (Skaper and Varon, 1982).

Other investigations, particularly those using ciliary-ganglion cells (see Section 2.4), have now revealed that trophic and neurite-promoting effects may be exerted by *separate* factors (Varon *et al.,* 1979; Adler and Varon, 1980). Such investigations have led to the recognition of a new class of agents, designated as *polyornithine-binding neurite-promoting factors* (PNPFs) (Collins, 1978; Adler and Varon, 1980, 1981a,b; Manthorpe *et al.,* 1981b; Adler *et al.,* 1981, 1983; Lander *et al.,* 1982). We (Varon *et al.,* 1983c; Manthorpe *et al.,* 1983c) have recently demonstrated that a similar activity resides with purified *laminin,* a glycoprotein component of basal lamina that in nerve tissue is associated with Schwann cells. Both PNPFs and laminin are effective on CNS as well as PNS neurons (Barbin *et al.,* 1983; Manthorpe *et al.,* 1983c). Neurite-promoting agents that operate directly from medium to neurons also exist (Henderson *et al.,* 1981; Kligman, 1982a,b), in addition to those that operate after binding to cell surfaces or extracellular matrices (as PNPFs may do) or as original constituents of the latter structures themselves (as laminin is). Also recognized has been the presence of *neurite inhibitors* in the form of certain lectins (Adler *et al.,* 1983) or in some CMs (Manthorpe *et al.,* 1981b) and in several sera (Skaper *et al.,* 1983a,b).

The investigation of neurite-promoting and neurite-inhibitory agents has yet to be pursued with the vigor with which NTFs are currently being investigated. It is known that PNPFs, as well as laminin, are unable by themselves to secure neuronal survival *in vitro* and thus are not "neuronotrophic" agents (Adler and Varon, 1980; Manthorpe *et al.,* 1981b, 1983c). On the other hand, there have been indications that their presence on the culture substrate may contribute to optimal survival under the influence of a suitable NTF (Edgar and Thoenen, 1982; Longo *et al.,* 1982). Their principal effect, however, is clearly on *neuritic growth*. It is not yet known to what extent PNPFs operate by opposing neurite-inhibiting properties of the untreated substrate or by actually conferring to it neurite-stimulatory competence. Nor is it fully clarified whether these substrate-bound agents address neurite initiation only (Collins, 1978) or also neuritic elongation or stabilization or both.

Whatever the neuritic aspects affected by PNPFs, one can readily speculate that they may be involved in *directional guidance* as well as growth. Two general modes of neuritic guidance can be considered (cf. Varon and Adler, 1980). One mode is by *attraction,* imposed by signals that are released from the destination site and form a gradient to be recognized by the advancing growth cone. Guidance by attraction may well be mediated via NTFs released from the innervation territory (e.g., Letourneau, 1978; Gundersen and Barrett, 1979). The other mode, illustrated in Fig. 1, is by *restriction,* imposed by surface-bound agents that define preferential paths for neuritic growth. Guidance by restriction could be achieved either by generating a neurite-conducive roadway (e.g., with PNPFs or laminin) across neurite-opposing surfaces (e.g., Adler and Varon, 1981a,b; Collins and Garrett, 1980) or by blocking neurite-conducive surfaces (e.g., with lectins or serum-derived inhibitors) anywhere but along a selected pathway (e.g., Adler *et al.,* 1983; Skaper *et al.,* 1983a,b).

The physiological involvement of neurite-promoting and neurite-inhibiting agents *in vivo,* while still largely a matter for speculation (Varon and Adler, 1980; Adler and Varon, 1981a,b), appears highly probable. Several investigators have stressed the importance of cell surfaces or their glycoprotein constituents, or both, in neuritic guidance during early development *in vivo* (see Chapters 1, 2, 4, and 5). One may readily suspect that abnormalities in neurite-modulating agents may be the basis for defective morphogenesis (e.g., Chapter 2) or abnormal differentiation (e.g., Chapter 1). Also increasingly attractive is the spec-

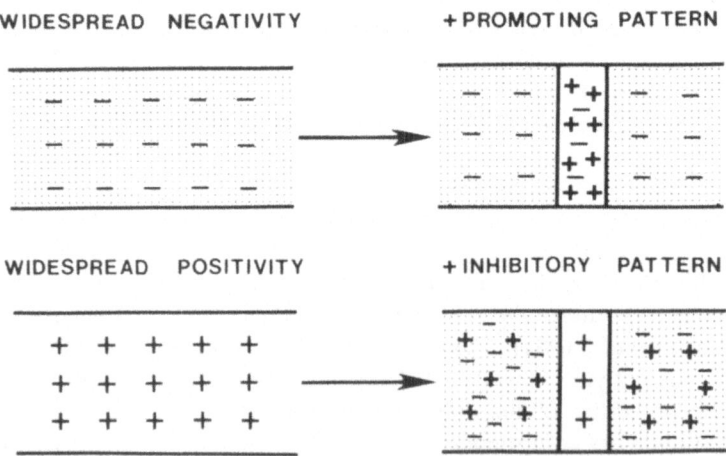

Figure 1. Guidance of neurites by the mode of restriction (see the text). Surface-bound agents could define preferential paths either by generating a neurite-conducive roadway across neurite-inhibiting surfaces *(top)* or by inhibiting a neurite-conducive surface everywhere except along a particular pathway *(bottom)*. Neurite promotion could be provided by agents such as PNPF or laminin, while the inhibition could be provided by certain lectins or serum-derived inhibitors.

ulation that cell-surface and extracellular-matrix properties may control, at least in part, the success or failure of axonal regeneration in both PNS and CNS (Varon and Lundborg, 1982; Varon and Manthorpe, 1983; Varon *et al.*, 1983d; Aguayo *et al.*, 1982; Manthorpe *et al.*, 1983c).

1.3. Test Systems and Methodology

Monolayer cultures of nerve cells, dissociated from the desired neural tissue, provide excellent tools for the detection, quantitative monitoring, and detailed investigation of both NTFs and NPFs, as well as the corresponding inhibitory agents (e.g., Varon and Raiborn, 1969, 1972; Burnham *et al.*, 1972; Barde *et al.*, 1978; Manthorpe *et al.*, 1981c; Varon *et al.*, 1974, 1981, 1983a). The source tissues are dissected and dissociated by brief exposure to trypsin followed by mechanical dispersion. The resulting cells are counted, diluted to the desired cell density, and seeded on substrates (e.g., polyornithine) that permit rapid and firm attachment of the still-dispersed cells. Procedures using 6-mm-diameter microwells have been developed to process a large number of samples economically (Manthorpe *et al.*, 1981c).

For *trophic* studies, the polyornithine-coated wells are also pretreated with PNPF, which causes early and extensive neuritic outgrowth and thus greatly simplifies the identification and direct counting of surviving neurons. The trophic material under investigation is serially diluted into the treated microwells, the test neurons are added, the cultures are fixed after 24–48 hr incubation, and the surviving neurons are counted over representative culture areas. These bioassays define as *1 trophic unit* (TU) the activity present in 1 ml culture medium that supports half-maximal survival of the test neurons; thus, the fold dilution required to achieve such an effect represents the trophic titer (TU/ml) of the original source material.

For *neurite-promoting* studies, the polyornithine-coated wells are treated with defined volumes of serially diluted samples of the material under investigation. The wells are then washed and supplied with test neurons and culture medium containing their NTF (to ensure maximal survival). After 24 hr incubation, the cultures are fixed and the proportion of neurons bearing neurites is determined by direct cell counts. Currently, *1 neurite-promoting unit* (NPU) is defined as the activity in 1 ml medium that, on presentation to the polyornithine substrate, confers to it half-maximal neurite-promoting competence. Similar approaches can be used for neurite-promoting agents that will not bind to the substrate, but rather operate directly from the culture medium. In such cases, untreated polyornithine serves as the culture substrate, the material under study is placed in the wells at serial dilutions, and the test neurons are added together with their trophic factor.

Inhibitory activities for either neuronal survival or neuritic growth and maintenance are tested in similar culture systems, in which both trophic and neurite-promoting agents already identified for a given test neuron are supplied at a fixed and optimal level (Manthorpe *et al.*, 1982c). The putative inhibitory material is presented at serial dilutions, and neuronal performance is evaluated 24 hr later. Neuronal survival or neuritic growth will not occur as long as the respective inhibitor is supplied at maximal concentrations, will be progressively unmasked as the inhibitor is further diluted, and will again be maximally expressed when the inhibitor has been adequately diluted out. Half-maximal performance designates the presence of *1 inhibitory unit* (IU) in 1 ml culture medium, and the dilution applied to achieve this effect defines the inhibitory titer of the original, undiluted material being tested.

Several features of such bioassays need careful consideration. *Neural dissociates* may comprise both nonneuronal cells (glia, fibroblasts) and neuronal cells, and among the latter, several subsets of neurons. Nonneurons (a potential source of both trophic and neurite-promoting agents) can be reduced to a minimum by appropriate dissociation or preplating steps (e.g., Varon and Raiborn, 1969, 1972; Varon *et al.*, 1973; Manthorpe *et al.*, 1980). Even so, some neurons may serve as potential innervation targets for other neurons in the same culture (hence, as potential sources of trophic and other factors), and it is critical to minimize the number of cells seeded per well to dilute out potential interactions (Manthorpe *et al.*, 1982c; Barbin *et al.*, 1983). Last, heterogeneity of the neuronal population used as test cells may not permit direct recognition of different neuronal subsets within it and may thus prevent the evaluation of exogenous agents that act selectively on one such subset. Given optimal choice and preparation of the test cells, two other aspects require constant attention (cf. Varon *et al.*, 1983a). One is the choice of the *substrate* and its precoating treatments, as well as the recognition that its properties will continuously be modified by exposure to the medium and to products of the cultured cells themselves. The other aspect is the choice of the *basal medium* as well as of the supplements added to it (besides the very material under investigation), in particular the negative as well as positive contributions that sera may make to both the medium and the substrate of the culture.

Another cogent concern is whether short-term survival (1–2 days) *in vitro* yields meaningful information with regard to longer-term neuronal maintenance *in vitro,* and eventually *in vivo.* Clearly, if survival fails in the short term, it will also fail in the longer term, and extrinsic conditions that ensure neuronal survival in the short term are *necessary* to permit it for longer times. Such conditions, however, may prove *insufficient* for the longer term for a variety of reasons, for example: (1) the trophic or neurite-promoting agents or critical basal-medium components supplied to the culture are not stable enough and may require a more sustained administration; (2) noxious materials are introduced or generated in the culture, accumulate in the medium or on the substrate, and must be constantly removed; or, most important, (3) survival requirements of the neurons may vary with developmental age *in vitro* as well as *in vivo,* as already demonstrated in several ganglionic systems (e.g., Barde *et al.,* 1978; Selak *et al.,* 1983a).

Last, one cannot overstress the limitations imposed by the insufficient knowledge of the molecular nature of the agents under study. It remains to be determined in most cases whether (1) trophic activities present in the same source material for different test neurons reside on the same or on distinct molecules (e.g., Manthorpe *et al.,* 1982a), (2) the same activity for the same test cells is carried by the same or different molecular species in different source materials, or (3) all the neurons within a given test population respond in the same qualitative and quantitative manner to the same trophic, neurite-promoting, or inhibitory agent.

1.4. Factors That Address Cholinergic Neurons

The conceptual and methodological background established in the preceding sections will be used in the remainder of this chapter to describe current knowledge of agents that modulate survival or neuritic production by *cholinergic* neurons. Identification and individual visualization of cholinergic neurons—a necessary prerequisite for such investigations—is not readily made in cell cultures derived from most PNS or CNS tissues known to comprise such nerve cells. Hence, most information has thus far been obtained from cultures in which the *majority* of neurons are cholinergic (and thus do not require individual appraisal). We shall review, here, studies concerned with (1) 8-day chick-embryo *ciliary-ganglion* cells, (2) 4-day chick-embryo *lumbar-cord* cells, and—to a much lesser extent—(3) 18-day fetal-rat *septal and striatal* cells.

2. CHOLINERGIC NEURONS IN THE PERIPHERAL NERVOUS SYSTEM

2.1. "Target"-Derived Ciliary Neuronotrophic Factors

Ciliary ganglia (CGs) contain two sets of neurons, both sets being cholinergic as well as cholinoceptive. One set, the choroid neurons, innervates the smooth-muscle cells in the eye choroid coat; the other set, the ciliary neurons, innervates the striated-muscle cells in the eye ciliary body and iris. In the chick embryo (Landmesser and Pilar, 1978), both neuronal sets are postmitotic cells that receive preganglionic inputs by embryonic day 5 (E5), extend their axons to their intraocular targets between E6 and E11, and lose about half their cell population over that age period. This last event is a typical example of the developmental neuronal death phenomenon that has been observed in both PNS and CNS tissues and has been attributed to the encounter and interaction between the axonal terminals and the innervation target cells (cf. Cowan, 1973; Varon and Adler, 1980). The main hypothesis has been that (1) before reaching their innervation target, developing neurons do not depend for their survival and growth on their target cells; and (2) innervation-target territories produce and release for their presynaptic neurons NTFs that will rescue from death those neurons that establish functional connections.

Neurons dissociated from CGs of the 8-day chick embryo (cCG8) will die if cultured without appropriate trophic supplementation (Helfand et al., 1976). We thought that the aforestated hypothesis might be validated using cCG8 neuronal cultures if one could demonstrate the existence of ciliary NTFs (CNTFs) preferentially located in the intrinsic eye-muscle tissues at the correct developmental stages, as well as an age-related dependence on such CNTFs by the CG neurons. The 8 day ganglia were dissociated and cultured in the presence of various chick-embryo extracts, under conditions in which no neurons would survive in the absence of appropriate medium supplementation (Varon et al., 1979). As illustrated in Fig. 2, CNTF activity was present in whole-embryo extract from E8 chick (CE$_8$), but greatly enriched in the eye extract (EYE$_8$) and even more so in extracts prepared from the very intraocular target tissues (CIPE$_8$) that contain the muscle cells innervated by the CG (Adler et al., 1979). Moreover, specific CNTF activity in the target tissue increased from E8 on (Landa et al., 1980), reaching a plateau by E15. The age dependence of CG neurons on target-derived CNTF was also verified both in dissociated cell cultures (Manthorpe et al., 1981a) and in ganglia maintained in suspension to preserve cell composition and intracellular arrangements (Adler and Varon, 1982). In both situations, E5 neurons survived even without CNTF addition to the culture medium, while E8 neurons died unless the CNTF was made available to them.

The intraocular target tissue of E15 chicks (CIPE$_{15}$) was chosen as the source material for purification of the eye CNTF molecule (Fig. 2). A 150-fold (from 100 to 15,000 TU/mg) increase in specific activity was obtained by this choice of E15 chick eye and the subdissection steps, and another 4-fold increase was achieved by fractionating the extract on a DEAE ion exchanger and an XM100 ultrafiltration membrane [CNTF-1 preparation (Manthorpe et al., 1980)]. Further fractionation of this material on sucrose-density gradients yields a 20,000-dalton product (CNTF-2) that is active at 3×10^5 TU/mg protein and can be prepared in batches of 10^6 TU (3 mg protein) at a time (Barbin et al., 1981; Manthorpe et al., 1983d). Isoelectric-focusing-gel electrophoresis, at the analytical level, still reveals a number of silver-stainable bands, with the CNTF restricted to the pH 5.0 region (Manthorpe et al., 1982a,b). The CNTF elutable from this region [Fig. 2 (CTNF-3)] has been calculated to be active at greater than 10^6 TU/mg protein, a range comparable to or better than the most highly purified NGF.

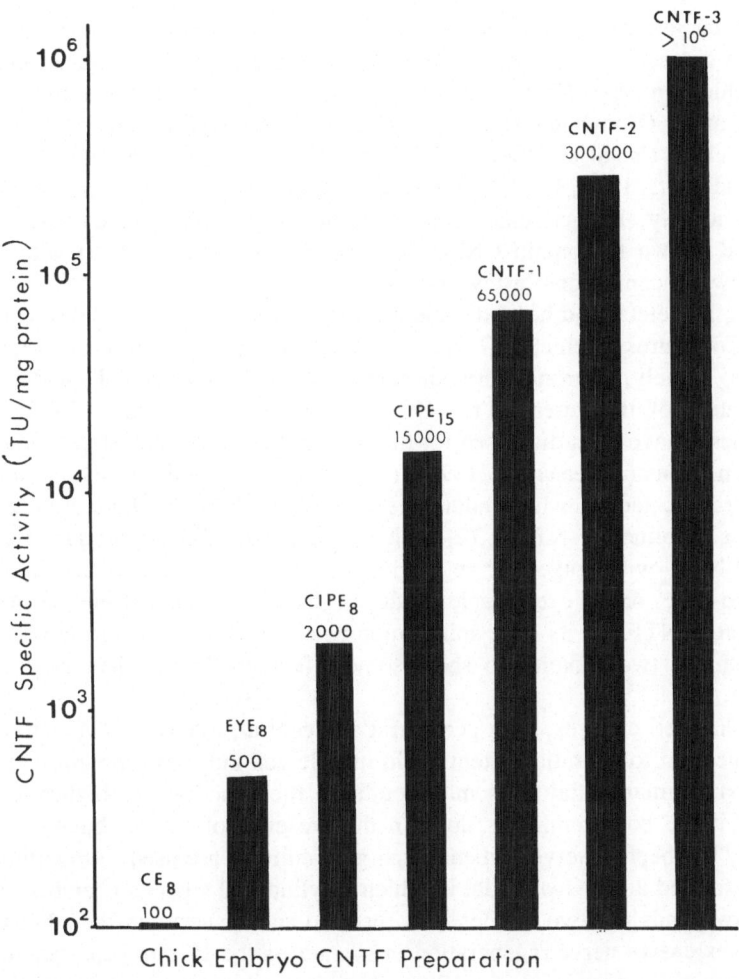

Figure 2. CNTF specific activity in different extract preparations. Aqueous high-speed supernatants were generated from 8-day whole chick embryos (CE_8), whole eyes (EYE_8), or the intraocular innervation territory of the CG, which includes the choroid and iris–ciliary body as well as the pigment epithelium ($CIPE_8$). Extracts were also prepared from 15-day embryo CIPE ($CIPE_{15}$) and fractions collected after submitting the extract to ion-exchange chromatography (CNTF-1) followed by sucrose-density centrifugation (CNTF-2) and isoelectric focusing (CNTF-3).

2.2. Other Sources of Ciliary Neuronotrophic Factors

CNTF activity has been observed in a variety of other source materials, though never, thus far, in the abundance and concentration available with the intraocular target tissues. One such source is *muscle* tissue. Extracts of chick-embryo carcass, rich in skeletal muscle (Adler *et al.*, 1979), and heart tissue from chick embryo (Ebendal *et al.*, 1979), adult rat (McLennan and Hendry, 1980), or ox (Bonyhady *et al.*, 1980) also contain CNTF activity. CMs from both skeletal (Nishi and Berg, 1977; Bennett and Nurcombe, 1979) and heart (Helfand *et al.*, 1976, 1978; Varon *et al.*, 1979) muscle-cell cultures also display CNTF activity. In fact, chick-heart-conditioned medium (HCM) was the first material shown to contain CNTF activity (Helfand *et al.*, 1976), although at very low concentrations (a few units/ml).

Both skeletal and heart muscle are innervation targets for cholinergic neurons. Cocultures of chick CG and chick skeletal muscle display two-way interactions, namely, neuronal survival supported by the muscle cells and synaptic innervation of the latter by the CG neurons (Nishi and Berg, 1977). Three hypotheses have been discussed for the neuronotrophic relationship from muscle to CG neurons (Adler *et al.*, 1979): (1) both eye and body muscle produce the same trophic factor, which addresses indiscriminately all *cholinergic* neurons (truly a "cholinergic NTF"); (2) both eye and body muscle produce the same *ciliary* NTF, with epigenetic regulation favoring production at the right target site and time; or (3) eye muscle produces CNTF and skeletal muscle produces a different NTF for its own spinal motor neurons, with enough cross-activity between the two proteins to show some effect on the "unphysiological" test neuron.

Glial-cell cultures, both peripheral and central, release CNTF activity into their medium to a similar extent as do muscle-cell cultures (unpublished observations). Primary glial cells may condition their medium to higher levels of CNTF if the conditioning is done in the presence of serum, but little CNTF activity has been observed when clonal glial cultures are used. Furthermore, we had described an *in vivo* model in which a cylindrical silicone chamber receives at its two ends the two stumps of a resected sciatic nerve in the adult rat and allows extensive nerve regeneration *in situ* across the 10-mm gap separating the two stumps (Lundborg *et al.*, 1982a; Varon and Lundborg, 1982; Williams *et al.*, 1983a). High levels (several hundred units/ml) of CNTF activity are found in the *wound fluid* that collects inside the chamber (Longo *et al.*, 1983a), and high CNTF activity can be extracted from the sciatic *nerve tissue* itself (Williams *et al.*, 1983b). Last, wound fluid can also be collected from lesions established in the *rat brain* and shown to display considerable CNTF activity (Manthorpe *et al.*, 1983b; Nieto-Sampedro *et al.*, 1982, 1983), as do extracts of the lesioned brain tissue itself (Nieto-Sampedro *et al.*, 1983).

Nothing is known at present about similarities and dissimilarities between muscle-derived and neural-tissue-derived CNTF molecules. The same considerations apply to neural CNTFs as to those raised for CNTFs from different muscle sources. One can therefore speculate as to whether neuronotrophic activity measured with CG neurons may be relevant to CNS cholinergic neurons as well.

2.3. Other Agents That Influence Ciliary-Ganglionic-Neuronal Survival

We have mentioned in Section 1.3 that other medium components, besides the NTF, can considerably affect neuronal survival *in vitro*. Serum has been for many years a necessary supplement to nearly all cell cultures, and it is only recently that Sato and his collaborators have shown that proliferative growth of many cells *in vitro* can be equally well supported in serum-free media receiving chemically defined supplements (Sato *et al.*, 1982). One such supplement (designated N1 or N2)—which comprises insulin, transferrin, putrescine, progesterone, and selenite—effectively promoted proliferation of neuroblastoma clonal cells (Bottenstein and Sato, 1979) and was later shown to support equally well the survival of primary neurons (Skaper *et al.*, 1979; Bottenstein *et al.*, 1980), always provided that the appropriate NTF was also supplied. The extent to which the N1 suplement replaces serum for neuronal survival and what individual N1 constituents are required for it can vary with the type of neural tissue (Selak *et al.*, 1983a; Skaper *et al.*, 1982) and the developmental age of the cultured neurons (Selak *et al.*, 1983a). The CG cultures have recently added another basis for differential N1 competence (Hewitt *et al.*, 1983; Skaper *et al.*, 1983c). In these studies, N1 supplementation was fully adequate with certain basal media, only partially adequate with others, and totally inadequate with yet others. A detailed definition of which basal-medium components are responsbile for such differences will greatly help in understanding (1) the requirements of these CG neurons and (2) the biochemical mechanisms addressed by the relevant N1 constituents.

2.4. Neurite-Promoting and Neurite-Inhibiting Factors

CG neurons have been a useful instrument through which substrate-binding neurite-modulating agents have recently been investigated. Given optimal trophic support (selected basal medium + chemically defined supplement + CNTF), CG neurons can be cultured with or without neuritic growth depending on the choice of substrate and special medium additives (Varon *et al.*, 1979). Using polyornithine-coated tissue-culture plastic (PORN) as substrate and HCM as the additive, Collins (1978) showed that HCM contains a PORN-binding material that promotes neuritic initiation very soon after seeding the CG neurons. Adler and Varon (1980) demonstrated that HCM contains two separate agents: (1) a

CNTF that does not bind to the substratum (see Section 2.2) and (2) a PORN-binding PNPF that has no survival-supportive activity but confers to the PORN substrate neurite-promoting or neurite-permissive qualities. Most CMs examined, including those derived from glial-cell cultures, contain such PNPFs (Adler *et al.*, 1981). Using schwannoma-CM, we have partially purified a large, acidic, carbohydrate-containing protein factor that can be presented to the PORN at 50–100 ng protein/ml medium to confer half-maximal neurite-promoting activity to the treated substrate (Manthorpe *et al.*, 1981b). An agent with similar PNPF properties for sympathetic neurons and proteoglycan like characteristics has been obtained from medium conditioned with bovine corneal endothelial-cell cultures (Lander *et al.*, 1982).

PNPF-treated PORN enchances neuritic outgrowth from all other ganglionic neurons tested (Adler *et al.*, 1981) and from CNS neurons as well (Longo *et al.*, 1982; Barbin *et al.*, 1983; Manthorpe *et al.*, 1983c). Both effects are illustrated in Fig. 3. The neurite-enhancing property of PNPF-treated PORN can be suppressed by treatment with concanavalin A (a lectin that binds to the PNPF

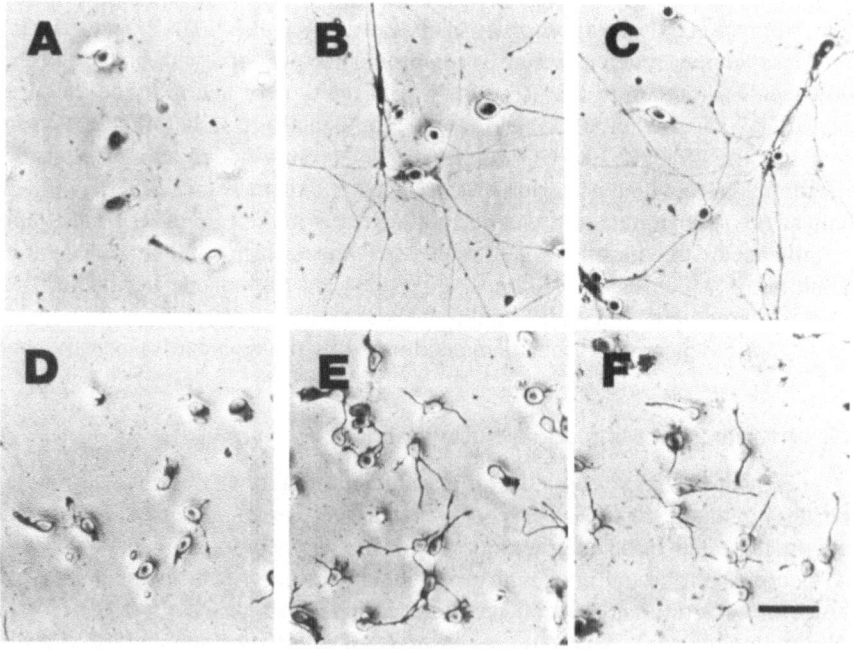

Figure 3. Neurite-promoting influence of substratum-bound agents. PORN dishes were left unexposed (A, D) or exposed to rat laminin (B, E) or rat PNPF (C, F). After the treated dishes were washed, cCG8 neurons (A–C) or 18-day fetal-rat striatal neurons (D–F) were added for 24 hr. Note the similar effects on neurite regeneration of the two treated substrata. Scale bar: 50 μm.

molecule) and restored by subsequent treatment with α-methylmannoside, a sugar specific for this lectin (Adler *et al.*, 1983). Also shown in Fig. 3 are the effects on both CG and CNS neurons on a PORN substrate that was pretreated with *laminin* (Varon *et al.*, 1983c; Manthorpe *et al.*, 1983c). Laminin is a glycoprotein constituent of basal laminae and is associated with Schwann cells in nerve tissue and in monolayer culture (Roufa *et al.*, 1982; Cornbrooks *et al.*, 1982). Purified rat or mouse laminin confers half-maximal support to PORN when presented at less than 25 ng/ml, making the neurite-promoting assay the most sensitive test for laminin detection and measurment yet described. The neurite-promoting effects of rat laminin are blocked by exposure of the laminin-treated PORN to antibody against rat laminin. The same antibody fails to block the activity of PNPF-treated PORN (despite the rat-cell origin of the PNPF), demonstrating molecular differences between these two agents and proposing the existence of a whole family of substrate-binding NPFs (Manthorpe *et al.*, 1983c).

When CGs are placed in explant cultures on PORN, the neurites growing out of the explants proceed radially for only a short distance and then assume a circular pattern, suggesting the release and short-range capture by the PORN of PNPFs generated directly by the ganglionic tissue (Adler and Varon, 1981a). Explant cultures of CGs on collagen, on the other hand, fail to display a neuritic outgrowth until nonneuronal cells have migrated out of the explant and then remain essentially confined to the nonneuronal-cell surface (Adler and Varon, 1981b). These and other observations have prompted the speculation (cf. Adler *et al.*, 1981; Varon and Adler, 1981) that PNPFs may be *cell-surface constituents* responsible *in vivo* as well as *in vitro* for axonal elongation and its directional guidance (cf. Section 1.2). *In vitro*, such surface constituents are released (in intact or fragmented forms) into the culture medium and need to reanchor to a substrate to exert their neurite-promoting activity. *In vivo*, they may define preferential roadways either by residing on selected cells [e.g., radial glia (cf. Chapter 2)] or by being released and packaged into selected extracellular structures (e.g., basal lamina), as discussed in a previous section. The recent observation that both PNPF and laminin are also active on CNS neurites raises the intriguing speculation that the presence or absence of such agents may play a role in axonal regeneration *in vivo* (cf. Varon *et al.*, 1983d; Manthorpe *et al.*, 1983c).

Monolayer cultures of CG neurons are now being used to investigate other aspects of neurite regulation by culture substrates (Hewitt *et al.*, 1983; Skaper *et al.*, 1983d). CG neurons display little if any neuritic growth when cultured in the presence of serum on collagen, as well as on PORN, substrates. If the medium used is serum-free and appropriately supplemented (see Section 2.3), considerable neuritic growth is displayed on both substrates. Serum-containing media, on the other hand, do not substantially reduce neuritic growth on PNPF- or laminin-pretreated PORN. It remains to be determined whether (1) serum interferes with neuritic growth by altering the substrate and (2) the serum components that interfere with neuritic expression of GC neurons are the same as those found to interfere with neuritic growth in other systems (Skaper *et al.*, 1983a,b).

3. CHOLINERGIC NEURONS IN THE SPINAL CORD

3.1. 4-Day Chick-Embryo Lumbar-Cord Culture System

In the spinal cord, the cholinergic motor neurons are accompanied by several other, noncholinergic neuronal subpopulations. Monolayer cultures of dissociated spinal cord cells, therefore, will ordinarily comprise a number of neuronal subsets, ony two of which (somatomotor and visceromotor neurons) will be cholinergic. The occurrence of cholinergic neurons in such cultures can be verified by their content of choline acetyltransferase (ChAT), the acetylcholine-synthesizing enzyme, but evaluation of the *number* of cholinergic neurons present requires their individual visualization—reagents for which (e.g., anti-ChAT antibodies) are not yet readily available (cf. Varon *et al.*, 1982). One way to minimize this problem is to start from a neural tissue the neuronal population of which is largely cholinergic and view its derivative culture as a first approximation of a cholinergic one. We have selected for this purpose (Longo *et al.*, 1982) the 4-day chick-embryo lumbar-cord (cLC4) segment, day 4 being the time at which most of the postmitotic cord cells are motor neurons. The monolayer cultures obtainable from cLC4 dissociates contain all the ChAT activity recovered in the dissociate itself and also display ChAT increases with time *in vitro*, demonstrating retention and maturation of the cultured cholinergic neurons (Manthorpe *et al.*, 1983a).

We had previously shown that E8 chick spinal neurons would survive in serum-free, N1-supplemented medium if seeded on collagen at high cell densities (Skaper *et al.*, 1979). The cLC4 neurons also did well under similar culture conditions. Furthermore, they displayed much better neuritic growth if seeded on polyornithine substrates that had been pretreated with PNPF (Longo *et al.*, 1982) or laminin (Manthorpe *et al.*, 1983c). The PNPF–PORN substrate and the serum-free, N1-supplemented medium were adopted for all subsequent studies (Longo *et al.*, 1982; Manthorpe *et al.*, 1982c, 1983a). No attempts have yet been made with the cLC4 neurons to define (1) optimal basal madia, (2) specific requirements for the N1 constituents, or (3) neurite sensitivity to serum-derived inhibitors.

When seeded at relatively high cell densities (e.g., 30,000 cells/cm^2), cLC4 neurons survived in large numbers even without the addition to the medium of putative trophic agents (Manthorpe *et al.*, 1982c). Low-density-seeded cultures

(3000 or fewer cells/cm^2), on the other hand, displayed very little neuronal survival unless their media were supplemented with (1) medium conditioned over high-density cLC4 cultures or (2) courses of putative trophic agents (see Section 3.2). Thus, the spinal neuronal cultures themselves produce and release trophic agents that they can utilize for their own neuronal survival. Figure 4 illustrates high-density cLC4 cultures at 1 and 5 days *in vitro*. Supplementation of these cultures with schwannoma-CM did not increase neuronal survival at 1 day and had, in fact, very toxic effects on the longer term (Manthorpe *et al.*, 1983a).

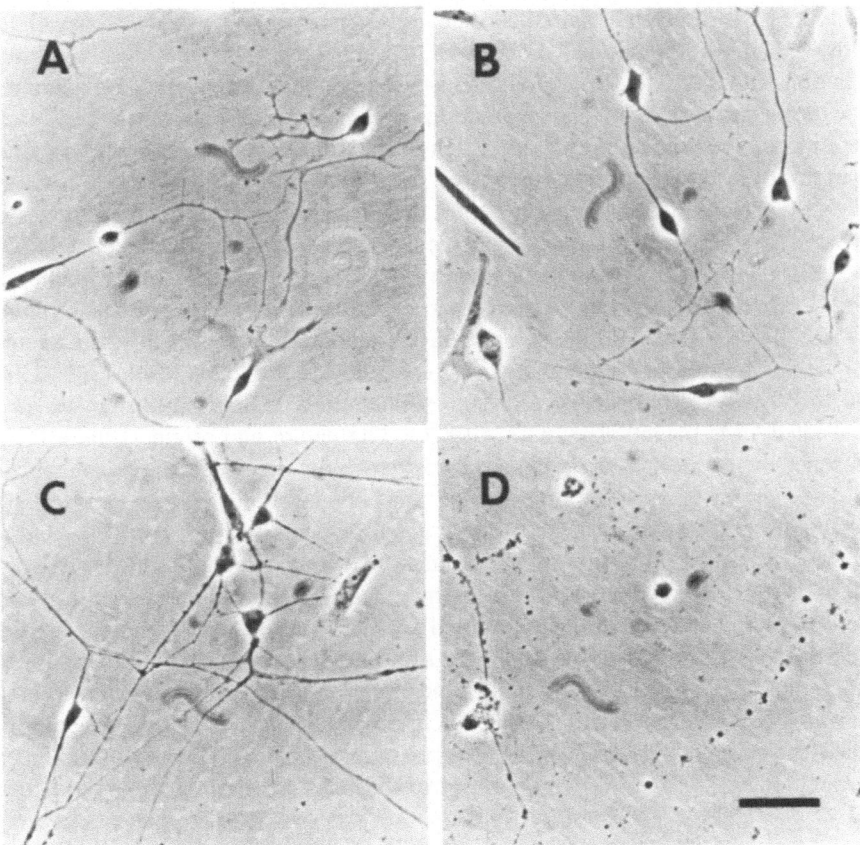

Figure 4. High-density cLC4 cultures on PORN. Supplementation of these cultures with rat RN22 schwannoma-CM (B, D) did not increase neuronal survival after 1 day *in vitro* (A, B), but had toxic influences after 5 days *in vitro* (C, D). Scale bar: 50 μm.

3.2. Trophic and Toxic Agents for Lumbar-Cord Neurons

Several sources of trophic agents for these CNS cholinergic neurons have been reported (cf. Varon *et al.*, 1982). Little is known as yet about tissue *extracts,* potentially the only type of source materials from which sizable amounts of trophic factors for these neurons could be expected. Most muscle or brain extracts examined also showed high toxicity for these test cells (Manthorpe *et al.*, 1983b; Nieto-Sampedro *et al.*, 1983).

Muscle- and glia-*CMs* contain variable levels of trophic activity for cLC4 neurons, up to 40 TU/ml (Manthorpe *et al.*, 1982c). In all cases, this trophic activity (1) passed through dialysis membranes and PM10 ultrafiltration membranes and (2) resisted inactivation by heat or trypsin. Thus, the trophic agents of these CMs directed to spinal cord neurons appear to be low-molecular-weight (lower than 10,000 daltons), non-protein compounds. When tested at low or no dilution, all CMs also displayed noxious effects on the survival of cLC4 neurons; the term *toxic* seems to be appropriate for such negative effects, since they could be exerted even on self-supported, high-density cLC4 cultures (cf. Fig. 4). The nature of these toxic agents has not yet been explored.

Both trophic and toxic agents for cLC4 neurons can be demonstrated to exist *in vivo,* as well as *in vitro,* specifically in *wound fluids* collected from PNS and CNS lesions in the adult or young rat (Longo *et al.*, 1983a,b; Manthorpe *et al.*, 1982c, 1983b; Nieto-Sampedro *et al.*, 1982, 1983). *Toxic* agents of these wound fluids affected all the spinal neurons supported by the CM trophic agents (or self-supported in high-density cultures) and were present at levels (100–2000 IU/ml) that were two or more orders of magnitude higher than in the *in vitro* collected CMs (Manthorpe *et al.*, 1982c). The nature and origin of these toxic substances, as well as their potential relevance for neural damage or repair or both *in vivo,* remain to be investigated. PNS wound fluids also contain *trophic* agents for cLC4 neurons, as they do for ganglionic ones (Longo *et al.*, 1983a,b). Comparison of the titration curves from wound fluids and CMs, however, suggests that (1) the wound-fluid agents may address only a subset (one third or lower) of the cLC4 neurons supported by CMs and (2) trophic titers for the responsive neurons could be 10- to 100-fold higher in wound fluid than in CM (Manthorpe *et al.*, 1982c). Wound fluid from CNS lesions had similar trophic features (Manthorpe *et al.*, 1983b; Nieto-Sampedro *et al.*, 1983). In addition, the trophic agents for cLC4 neurons present in both fluids displayed properties typical of *protein* factors, unlike those of the CMs (Manthorpe *et al.*, 1982c).

3.3. *In Vivo* Model for Regeneration of Spinal Motor Neurons

Both motor and sensory components of peripheral nerve are capable of regenerating after nerve lesions in the adult mammal. Peripheral motor and sensory components are also involved in a number of axonopathies and neuronopathies of unknown as well as toxic origins (cf. Varon *et al.*, 1982; Spencer and Schaumburg, 1980). Both the pathological damage and the potential repair of spinal-cord motor neurons may involve neuronotrophic, neuronotoxic, or neurite-modulating agents that can be studied *in vitro* and, eventually, *in vivo*. An *in vivo* model that provides several experimental opportunities is currently under investigation in our laboratory (Lundborg *et al.*, 1982a,b; Varon and Lundborg, 1982; Williams *et al.*, 1983a).

4. CHOLINERGIC NEURONS IN BRAIN TISSUE

4.1. *In Vivo* Models for Maintenance and Repair of Intrinsic Cholinergic Neurons in the Central Nervous System

A CNS system that has been extensively investigated with regard to both development and potential regeneration is the septohippocampal system. In the adult rat, transection of the fimbria–fornix tract of septohippocampal fibers leads to the degeneration of most or all of the cholinergic septal neurons. Two situations have been reported, however, in which regeneration of some cholinergic efferents from the septum (hence, survival of some cholinergic septal neurons) can take place. Kromer *et al.* (1981) have implanted in the fimbria–fornix lesion itself a graft of *fetal rat hippocampus*, which becomes invaded by host septal cholinergic fibers. One may speculate that the graft may have served as a source of NTFs (consistent with its relationship as the innervation territory for septal neurons), as a neurite-promoting terrain (consistent with its ability to sustain and possibly guide septal cholinergic fibers during normal development), or as both. Rothman–Schonfeld *et al.* (1982) have implanted in the fimbria–fornix lesion *fetal rat iris* tissue as a potential acceptor of cholinergic afferents and have applied HCM to the host septum as a potential source of cholinergic NTF. They reported a considerable increase in cholinergic fibers and ChAT activity in the iris graft up to 16 days of treatment.

Another approach to septohippocampal investigations has been to attempt replacing the host septum with fetal rat brain grafts containing cholinergic neurons, as potential sources for cholinergic reinnervation of the deafferented host hippocampus. Björklund and co-workers (Björklund and Stenevi, 1977; Schmidt *et al.*, 1981) have been successful both with *septal* implants and with the introduction of dissociated septal cells. We are currently exploring with them the possibility of improving survival and performance of the implanted cholinergic neurons by the concurrent and sustained intracerebral administration of trophic and other agents. Cotman and co-workers (Lewis and Cottman, 1980), have used *striatal* as well as septal fetal tissue as implants into cavities formed in the entorhinal cortex of the rat host and have examined their success in supplying cholinergic fibers to the deafferented host hippocampus. They have observed, by acetylcholinesterase (AChE) staining, that survival of cholinergic neurons in the graft and growth of cholinergic fibers into the host hippocampus were very low if implantation was carried out at lesion time, but increased considerably if it was delayed for increasing time periods. In collaboration with Cotman's group, we reported that success of the cholinergic graft increases with time postlesion in precise correlation with the temporal accumulation of CNTF (and other trophic) activities in both the lesion fluid and the lesioned tissue (Manthorpe *et al.*, 1983b; Nieto-Sampedro *et al.*, 1983). These observations encourage the speculation that (1) trauma in neural tissue triggers the accumulation of NTFs and (2) these factors may be important in the subsequent evolution of the neural wound.

4.2. Septal-Cell and Striatal-Cell Cultures

Septal and striatal tissues represent choice sources of test cells for the investigation *in vitro* of trophic and neurite-promoting agents directed to intrinsic brain cholinergic neurons. No such studies have yet been reported, despite the description for other purposes of both septal-cell (Jirikowski *et al.*, 1981) and striatal-cell (Messer, 1981; Panula *et al.*, 1979) cultures.

We have now defined, along the lines used for spinal-cord neurons, cell-culture systems from 18-day fetal rat septum and striatum that are suitable tools for the study of trophic and neurite-modulating agents (Barbin *et al.*, 1983). The cell dissociates are seeded at low densities on PNPF-coated polyornithine substrates and in serum-free, N1-supplemented medium (cf. Fig. 3). Under such conditions, no neurons survive unless exogenous trophic agents are supplied. High-density cultures condition their own medium and thus support neuronal survival without addition of exogenous trophic agents. Up to 60% of the seeded cells survive in either high-density or properly supplemented low-density conditions, and do so for at least several days. Better than 95% of the cells surviving at 24 hr are recognizable neurons, both by morphological criteria and by their tetanus-toxin-binding competence. We do not know, however, whether or to what extent the surviving neurons include *cholinergic* ones. Cholinergic neurons constitute but a small proportion of either septal or striatal neurons, and their cholinergic properties develop mainly postnatally (Honegger *et al.*, 1979; Hefti, personal communication). Thus, our low-seeded cultures are unlikely to permit ChAT measurements, and even AChE visualization of cholinergic neurons may not be possible until after 1–2 weeks *in vitro*. The following section on trophic regulation of septal and striatal neuronal cultures, therefore, may or may not prove relevant with regard to the cholinergic subset of these tissues.

4.3. Trophic Agents for Septal and Striatal Neurons

Neuronal survival in the low-density cultures is not supported by NGF or by the eye-derived CNTF. It is supported by a variety of *glia-CMs* (Barbin *et al.*, 1983), by *CNS wound fluid* (Manthorpe *et al.*, 1983e), and by rat and beef *brain extracts* (Selak *et al.*, 1983b). The trophic activity for septal and striatal neurons observed in the CMs is associated with low-molecular-weight, non-protein agents and therefore can be segregated from the CM activities for ciliary and other ganglionic neurons (Varon *et al.*, 1983b). In contrast, the trophic activity for septal and striatal neurons present in CNS wound fluid appears to be associated with protein agents and thus is not distinguishable at present from those observed to support the cholinergic test neurons from CGs (Manthorpe *et al.*, 1983e).

5. CONCLUSIONS AND PROJECTIONS

Cholinergic neurons are important subsystems for both the normal performance of several peripheral and central neural functions and a variety of human pathological situations (cf. Appel, 1981; Varon *et al.*, 1982). Understanding of both normal and abnormal cholinergic behaviors will be greatly advanced by increasing our knowledge of the biological agents or situations or both that can affect maintenance and performance of cholinergic neurons. *In vitro* cultures of dissociated neurons are well-recognized, excellent tools for the investigation of neuronal regulation by extrinsic signals from the microenvironment in which neurons must operate. The use of CG neurons has already made it possible to isolate the first trophic factor for cholinergic neurons and to identify a number of potential sources for additional ones. Cultures of CNS neurons that include cholinergic elements are also increasingly employed for the investigation of such trophic agents. With both peripheral and central test neurons, efforts are being made to identify and investigate other agents that specifically address the regulation of neuritic elongation and, perhaps, maintenance or repair. *In vivo* models have been developed to study the potential involvement of such agents in normal and pathological situations *in situ*. Most encouraging of all may be the perception, from the findings in several laboratories, that there may be relatively little species specificity concerning the acceptance by neurons of appropriate trophic and neurite-promoting agents. Two important obstacles to future progress are (1) the difficulty in visualizing cholinergic neurons in heterogeneous cultures or tissue sections and (2) the very small amounts of factors that can be collected in biological sources known at present. Nevertheless, it can be hoped that both problems may be corrected in the not-too-distant future by the ongoing technological advances in the fields of monoclonal antibodies and genetic engineering, respectively.

ACKNOWLEDGMENTS. Part of this work was supported by Grant NS-16349 from the National Institute of Neurological and Communicative Disorders and Stroke and by a past grant from the Muscular Dystrophy Association.

REFERENCES

Adler, R., and Varon, S., 1980, Cholinergic neuronotrophic factors: V. Segregation of survival- and neurite-promoting activities in heart conditioned media, *Brain Res.* **188**:437–448.

Adler, R., and Varon, S., 1981a, Neuritic guidance by polyornithine attached materials of ganglionic origin, *Dev. Biol.* **81**:1–11.

Adler, R., and Varon, S., 1981b, Neuritic guidance by nonneuronal cells of ganglionic origin, *Dev. Biol.* **86**:69–80.

Adler, R., and Varon, S., 1982, Neuronal survival in intact ciliary ganglia *in vivo* and *in vitro:* CNTF as a target surrogate, *Dev. Biol.* **92**:470–475.

Adler, R., Landa, K.B., Manthorpe, M., and Varon, S., 1979, Cholinergic neuronotrophic factors. II. Selective intraocular distribution of soluble trophic activity for ciliary ganglionic neurons, *Science* **204**:1434–1436.

Adler, R., Manthorpe, M., Skaper, S.D., and Varon, S., 1981, Polyornithine-attached neurite-promoting factors (PNPFs): Culture sources and responsive neurons, *Brain Res.* **206**:129–144.

Adler, R., Manthorpe, M., and Varon, S., 1983, Lectin reactivity of PNPF, a polyornithine-binding neurite promoting factor, *Dev. Brain Res.* **6**:69–75.

Aguayo, A., David, S., Richardson, P., and Bray, G., 1982, Axonal elongation in peripheral and central nervous system transplants, *Adv. Cell. Neurobiol.* **3**:215–234.

Appel, S.H., 1981, A unifying hypothesis for the cause of amyotrophic lateral sclerosis, parkinsonism, and Alzheimer's disease, *Ann. Neurol.* **10**:499–505.

Barbin, G., Manthorpe, M., and Varon, S., 1981, Molecular behaviors of ciliary neuronotrophic factor, *Soc. Neurosci. Abstr.* **7**:554.

Barbin, G., Selak, I., Manthorpe, M., and Varon, S., 1983, Use of central neuronal cultures for the detection of neuronotrophic agents, *Neurosci.* (in press).

Barde, Y.-A., Lindsay, R.M., Monard, D., and Thoenen, H., 1978, New factor released by cultured glioma cells supporting survival and growth of sensory neurones, *Nature (London)* **274**:818.

Bennett, M.R., and Nurcombe, V., 1979, The survival and development of cholinergic neurons in skeletal muscle conditioned media, *Brain Res.* **173**:543–548.

Björklund, A., and Stenevi, U., 1977, Reformation of severed septohippocampal cholinergic pathway in the adult rat by transplanted septal neurons, *Cell Tissue Res.* **185**:289–302.

Bonyhady, R.E., Hendry, I.A., Hill, C.E., and McLennan, I.S., 1980, Characterization of a cardiac muscle factor required for the survival of cultured parasympathetic neurones, *Neurosci. Lett.* **18**:197–201.

Bottenstein, J.E., and Sato, G., 1979, Growth of a rat neuroblastoma cell line in serum-free supplemented medium, *Proc. Natl. Acad. Sci. U.S.A.* **76**: 514–517.

Bottenstein, J.E., Skaper, S.D., Varon, S., and Sato, G.H., 1980, Selective survival of neurons from chick embryo sensory ganglionic dissociates using defined, serum-free supplemented medium, *Exp. Cell Res.* **125**:183–190.

Burnham, P., Raiborn, C., and Varon, S., 1972, Replacement of nerve growth factor by ganglionic non-neuronal cells for the survival *in vitro* of dissociated ganglionic neurons, *Proc. Natl. Acad. Sci. U.S.A.* **69**:3556–3560.

Campenot, R.B., 1977, Local control of neurite development by nerve growth factor, *Proc. Natl. Acad. Sci. U.S.A.* **74**:4516–4519.

Campenot, R.B., 1982a, Development of sympathetic neurons in compartmentalized cultures. I. Local control of neurite growth by nerve growth factor, *Dev. Biol.* **93**:1–12.

Campenot, R.B., 1982b, Development of sympathetic neurons in compartmentalized cultures. II. Local control of neurite survival by nerve growth factor, *Dev. Biol.* **93**:13–21.

Collins, F., 1978, Induction of neurite outgrowth by a conditioned medium factor bound to culture substratum, *Proc. Natl. Acad. Sci. U.S.A.* **75**:5210–5213.

Collins, F., and Garrett, J.E., 1980, Elongating nerve fibers are guided by a pathway of material released from embryonic nonneuronal cells, *Proc. Natl. Acad. Sci. U.S.A.* **77:**6226–6228.

Cornbrooks, C., Carey, D., Timpl, R., McDonald, J., and Bunge R., 1982, Immunohistochemical visualization of fibronectin and laminin in adult rat peripheral nerve and peripheral nerve cells in culture, *Soc. Neurosci. Abstr.* **8:**240.

Coughlin, M.D., Dibner, M.D., Boyer, D.M., and Black, I.B., 1978, Factors regulating development of an embryonic mouse sympathetic ganglion, *Dev. Biol.* **66:**513–528.

Cowan, W.M., 1973, Neuronal death as a regulative mechanism in the control of cell number in the nervous system, in: *Development and Aging in the Nervous System* (M. Rockstein and M.L. Sussman, eds.), Academic Press, New York, pp. 19–41.

Ebendal, T., Jordell-Kylberg, A., and Soderstrom, S., 1978, Stimulation by tissue explants on nerve fiber outgrowth in culture, *Zoon* **6:**235–243.

Ebendal, T., Belew, M., Jacobson, C.-O., and Porath, J., 1979, Neurite outgrowth elicited by embryonic chick heart: Partial purification of the active factor, *Neurosci. Lett.* **14:**91–95.

Edgar, D., and Thoenen, H., 1982, Modulation of NGF-induced survival of chick sympathetic neurons by contact with a conditioned medium factor bound to the substratum, *Dev. Brain Res.* **5:**89–92.

Greene, L.A., and Shooter, E.M., 1980, The nerve growth factor: Biochemistry, synthesis, and mechanism of action, *Annu. Rev. Neurosci.* **3:**352–402.

Gundersen, R.W., and Barrett, J.N., 1979, Neuronal chemotaxis: Chick dorsal root axons turn toward high concentrations of nerve growth factor, *Science* **206:**1079–1080.

Helfand, S.L., Smith, G., and Wessells, N.K., 1976, Survival and development in culture of dissociated parasympathetic neurons from ciliary ganglia, *Dev. Biol.* **50:**541–547.

Helfand, S.L., Riopelle, R.J., and Wessells, N.K., 1978, Nonequivalence of conditioned medium and nerve growth factor for sympathetic, parasympathetic and sensory neurons, *Exp. Cell Res.* **113:**39–45.

Henderson, C.E., Huchet, M., and Changeux, J.-P., 1981, Neurite outgrowth from embryonic chicken spinal neurons is promoted by media conditioned by muscle cells, *Proc. Natl. Acad. Sci. U.S.A.* **78:**2625–2629.

Hewitt, E., Manthorpe, M., Skaper, S.D., Davis G.E., and Varon, S., 1983, Influences of culture media on the survival and neuritic growth of chick embryo ciliary ganglion neurons in monolayer cultures, *In Vitro* **19:**280.

Honegger, P., Lenoir, D., and Favrod, P., 1979, Growth and differentiation of aggregating fetal brain cells in a serum-free, defined medium, *Nature (London)* **282:**305–308.

Jirikowski, G., Reisert, I., and Pilgrim, C.H., 1981, Neuropeptides in dissociated cultures of hypothalamus and septum: Quantitation of immunoreactive neurons, *Neuroscience* **6:**1953–1960.

Kligman, D., 1982a, Isolation of a protein from bovine brain which promotes neurite extension from chick embryo cerebral cortex neurons in defined medium, *Brain Res.* **250:**93–100.

Kligman, D., 1982b, Neurite outgrowth from cerebral cortical neurons is promoted by medium conditioned over heart cells, *J. Neurosci. Res.* **8:**281–287.

Kromer, L., Björklund, A., and Stenevi, U., 1981, Innervation of embryonic hippocampal implants by regenerating axons of cholinergic septal neurons in the adult rat, *Brain Res.* **210:**153–172.

Landa, K.B., Adler, R., Manthorpe, M., and Varon, S., 1980, Cholinergic neuronotrophic factors. III. Developmental increase of trophic activity for chick embryo ciliary ganglionic neurons in their intraocular target tissues, *Dev. Biol.* **74:**401–408.

Lander, A.D., Fujii, D.K., Gospodarowicz, D., and Reichardt, L.F., 1982, Characterization of a factor that promotes neurite outgrowth: Evidence linking activity to a heparan sulfate proteoglycan, *J. Cell Biol.* **94:**574–585.

Landmesser, L., and Pilar, G., 1978, Interactions between neurons and their targets during *in vivo* synaptogenesis, *Fed. Proc. Fed. Am. Soc. Exp. Biol.* **37:**2016–2022.

Letourneau, P.C., 1978, Chemotactic response of nerve fiber elongation to nerve growth factor, *Dev. Biol.* **66:**183–196.

Levi-Montalcini, R., and Angeletti, P.U., 1968, Nerve growth factor, *Physiol. Rev.* **48:**534–569.

Lewis, E., and Cotman, C.W., 1980, Mechanisms of septal lamination in the developing hippocampus revealed by outgrowth of fibers from septal implants. II. Absence of guidance by degenerative debris, *J. Neurosci.* **2:**66–77.

Longo, F.M., Manthorpe, M., and Varon, S., 1982, Spinal cord neuronotrophic factors (SCNTFs). I. Bioassay of schwannoma and other conditioned media, *Dev. Brain Res.* **3:**277–294.

Longo, F. M., Manthorpe, M., Skaper, S. D., Lundborg, G., and Varon, S., 1983a, Neuronotrophic activities accumulate *in vivo* within silicone nerve regeneration chambers, *Brain Res.* **261:**109–117.

Longo, F.M., Skaper, S.D., Manthorpe, M., Williams, L.R., Lundborg, G., and Varon, S., 1983b, Temporal changes of neuronotrophic activities accumulating *in vivo* within nerve regeneration chambers, *Exp. Neurol.* (in press).

Lundborg, G., Dahlin, L.B., Danielsen, N., Gelberman, R.H., Longo, F.M., Powell, H.C., and Varon, S., 1982a, Nerve regeneration in silicone chambers: Influence of gap length and presence of distal stump, *Exp. Neurol.* **76:**361–375.

Lundborg, G., Longo, F.M., and Varon, S., 1982b, Nerve regeneration model and trophic factors *in vivo*, *Brain Res.* **232:**157–161.

Manthorpe, M., Skaper, S.D., Adler, R., Landa, K.B., and Varon, S., 1980, Cholinergic neuronotrophic factors. IV. Fractionation properties of an extract from selected chick embryonic eye tissues, *J. Neurochem.* **34:**69–75.

Manthorpe, M., Adler, R., and Varon, S., 1981a, Cholinergic neuronotrophic factors. VI. Age-dependent requirements by chick embryo ciliary ganglionic neurons, *Dev. Biol.* **85:**156–163.

Manthorpe, M., Varon, S., and Adler, R., 1981b, Neurite-promoting factor (NPF) in conditioned medium from RN22 schwannoma cultures: Bioassay, fractionation and other properties, *J. Neurochem.* **37:**759–767.

Manthorpe, M., Skaper, S.D., and Varon, S., 1981c, Neuronotrophic factors and their antibodies: *In vitro* microassays for titration and screening, *Brain Res.* **230:**295–306.

Manthorpe, M., Skaper, S.D., Barbin, G., and Varon, S., 1982a, Cholinergic neuronotrophic factors (CNTF's). VII. Concurrent activities on certain nerve growth factor–responsive neurons, *J. Neurochem.* **38:** 415–421.

Manthorpe, M., Barbin, G., and Varon, S., 1982b, Isoelectric focusing of the chick eye ciliary neuronotrophic factor, *J. Neurosci. Res.* **8:**233–239.

Manthorpe, M., Longo, F.M., and Varon, S., 1982c, Comparative features of spinal neuronotrophic factors in fluids collected *in vitro* and *in vivo*, *J. Neurosci. Res.* **8:**241–250.

Manthorpe, M., Luyten, W., Longo, F.M., and Varon, S., 1983a, Endogenous and exogenous factors support neuronal survival and choline acetyltransferase activity in embryonic spinal cord cultures, *Brain Res.* **267:**57–66.

Manthorpe, M., Nieto-Sampedro, M., Skaper, S.D., Lewis, E.R., Barbin, G., Longo, F.M., Cotman, C.W., and Varon, S., 1983b, Neuronotrophic activity in brain wounds in the developing rat: Correlation with implant survival in the wound cavity, *Brain Res.* **267:**47–56.

Manthorpe, M., Engvall, E., Ruoslahti, E., Longo, F.M., Davis, G.E., and Varon, S., 1983c, Laminin promotes neuritic regeneration from cultured peripheral and central neurons, *J. Cell Biol.* (in press).

Manthorpe, M., Barbin, G., and Varon, S., 1983d, Ciliary neuronotrophic factor (CNTF): Biochemical and biological properties, *J. Neurochem. (in press)*.

Manthorpe, M., Williams, L.R., Barbin, G., Selak, I., and Varon, S., 1983e, Trophic and toxic factors for central neurons in rat brain wound fluid (in prep.).

McLennan, I.S., and Hendry, I.A., 1980, Influences of cardiac extracts on cultured ciliary ganglia, *Dev. Neurosci.* **3:**1–10.

Messer, A., 1981, Primary monolayer cultures of the rat corpus striatum: Morphology and properties related to acetylcholine and γ-aminobutyrate, *Neuroscience* **6**:2677–2687.

Nieto-Sampedro, M., Lewis, E.R., Cotman, C.W., Manthorpe, M., Skaper, S.D., Barbin, G., Longo, F.M., and Varon, S., 1982, Brain injury causes a time-dependent increase in neuronotrophic activity at the lesion site, *Science* **217**:860–861.

Nieto-Sampedro, M., Manthorpe, M., Barbin, G., Williams, L.R., Varon, S., and Cotman, C.W., 1983, Injury-induced neuronotrophic activity in adult brain: Correlation with survival of delayed implants in the wound cavity, *J. Neuroscience* (in press).

Nishi, R., and Berg, D.K., 1977, Dissociated ciliary ganglion neurons *in vitro:* Survival and synapse formation, *Proc. Natl. Acad. Sci. U.S.A.* **74**:5174–5175.

Panula, P., Rechardt, L., and Hervonen, H., 1979, Observations on the morphology and histochemistry of the rat neostriatum in tissue culture, *Neuroscience* **4**:235–248.

Peterson, E.R., and Crain, S.M., 1982, Nerve growth factor attenuates neurotoxic effects of taxol on spinal cord-ganglion explants from fetal mice, *Science* **217**:377–379.

Rothman-Schonfeld, A., Thal, L.J., Horowitz, S.G., and Katzman, R., 1981, Heart conditioned medium promotes cholinergic regeneration *in vivo*, *Brain Res.* **229**:541–546.

Roufa, D., Cornbrooks, C., Johnson, M., and Bunge, M., 1982, Preparation of homogeneous populations of immature Schwann cells from early embryonic sympathetic ganglia, *Soc. Neurosci. Abstr.* **8**:240.

Sato, G.H., Pardee, A.B., and Sirbasku, D.A. (eds.), 1982, *Growth of Cells in Hormonally Defined Media*, Vol. 9, Cold Spring Harbor Conference on Cell Proliferation, Cold Spring Harbor Laboratory, New York.

Schmidt, R.H., Björklund, A., and Stenevi, U., 1981, Intracerebral grafting of dissociated CNS tissue suspensions: A new approach for neuronal transplantation to deep brain sites, *Brain Res.* **218:** 347–356.

Selak, I., Skaper, S.D., and Varon, S., 1983a, Age-dependent requirements of sympathetic neurons in serum-free cultures, *Dev. Brain Res.* **7**:171–179.

Selak, I., Barbin, G., Manthorpe, M., and Varon, S., 1983b, Trophic and toxic factors for CNS neurons in beef brain extracts (in prep.).

Skaper, S.D., and Varon, S., 1982, Ionic behaviors and nerve growth factor dependence in developing chick ganglia. II. Studies with purified neurons of dorsal root ganglia, *Dev. Biol.* **98:** 257–264.

Skaper, S.D., Adler, R., and Varon, S., 1979, A procedure for purifying neuron-like cells in cultures from central nervous tissues with a defined medium, *Dev. Neurosci.* **2**:233–237.

Skaper, S.D., Selak, I., and Varon, S., 1982, Molecular requirements for survival of cultured avian and rodent dorsal root ganglionic neurons responding to different trophic factors, *J. Neurosci. Res.* **8**:251–261.

Skaper, S.D., Selak, I., and Varon, S., 1983a, Serum- and substratum-dependent modulation of neuritic growth, *J. Neurosci. Res.* **9**:359–369.

Skaper, S.D., Selak, I., and Varon, S., 1983b, Characterization and partial purification of a neurite inhibitory factor from fetal calf serum (submitted).

Skaper, S.D., Selak, I., Manthorpe, M., and Varon, S., 1984, Chemically defined requirements for the survival of cultured 8-day chick embryo ciliary ganglionic neurons, *Brain Res.* (in press).

Skaper, S.D., Davis, G.E., Moonen, G., Manthorpe, M., and Varon, S., 1983d, Serum inhibits neuritic growth from primary ciliary ganglionic neurons (in prep.).

Spencer, P.S., and Schaumburg, H.H. (eds.), 1980, *Experimental and Clinical Toxicology*, Williams and Wilkins, Baltimore.

Thoenen, H., and Barde, Y.-A., 1980, Physiology of nerve growth factor, *Physiol. Rev.* **60:** 1284–1335.

Varon, S., 1975a, Nerve growth factor and its mode of action, *Exp. Neurol.* **48**:75–92.

Varon, S., 1975b, *In vitro* approaches to the study of neural tissue aging, in: *Survey of the Aging Nervous System* (G. Maletta, ed.), DHEW Publication No. NIH 74-296, pp. 59–76.

Varon, S., 1977, Neural growth and regeneration: A cellular perspective, *Exp. Neurol.* **54:**1–6.

Varon, S., and Adler, R., 1980, Nerve growth factors and control of nerve growth, *Curr. Top. Dev. Biol.* **16:**207–252.

Varon, S., and Adler, R., 1981, Trophic and specifying factors directed to neuronal cells, *Adv. Cell. Neurobiol.* **2:**115–163.

Varon, S., and Bunge, R.P., 1978, Trophic mechanisms in the peripheral nervous system, *Annu. Rev. Neurosci.* **1:**327–361.

Varon, S., and Lundborg, G., 1983, *In vivo* model for peripheral nerve regeneration and the presence of neuronotrophic factors, in: *Nervous System Regeneration* (A.M. Giuffridastella, B. Haber, G.A. Hashim, and J.R. Perez-Polo eds.), Alan R. Liss, New York **19:**221–240.

Varon, S., and Manthorpe, M., 1982, Schwann cells: An *in vitro* perspective, *Adv. Cell. Neurobiol.* **3:**35–95.

Varon, S., and Manthorpe, M., 1983, *In vitro* models for neuroplasticity and repair, in: *Neuroplasticity and Repair in the Central Nervous System* (C.L. Bolis and F.E. Bloom, eds.), WHO Study Group Geneva (in press).

Varon, S., and Raiborn, C.W., Jr., 1969, Dissociation, fractionation and culture of embryonic brain cells, *Brain Res.* **12:**180–199.

Varon, S., and Raiborn, C., 1972, Dissociation, fractionation and culture of chick embryo sympathetic ganglionic cells, *J. Neurocytol.* **1:**211–221.

Varon, S., and Skaper, S.D., 1983a, The Na^+, K^+-pump may mediate the control of nerve cells by nerve growth factor, *Trends Biochem. Sci.* **8:**22–25.

Varon, S., and Skaper, S.D., 1983b, Nerve growth factor and neuronotrophic functions, in: *Somatic and Autonomic Nerve–Muscle Interactions* (G. Burnstock, G. Urbova, and R. O'Brien eds.), Elsevier-North-Holland, Amsterdam pp. 213–251.

Varon, S., and Somjen, G., 1979, Neuron–glia interactions, *Neurosci. Res. Program Bull.* **17:**1–239.

Varon, S., Nomura, J., Perez-Polo, J.R., and Shooter, E.M., 1972, The isolation and assay of the nerve growth factor proteins, in: *Methods and Techniques of Neurosciences* (R. Fried, ed.), Marcel Dekker, New York, pp. 203–229.

Varon, S., Raiborn, C., and Tyszka, E., 1973, *In vitro* studies of dissociated cells from newborn mouse dorsal root ganglia, *Brain Res.* **54:**51–63.

Varon, S., Raiborn, C., and Norr, S., 1974, Association of antibody to nerve growth factor with ganglionic non-neurons (glia) and consequent interference with their neuron-supportive action, *Exp. Cell Res.* **88:**247–256.

Varon, S., Manthorpe, M., and Adler, R., 1979, Cholinergic neuronotrophic factors. I. Survival, neurite outgrowth and choline acetyltransferase activity in monolayer cultures from chick embryo ciliary ganglia, *Brain Res.* **173:**29–45.

Varon, S., Skaper, S.D., and Manthorpe, M., 1981, Trophic activities for dorsal root and sympathetic ganglionic neurons in media conditioned by Schwann and other peripheral cells, *Dev. Brain Res.* **1:**73–87.

Varon, S., Manthorpe, M., and Longo, F.M., 1982, Growth factors and motor neurons, in: *Advances in Neurology, Human Motor Neuron Diseases* (L.P. Rowland, ed.), Raven Press, New York **36:**453–472.

Varon, S., Adler, R., Manthorpe, M., and Skaper, S.D., 1983a, Culture strategies for trophic and other factors directed to neurons, in: *Neuroscience Approached through Cell Culture,* (S.E. Pfeiffer, ed.), CRC Press, Boca Raton, Florida, **2:**53–77.

Varon, S., Skaper, S.D., Barbin, G., Selak, I., and Manthorpe, M., 1983b, Low molecular weight agents for the survival of cultured CNS neurons *J. Neuroscience* (in press).

Varon, S., Engvall, E., Ruoslahti, E., Longo, F.M., and Manthorpe, M., 1983c, Laminin promotes axonal regeneration from cultured PNS and CNS neurons, *Trans. Am. Soc. Neurochem.* **14**:225.

Varon, S., Manthorpe, M., Longo, F.M., and Williams, L.R., 1983d, Growth factors in regeneration of neural tissue, in: *Nerve, Organ, and Tissue Regeneration: Research Perspectives,* (F.J. Seil, ed.) Academic Press, New York pp. 127–155.

Williams, L.R., Longo, F.M., Powell, H.C., Lundborg, G., and Varon, S., 1983a, Spatial–temporal progress of peripheral nerve regeneration within a silicone chamber: Parameters for a bioassay, *J. Comp. Neurol.* **218**:460–470.

Williams, L.R., Manthorpe, M., Nieto-Sampedro, M., Cotman, C.W., and Varon, S., 1983b, Ciliary neuronotrophic activity in high in peripheral nerve and root extracts *J. Neurosci. Res.* (in press).

V

Molecular Biology of Neural Development and Function

Session Chairman: Xandra O. Breakefield

Expression of Opioid Peptide Genes in Different Species

Edward Herbert, Michael Comb, Haim Rosen, and Gerard Martens

1. INTRODUCTION

In the past 15 years, a number of small peptides that mediate specific kinds of behavior have been isolated. For example, hypothalamic releasing factors ranging from 3 to 41 amino acids in length regulate the production and release of anterior pituitary hormones that mediate growth, reproductive cycles, and diurnal light and temperature cycles. Substance P and opioid peptides regulate the perception of pain in the animal by serving as neurotransmitters and neuromodulators in the nervous system. Other peptides have been implicated in regulating sleep and temperature (bombesin). The train of responses involved in egg-laying behavior in *Aplysia* has been shown to be triggered by a small group of related peptides. Finally, peptide pheromones (a and α factors) mediate the union of yeast cells during conjugation.

For the past five years, we have been interested in the regulation of expression of a class of neuroactive peptides known as opioid peptides. These peptides mimic the effects of morphine when injected into animals except that they are much more potent than morphine.

The simplest of the opioid peptides are the pentapeptides leucine-enkephalin (Leu-enkephalin) and methionine-enkephalin (Met-enkephalin), first isolated from

EDWARD HERBERT, MICHAEL COMB, HAIM ROSEN, and GERARD MARTENS ● Department of Chemistry, University of Oregon, Eugene, Oregon 97403.

brain tissue by Hughes *et al.* (1975) (Fig. 1). As shown Fig. 1, the other opioid peptides are C-terminal extensions of either Leu- or Met-enkephalin. β-Endorphin, for example, is a C-terminal 26-amino-acid extension of Met-enkephalin (Bradbury *et al.*, 1976; Li and Chung, 1976) and dynorphin is a 12-amino-acid C-terminal extension of Leu-enkephalin (Goldstein *et al.*, 1979, 1981). Peptide E, which contains both a Met- and a Leu-enkephalin sequence, is one of a group of large enkephalin-containing peptides found in the bovine adrenal medulla (Kilpatrick *et al.*, 1981). Some of these peptides could be intermediates in the processing of high-molecular-weight enkephalin precursors. Peptide E, however, is thought to be an end product of processing because of its high potency as an opiate (much higher potency than enkephalins). The integrity of the enkephalin moiety is essential for activity of all the opioid peptides.

The structural relationships of the opioid peptides suggested initially that the larger ones such as β-endorphin and dynorphin might be precursors to Met- and Leu-enkephalin. However, it has become clear in the past three years by the use of recombinant DNA technology that the synthesis of the opioid peptides is controlled by at least three different genes. One gene codes for proenkephalin, which contains six Met-enkephalin sequences and one Leu-enkephalin sequence (Comb *et al.*, 1982; Gubler *et al.*, 1982; Noda *et al.*, 1982). A second gene codes for pro-opiomelanocortin (POMC), from which β-endorphin, ACTH, and melanocyte-stimulating hormones (α-, β-, and γMSH) are derived (Roberts *et al.*, 1977; Mains *et al.*, 1977; Nakanishi *et al.*, 1979). A third gene codes for the precursor to both dynorphin and neoendorphins (Kakidani *et al.*, 1982). The latter gene contains three copies of Leu-enkephalin in mammals.

The precursors to the opioid peptides belong to a growing class of proteins called *polyfunctional proteins* or *polyproteins* because they give rise to two or more biologically active peptides. Some remarkable similarities in the structure of the polyfunctional opioid peptide precursors are revealed in Fig. 2: (1) The precursors are about the same length and the sequences of almost all the biologically active peptides are located in the C-terminal half of the molecules. (2) The biologically active domains are flanked on both sides by pairs of basic amino acid residues, creating potential cleavage sites for trypsinlike enzymes. (3) The concentration of cysteine residues in the N-terminal region of each protein suggests that disulfide bridges are necessary for stabilizing the proteins in conformations required for correct processing.

2. ISOLATION AND CHARACTERIZATION OF OPIOID PEPTIDE GENES

The similarity in structure of the precursors to the opioid peptides suggests that there is a close evolutionary relationship among these peptides. This relationship should be reflected in the structure of the genes that code for these peptides.

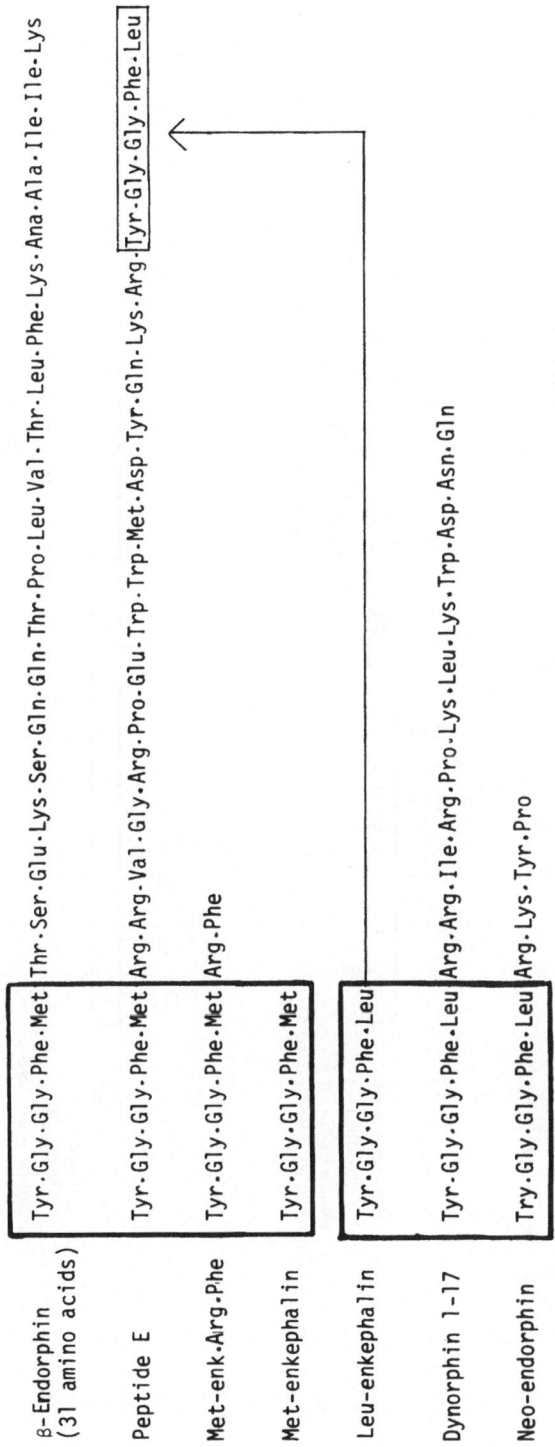

Figure 1. Structure of the opioid peptides depicting common functional enkephalin sequences. Only 25 of the 31 amino acids of β-endorphin are shown.

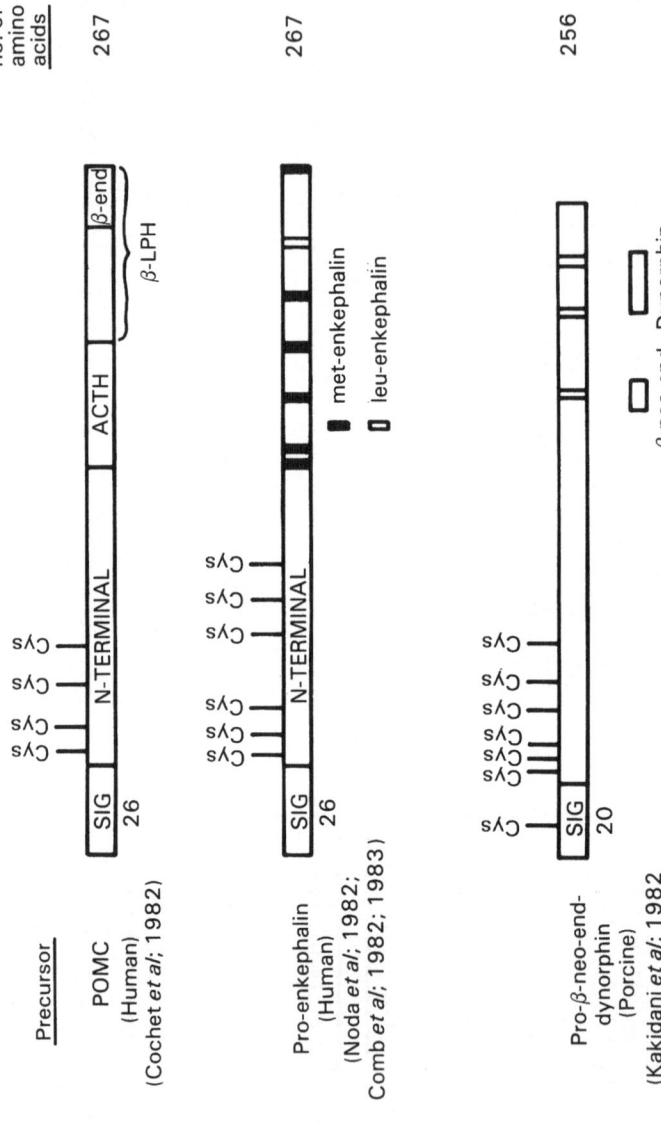

Figure 2. Structure of precursors to the opioid peptides. Schematic representation of the three opioid precursors determined from the amino acid sequences. (SIG) Signal peptide; (Cys) cysteine residue. The amino acid sequences are derived from nucleotide sequences of cloned complementary DNA.

The genes of eukaryotic organisms contain coding regions separated by noncoding regions (introns). The arrangement of these regions can give a clue as to the evolutionary relationship of different genes. The genes that code for human POMC (Chang *et al.*, 1980; Whitfeld *et al.*, 1982) and for proenkephalin (Noda *et al.*, 1982; Comb *et al.*, 1983) have recently been isolated from lambda gene libraries and sequenced by recombinant DNA techniques. The similarity in the structure of the two genes is quite remarkable, as shown in Fig. 3. Both genes contain large 3' exons that code for more than 80% of the protein. A large intron on the 3' side of the major exon [3 kilobases (kb) in length] separates this exon from a smaller exon that codes for the signal sequence of each polyprotein. A second intron separates the signal sequence exon from an exon that codes for the 5' untranslated region of the messenger RNA (mRNA). Therefore, a single large exon codes for all the known biologically active peptides derived from proenkephalin and POMC. Structural similarity at the level of the gene supports the idea of a close evolutionary relationship of POMC and proenkephalin. It will be very interesting from an evolutionary point of view to compare the structure of the prodynorphin gene with the genes shown in Fig. 3. Studies of the prodynorphin gene are under way in Numa's laboratory in Japan and in our own laboratory.

2.1. Chromosomal Locations of Opioid Peptide Genes

Another important question concerning the interrelationship of the opioid peptide genes is how closely linked they are in the genome. POMC has been shown to be located on chromosome 2 in the human (Owerbach *et al.*, 1981) and chromosome 12 in the mouse (Uhler *et al.*, 1983). We have been trying to determine the location of the proenkephalin and prodynorphin genes in the human genome in collaboration with Drs. Frank Ruddle and Peter Barker at Yale University, using the hybridoma technique. The results indicate that the proenkephalin gene is located on chromosome 12 (Comb *et al.*, 1983). Thus, the POMC and proenkephalin genes are not on the same chromosome in the human. We do not yet have any information about the location of the prodynorphin gene.

2.2. Detailed Structure of the Human Proenkephalin Gene—Methylation Sites and Regulation

Southern blot analysis of human DNA using cloned proenkephalin complementary DNA (cDNA) as a hybridization probe indicates that there is only one human proenkephalin gene (Comb *et al.*, 1983). A detailed structural analysis of this gene and the regions flanking this gene (isolated from a lambda gene library provided by Maniatis) has been carried out by cloning and sequencing techniques.

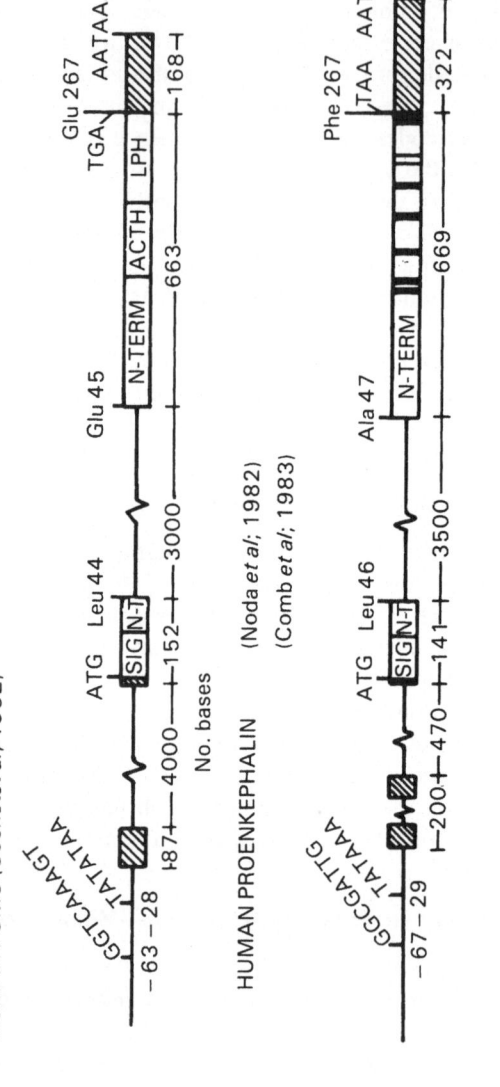

Figure 3. Structural comparison of the human POMC and proenkephalin genes. The data are taken from Comb *et al.* (1983). The lines indicate introns (nontranscribed sequences) and the boxes indicate exons (transcribed sequences). ATG positions −63, −28 for POMC, and −67, −29 for proenkephalin sequences are common to the majority of the eukaryotic genes. (ATG) Translation start site; (AATAAA) poly(A) addition site. Note the presence of N-terminal (N-T) sequences on the exon carrying the signal (SIG) peptide.

The results of this analysis for a 10-kb segment of human DNA are shown in Fig. 4. The main exon is in the middle of the gene segment. The site for initiation of transcription (cap site) is close to the 5′ border of the gene. Several kinds of repeat sequences are present in the intronic regions. These repeat sequences are observed in many other genes, but their significance is not very well understood. The AATAAA sequence at the 3′ end of the main exon is a poly(A) addition site that is found in other genes that code for mRNA that have a poly(A) tail.

A particularly interesting feature of the proenkephalin gene that is shared with the POMC gene is the highly asymmetric distribution of guanine (G) and cytosine (C) residues. The 5′ flanking sequences and sequences from the 1st exon (closest to the 5′ end on the left) to the 3rd exon exhibit high C_pG content (67% of total bases). Several hundred bases on the 3′ side of the 3rd exon, there is a dramatic drop in GtC content and the GtC content remains low through the 4th exon (the main exon). The C_pG content rises again 2 kb on the 3′ side of the main exon. The frequency of the relatively rare dinucleotide C_pG parallels the C_pG distribution, exhibiting clustering in the 5′ region of the gene (exons I, II, and III). C_pG accounts for 10% of the total dinucleotide sequences in this region, which is 10 times the frequency of C_pG in the genome as a whole. Clustering of C_pG also occurs in the 3′ untranslated region of the gene, where it accounts for 8% of the total dinucleotide sequences (Fig. 4).

The clustering of C_pG sequences in the 5′ and 3′ untranslated regions is of

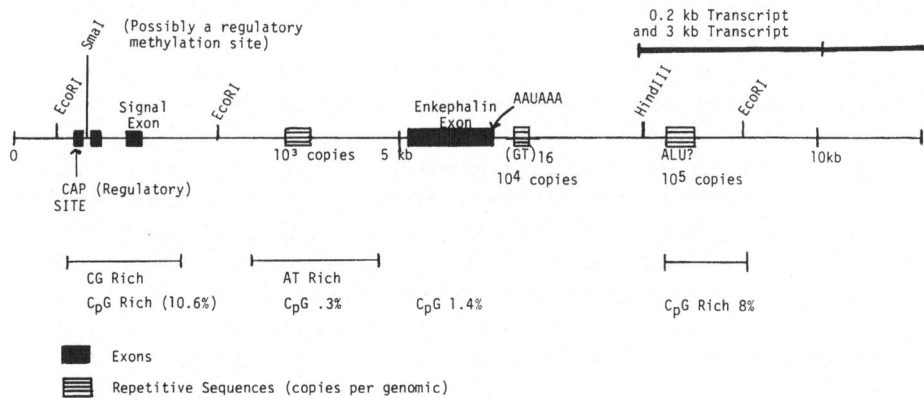

Figure 4. Organization of a 10-kb segment of human DNA containing the proenkephalin gene. The left-hand side of the gene is the 5′ end. The exons (■) are referenced beginning with the first one closest to the 5′ end. The transcripts referred to in the upper right-hand corner (0.2 and 3 kb) have been detected in poly(A) RNA isolated from bovine adrenal medulla by the use of DNA probes to the 5′ region of the gene (Comb *et al.*, 1983). As yet, we do not know the function of these transcripts. (|) Sites digested by restriction endonucleases *Eco*RI, *Sma*I, and *Hind*III.

particular interest because methylation of C residues in this sequence has been implicated in the control of gene expression during development. An inverse relationship has been observed between the level of methylation of C in C_pG sequences in the neighborhood of a gene and the transcriptional activity of that gene (Felsenfeld and McGhee, 1982).

2.3. Methylation of Specific C_pG Sites in DNA from Different Tissues and Relationship to Regulation of Expression

The action of several restriction enzymes (*SmaI*, *HpaII*, and *HhaI*) is inhibited when the C residue in C_pG within the restriction site is methylated. Thus, the failure of a methylation-sensitive enzyme to cleave DNA at a known restriction site indicates the presence of 5-methylcytosine at that site. The methylated sites can be detected by performing Southern blot hybridizations of DNA that has been digested with the methylation-sensitive restriction enzymes. This type of analysis has been performed with DNA isolated from a number of human tissues, some of which are known to express enkephalin peptides, such as the adrenal, and other tissues that do not express these peptides, including leukocytes, thymus, pituitary, and probably placenta. It was hoped that we would detect specific C_pG sites that are less methylated in the DNA from adrenal than in the DNA from tissues that do not express enkephalins. Although the analysis is not complete, this hope appears to have been realized with regard to the C_pG sequence in the *SmaI* site just 3' to the cap site (Fig. 4).

How can one determine whether the aforementioned *SmaI* methylation site is involved in regulation of expression of the enkephalin gene? One strategy we are using to attempt to answer this question is the gene-transfer approach. The human proenkephalin gene has been inserted into viral vectors that have the capbility of entering cells and transforming them with high efficiency. Since the viral vectors contain antibiotic-resistance markers, the transformed cells can be selected for by their ability to grow on certain antibiotics. The ability of these vectors to transform various cell lines is being determined. Once transformed lines that express the proenkephalin gene are established, we will attempt to alter the structure of the methylation sites (demethylate or methylate) and other possible regulatory sites and determine the effect of these alterations on the level of expression of proenkephalin mRNA. The proenkephalin mRNA level will be measured by molecular hybridization with cDNA probes (Comb *et al.*, 1982).

3. COMPARATIVE STUDIES OF OPIOID PEPTIDE GENES

The opioid peptides and ACTH have been reported to occur in a wide range of species from man to *Tetrahymena* (LeRoith *et al.*, 1982). We were interested in knowing whether these peptides in the nonmammalian organisms are derived from precursors similar to those described above for man. Using human proenkephalin cDNA as a hybridization probe, we have been able to isolate a proenkephalin gene from a *Xenopus laevis* lambda gene library (kindly given to us by Dr. Igor Dawid of the National Institutes of Health). The major exon of the *X. laevis* gene has been almost completely sequenced, and a low-resolution model of the structure of this region of the proenkephalin gene from *X. laevis* and from the human is presented in Fig. 5. Some remarkable similarities are apparent: (1) There are seven copies of enkephalin in the amphibian. (2) The enkephalin sequences are arranged in precisely the same way as in the human gene; that is, the spacer regions between the enkephalins are the same length in the genes in the two species. (3) Six of seven enkephalin sequences are flanked by pairs of basic amino acid residues in the amphibian precursor as in the mammalian precursor, suggesting that the same kinds of enzymes are involved in cleaving these peptides out of the precursors in the two classes of animals. (4) It should be noted that the peptide Met-enkephalin-Arg-Gly-Leu in the mammalian precursor is replaced by Met-enkephalin-Arg-Gly-Tyr in the amphibian precursor. The Met-enkephalin-Arg-Gly-Tyr is a potent opioid peptide in mammals.

Perhaps the most interesting feature of the amphibian precursor is that it has Met-enkephalin in place of Leu-enkephalin in the peptide E sequence. Thus,

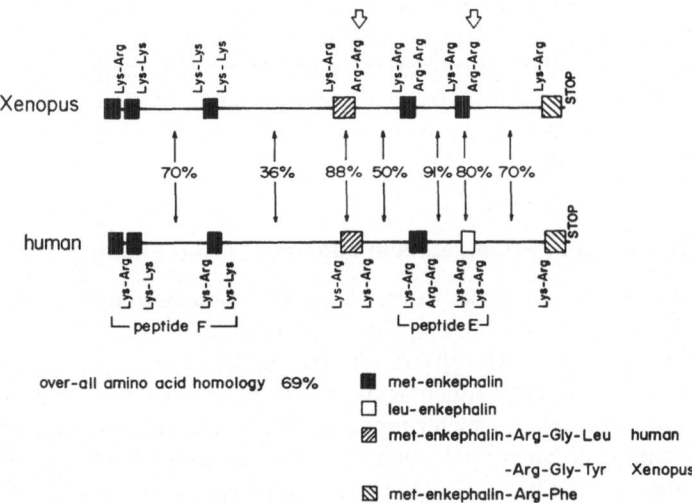

Figure 5. Comparison of human and *Xenopus laevis* proenkephalin genes showing homology of amino acid sequences in the major exon. The percentage homologies indicated between the genes are for the sequences in the spacer regions between copies of enkephalin. The one exception is 88% homology for the Met-enkephalin-Arg-Gly-Leu sequence.

X. laevis has no Leu-enkephalin in the portion of the precursor that has been sequenced to date. This is particularly interesting in view of the finding by Cone and Goldstein (1982) that immunoreactive dynorphin is present in the amphibian *Bufo marinus* (a toad). In mammals, dynorphin contains Leu-enkephalin. Hence, the question arises as to whether amphibian dynorphin contains Leu- or Met-enkephalin. Since the antiserum used to detect dynorphin in *B. marinus* does not require the Leu-enkephalin moiety for cross-reactivity, it is not clear whether amphibian dynorphin contains Leu- or Met-enkephalin.

It would be of interest to know when in the course of evolution the switch occurred from Met-enkephalin to Leu-enkephalin in peptide E and whether this switch occurred in the dynorphin gene as well as the enkephalin gene. This switch requires only a single base change, since there are six codons for Leu. Evolutionary divergence of mammals and amphibia occurred about 350 million years ago. It would be of great interest to determine the structure of the dynorphin precursor in Amphibia, particulary since this precursor also contains α-neo-endorphin, another Leu-enkephalin-containing opioid in mammals. If the amphibian dynorphin gene also contains Met-enkephalin in place of Leu-enkephalin, then we would like to know whether the switch from Met- to Leu-enkephalin occurred independently in the two genes during evolution. It would be necessary to determine the structure of the opioid peptide genes in other vertebrate species to answer this question.

Finally, it is important to point out that despite the similarity in length of the spacer regions between enkephalins, the sequences are not very well conserved in these regions except for peptide E, as shown in Fig. 5. The amino acid homologies in the spacer regions range from 91% (peptide E spacer) to 36% (for the longest spacer). Hence except for peptide E, most of the sequences in these regions do not appear to be important for processing of the precursor or for other functions.

3.1. Comparison of Proenkephalin and Yeast Pro-α-Factor

The precursor to α-factor has recently been sequenced by Kurjan and Herskowitz (1982). This precursor contains four identical copies of α-factor, which is 13 amino acids long. The spacer regions, unlike those in proenkephalin, are exactly the same length (9 amino acids) and have almost the identical sequence. Thus, the multiple-copy α-factor precursor (Fig. 6) appears to have evolved to provide for more efficient production of a secretory peptide. The variable nature of the spacer regions in proenkephalin is compatible with a more varied use of different regions of the precursor. Since enkephalins are expressed in many tissues, it is conceivable that different peptides are cut out of the precursor in different tissues. The spacer regions could specify differential tissue expression

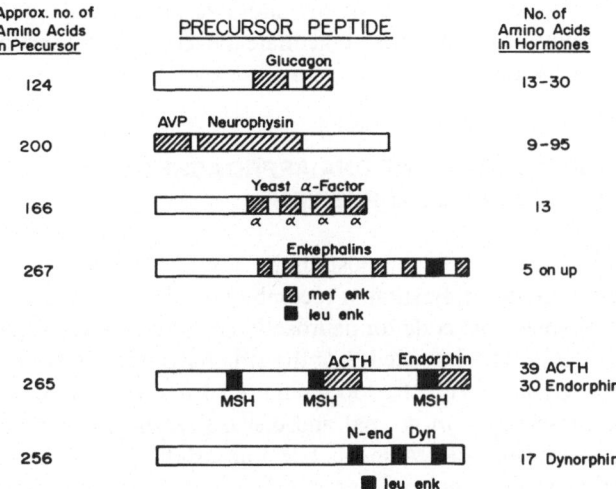

Figure 6. Polyprotein precursors of small bioactive peptides. (AVP) Arginine vasopressin.

of the precursor. In fact, differential processing of POMC has been observed in the anterior and intermediate lobes of mouse and rat pituitary.

4. POLYFUNCTIONAL PROTEINS

As pointed out in Section 1, the opioid peptide precursors belong to a growing class of proteins known as polyfunctional proteins or polyproteins. Models of the structure of some of the polyfunctional proteins that have been identified including precursors to the opioid peptides are shown in Fig. 6. In addition to the proteins shown in this Figure, it has very recently been found that the precursor to caerulein, an amphibian decapeptide with the same C-terminal structure as mammalian gastrin and cholecystokinin, contains two copies of caerulein (Hoffman *et al.*, 1983). Also, several peptides that control egg-laying behavior in *Aplysia* are synthesized in the same precursor protein (Scheller *et al.*, 1983). In some cases, polyproteins may serve to coordinate biochemical processes. For example, vasopressin and neurophysin II, the carrier or binding protein for vasopressin, are made in the same precursor (Land *et al.*, 1982). In other cases, polyproteins may coordinate complex behavioral responses. For example, ACTH, endorphin, and MSHs are released from POMC in the anterior pituitary in response to stress. ACTH stimulates the release from the adrenal cortex of glucocorticoids that speed up glucose metabolism. Endorphins may mediate the analgesic component of the stress response, and γMSH (derived from the N terminus of POMC) may potentiate the steroidogenic action of ACTH (Pedersen and Brownie, 1980).

5. IMPACT OF RECOMBINANT DNA APPROACHES ON STUDIES OF NEUROPEPTIDE GENE EXPRESSION

It is highly significant that almost all the advances described in this chapter came about through the application of recombinant DNA technology to the study of expression of genes that code for neuroactive peptides. These techniques have provided us not only with knowledge of the structures of the neuropeptide genes and precursors but also with the tools required for investigating regulation of expression of these genes in normal and disease states. DNA probes are now available for measuring mRNA levels and transcription rates of specific genes in different tissues. For example, the regulation of transcription of the POMC gene in the anterior and intermediate lobes of the rat pituitary by glucocorticoids has been studied in intact animals (Herbert *et al.*, 1981) using cloned POMC cDNA as a hybridization probe. *In situ* hybridization methods are being developed in several laboratories for measuring mRNA levels in specific cells in the endocrine and nervous systems. This technology will make it possible to study regulation of transcriptional activity of specific genes in individual neurons.

Perhaps the developments that will affect the future course of research on the nervous and endocrine systems the most are those that are taking place in the gene-transfer field. If the techniques of gene transfer can be refined sufficiently, they will allow us to determine the effect of specific alterations of gene structure on gene expression in the adult and in the developing organism. The impact of these techniques on developmental biology is likely to be very dramatic.

ACKNOWLEDGMENTS. This work is supported by the following grants (Edward Herbert, Principal Investigator): National Institute of Arthritis, Diabetes, Digestive and Kidney Diseases Grants AM16879 and AM30155 and National Institute of Drug Abuse Grant DA02736. Gerard Martens is supported by a postdoctoral fellowship from the Netherlands Organization for the Advancement of Pure Research (ZWO). Haim Rosen is supported by a Chaim Weizman postdoctoral fellowship. Michael Comb is supported by Molecular Biology Training Grant GM07759.

REFERENCES

Bradbury, A.F., Smyth, D.G., Snell, C.R., Birdsall, N.J.M., and Hulme, E.C., 1976, C fragment of lipotropin has a high affinity for brain opiate receptors, *Nature (London)* **260:**793–795.

Chang, A.C.Y., Cochet, M., and Cohen, S.N., 1980, Structural organization of human genomic DNA encoding the pro-opiomelanocortin peptide, *Proc. Natl. Acad. Sci. U.S.A.* **77**:4890–4894.

Cochet, M., Chang, A.C.Y., and Cohen, S.N., 1982, Characterization of the structural gene and putative 5¹ regulatory sequences for human pro-opiomelanocortin, *Nature*, **297**:335–339

Comb, M., Seeburg, P., Adelman, J., Eiden, L., and Herbert, E., 1982, Primary structure of the human Met- and Leu-enkephalin precursor and its mRNA, *Nature (London)* **295**:663–666.

Comb, M., Rosen, H., and Herbert, E., 1983, Structure of the human pro-enkephalin gene: Clustering of C_pG sequences and relationship to methylation, *J. DNA* **2**(3):278–290.

Cone, R.I., and Goldstein, A., 1982, A dynorphin like opioid in the central nervous system of an amphibian, *Proc. Natl. Acad. Sci. U.S.A.* **79**:3345–3349.

Felsenfeld, G., and McGhee, J., 1982, Methylation and gene control, *Nature (London)* **296**:602–605.

Goldstein, A., Tachibana, S., Lowney, L.I., Hunkapillar, M., and Hood, L., 1979, Dynorphin (1–13), an extraordinarily potent opioid peptide, *Proc. Natl. Acad. Sci. U.S.A.* **76**:6666–6670.

Goldstein, A., Fischli, W., Lowney, L.I., Hunkapillar, M., and Hood, L., 1981, Porcine pituitary dynorphin: Complete amino acid sequence of the biologically active heptadecapeptide, *Proc. Natl. Acad. Sci. U.S.A.* **78**:7219–7223.

Gubler, U., Seeburg, P., Hoffman, B.J., Gage, L.P., and Udenfriend, S., 1982, Molecular cloning establishes proenkephalin as precursor of enkephalin-containing peptides, *Nature (London)* **295**:206–208.

Herbert, E., Birnberg, N., Lissitsky, J-C., Civelli, O., and Uhler, M., 1981, Pro-opiomelanocortin: A model for regulation of expression of neuropeptides in pituitary and brain, *Neurosci. Newslett.* **12**:16–27.

Hoffman, W., Bach, T.C., Seliger, H., and Kreil, G., 1983, Biosynthesis of caerulein in the skin of *Xenopus laevis:* Partial sequences of precursors as deduced from cDNA clones, *Eur. Mol. Biol. Assoc. J.* **2**(1):111–114.

Hughes, J., Smith, J.W., Kosterlitz, H.W., Forthgill, L.A., Morgan, B.A., and Morris, H.R., 1975, Identification of two related pentapeptides from brain with potent opioid agonist activity, *Nature* **258**:557–579.

Kakidani, H., Furutani, Y., Takahashi, H., Noda, M., Morimoto, Y., Hirose, T., Asai, M., Inayama, S., Nakanishi, S., and Numa, S., 1982, Cloning and sequence analysis of cDNA for porcine β-neo-endorphin/dynorphin precursors, *Nature (London)* **298**:245–249.

Kilpatrick, D.L., Takashi, T., Jones, B.N., Stern, A.S., Shively, J.E., Hullihan, J., Kimura, S., Stein, S., and Udenfriend, S., 1981, A highly potent 3200-dalton adrenal opiate that contains both a Met- and Leu-enkephalin sequence, *Proc. Natl. Acad. Sci. U.S.A.* **78**:3265–3268.

Kurjan, J., and Herskowitz, I., 1982, Structure of a yeast pheromone gene (MFα): A putative α-factor precursor containing four tandem copies of mature α factor, *Cell* **30**:933–943.

Land, H., Schütz, G., Schmale, H., and Richter, D., 1982, Nucleotide sequence of clones cDNA encoding the bovine arginine vaso-pressin-neurophysin II precursor, *Nature (London)* **295**:299–303.

LeRoith, D., Liotta, A.S., Roth, J., Shiloach, J., Lewis, M.E., Pert, C.B., and Krieger, D.T., 1982, Corticotropin and β-endorphin like materials are native to unicellular organisms, *Proc. Natl. Acad. Sci. U.S.A.* **79**:2086–2090.

Li, C.H., and Chung, D., 1976, Isolation and structure of an untriakontapeptide with opiate activity from camel pituitary glands, *Proc. Natl. Acad. Sci. U.S.A.* **73**:1145–1148.

Mains, R.E., Eipper, B.A., and Ling, N., 1977, Common precursor to corticotropins and endotropins, *Proc. Natl. Acad. Sci. U.S.A.* **74**:3014–3018.

Nakanishi, S., Inoue, A., Kita, J., Kakamura, M., Chang, A.C.Y., Cohen, S.N., and Numa, S., 1979, Nucleotide sequence of cloned cDNA for bovine corticotropin -P- lipotropin precursor, *Nature* **278**:423–427.

Noda, M., Furutani, Y., Takahashi, H., Toyosato, M., Hirose, T., Inayama, S., Nakanishi, S., and Numa, S., 1982, Cloning and sequence analysis of cDNA for bovine adrenal preproenkephalin, *Nature (London)* **295**:202–206.

Owerbach, D., Rutter, W.J., Roberts, J.L., Whitfeld, P., Shine, J., Seeburg, P.H., and Shows, T.B., 1981, The proopiocortin (adrenocorticotropin/β-lipotropin) gene is located on chromosome 2 in humans, *Somat. Cell Genet.* **7**:359–369.

Pedersen, R.C., and Brownie, A.C., 1980, Adrenocortical response to corticotropin is potentiated by a part of the N-terminal region of pro-corticotropin/endorphin, *Proc. Natl. Acad. Sci. U.S.A.* **77**:2239–2243.

Roberts, J.E., and Herbert, E., 1977, Characterization of a common precursor to corticotropin and β-lipotropin: identification of β-lipotropin peptides and their arrangement relative to corticotropin, *Proc. Natl. Acad. Sci. U.S.A.* **74**:5300–5304.

Scheller, R.H., Jackson, J.F., McAllister, L.B., Rothman, B.S., Mayeri, E., and Axel, R., 1983, A single gene encodes multiple neuropeptides mediating a steroetyped behavior *Cell.* **32**:7–22.

Uhler, M., and Herbert, E., D'Eustachio, P., and Ruddle, F.O., 1983, The mouse genome contains two non-allellic proopiomelanocortin genes *J. Biol. Chem.* **258**:9444–9453.

Whitfeld, P.L., Seeburg, P.H., and Shine, J., 1982, The human pro-opiomelanocortin gene: Organization, sequence and interspersion with repetitive DNA, *J. DNA* **1**:133–144.

Chapter 19

Isolation and Characterization of DNA Sequences Coding for Mouse and Human β-Nerve Growth Factor

Axel Ullrich, Alane Gray, Cara Berman, and Thomas J. Dull

1. INTRODUCTION

Recombinant DNA techniques developed in recent years enable the efficient isolation and characterization of DNA sequences that code for specific genes. Isolated DNA fragments represent unique tools that permit the investigation of a diverse array of questions, such as determination of the size and coding potential of a specific messenger RNA (mRNA), elucidation of the structure and chromosomal location of a particular gene, determination of the relatedness of genes between different species, identification of sites of biosynthesis, regulation of gene expression, and, perhaps most important, identification of genetic alterations that may be responsible for genetic disease. Laboratory manipulation of gene sequences, employing chemical DNA synthesis and enzymatic methods, enables interspecies transfer of genetic information and even production of rare proteins of medical importance in organisms such as *Escherichia coli, Saccharomyces cerevisiae*, or mammalian cells in culture. We employed recombinant DNA techniques for the isolation and characterization of mouse and human DNA sequences that code for the biosynthetic precursor of the β-subunit of nerve growth factor (βNGF). Our strategy was to first clone a complementary DNA

AXEL ULLRICH, ALANE GRAY, CARA BERMAN, and THOMAS J. DULL ● Department of Molecular Biology, Genentech, Inc., South San Francisco, California 94080.

(cDNA) of mouse βNGF mRNA and subsequently use this cDNA as a tool to isolate the human gene that codes for the human βNGF protein, which has not yet been characterized.

2. ISOLATION OF COMPLEMENTARY DNA CLONES THAT CODE FOR MOUSE PRO-β NERVE GROWTH FACTOR

The cDNA cloning approach took advantage of the known 118-amino-acid-long sequence of the mouse βNGF subunit, employing synthetic oligonucleotide primers; the difference in NGF levels between male and female mouse submaxillary glands (MSGs) (Ishii and Shooter, 1975) was used as an additional means of identification. Three small portions of the βNGF amino acid sequence were chosen, and oligonucleotide pools complementary to all possible sequences that code for them were synthesized (Crea and Horn, 1980) (Fig. 1). Initial attempts to identify or isolate βNGF cDNA clones from an oligodeoxythymidylic acid [oligo(dT)]-primed, male MSG cDNA bank ($\approx$10,000 clones) failed, using the synthetic oligonucleotides as hybridization probes (Wallace *et al.*, 1981). This result indicated that while βNGF comprised approximately 0.1% of the protein in the male MSG (Harper and Thoenen, 1980), its mRNA was not of equal abundance. Therefore, the primer pool representing sequences closest to the carboxyl terminus of the protein [Fig. 1 (1)] was used to specifically prime reverse transcription of male MSG polyadenylic acid [poly(A)]-containing [poly(A)$^+$] RNA, to first enrich for βNGF-specific nucleotide sequences. Complementary DNA molecules greater than 200 base pairs (bp) in length were cloned into pBR322; 10,000 clones were screened using the 5'-^{32}P-labeled βNGF primer pool originally used in the cDNA priming as hybridization probe. Approximately. 8% of the clones gave a positive signal under high-stringency hybridization conditions. We assume that the remaining 92% of the primed cDNA bank resulted from self-priming as well as from priming by trace amounts of oligo(dT) eluted during preparation of poly(A)$^+$ MSG RNA. S1 nuclease treatment during the cloning procedure may also have removed some of the 3' terminal primer sequence, resulting in fewer detected positive clones.

Clones scored as positive in the first screen were rescreened using as hybridization probes radiolabeled primer pools (2) and (3), derived from sequences upstream from oligonucleotide pool (1) (Fig. 1). In addition, [^{32}P]-cDNA primed with pool (1) from male or female poly(A)$^+$ MSG RNA was used as probes on duplicate filters. Nine male-specific clones that hybridized with oligonucleotide pools (2) and (3) were identified. Restriction enzyme analyses demonstrated that all nine had common *Hae*III and *Hinf*I fragments. The clone that contained the longest DNA insert (pmβN-9G1; 716 bp) was sequenced in its entirety. The amino acid sequence deduced from the nucleotide sequence contained the ex-

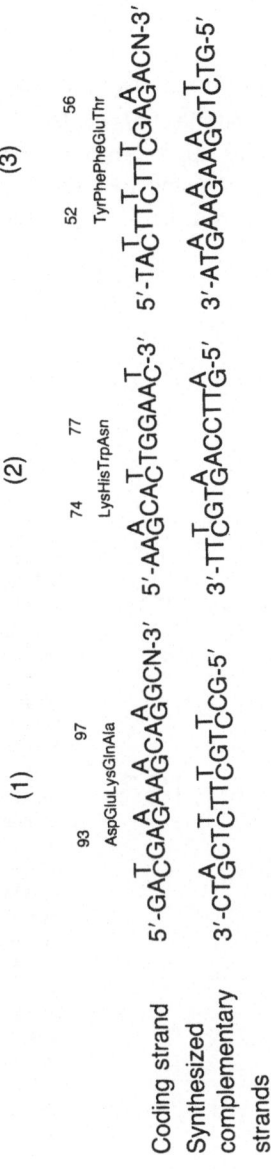

Figure 1. Synthetic oligodeoxyribonucleotides used for cDNA priming or colony hybridization or both. Primer pools (1) and (3) consist of 16 14-nucleotide-long oligonucleotides, which were synthesized separately (Crea and Horn, 1980) and are complementary to the only possible nucleotide sequences that code for the amino acid sequence between positions 93 and 97 or 52 and 56, respectively, of βNGF (only the first two nucleotides of Ala 97 were used). (N) A, T, G, or C. Oligodeoxynucleotide pool (2) was derived from position 74–77 in the βNGF amino acid sequence and was synthesized in two pools of four sequences.

pected βNGF sequence in one translational frame in addition to an amino-terminal pro-sequence (see Fig. 5).

To determine the size of the complete βNGF mRNA, we used Northern blot hybridization (Fig. 2) (Lehrach *et al.*, 1977) as well as primer extension analysis (Fig. 3). A 665-bp-long [³²P]-DNA fragment that included βNGF as well as propeptide sequences (pmβN-9G1 *Hga*I–*Hinc*II in Fig. 3) hybridized to a male-MSG-specific RNA species approximately 1300 nucleotides long (Fig. 2). A primer extension experiment using two short, double-stranded, 5′-end-labeled restriction fragments localized the 5′ end of the βNGF mRNA to about

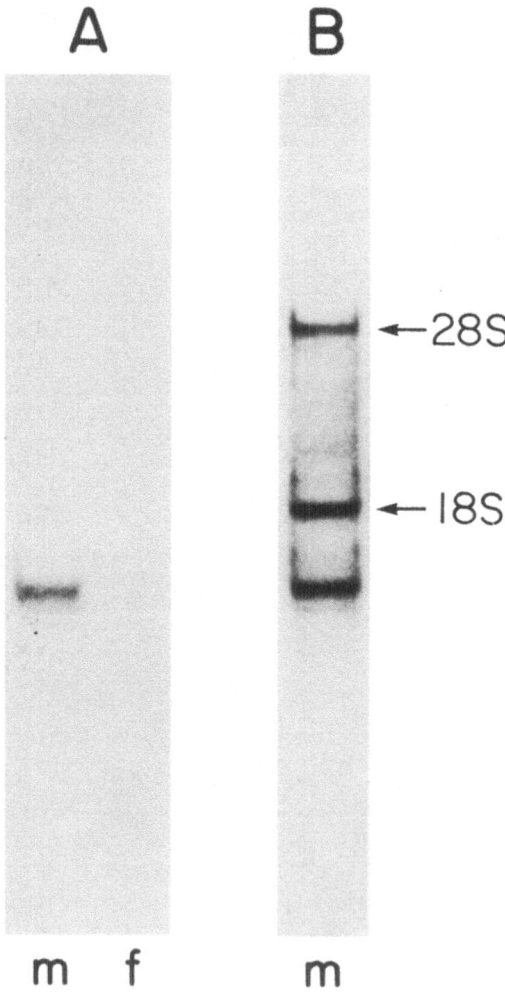

Figure 2. Northern blot hybridization analysis of male and female MSG poly(A)⁺ RNA. Male and female MSG poly(A)⁺ RNAs, purified as described (Ullrich *et al.*, 1977) from adult (≈30 g) Swiss Webster male or female mice, were separated on a 1.2% agarose–formaldehyde gel and transferred to a nitrocellulose filter for Northern blot analysis as described (Lehrach *et al.*, 1977). (A/m) Male MSG poly(A)⁺ RNA (2 μg); (A/f) female MSG poly(A)⁺ RNA (2 μg). A 665-bp *Hga*I–*Hinc*II fragment of clone pmβN-9G1 was radiolabeled (Taylor *et al.*, 1976) and used as a hybridization probe. (B) Hybridization with a mixture of the probe used in (A) and reverse-transcribed, radiolabeled rat ribosomal RNA. The ribosomal RNA reverse transcription used short pieces of calf thymus DNA as primer and was carried out with α-³²P-labeled nucleotides (Taylor *et al.*, 1976). The detected hybridization reflects the presence of small amounts of ribosomal RNA in our poly(A)⁺ RNA preparation and provides an internal standard. Rat ribosomal RNAs, 28 S and 18 S, are 4718 and 1874 nucleotides in length, respectively (I. Wool, personal communication).

230 bases upstream from the 5' end of the pmβN-9G1 cDNA fragment, leaving approximately 370 bases downstream from the 3' end of our clone (data not shown). All but approximately 30 nucleotides of the missing 5' sequences are contained in clones pmβN-16F7 and pmβN-21B5, which overlap each other, and clone pmβN-9G1, isolated as described in the Fig. 3 caption.

To obtain cloned cDNA that included the sequences from the 3' end of the primer sequence downstream to the 3' poly(A) sequence, we first enriched for βNGF mRNA by fractionating total poly(A)$^+$ male MSG RNA on a preparative urea–agarose gel (Lehrach et al., 1977). The largest-size fraction containing sequences that hybridized to a βNGF cDNA probe was used for oligo(dT)-primed cDNA synthesis and cloning. The screening of 4400 clones resulted in 4 positive hybridization signals. Nucleotide sequence analysis of clones pmβN-12E4 and pmβN-8B3 added 239 nucleotides to the 3' coding and untranslated sequences. Although oligo(dT)-primed, none of our clones contained the entire 3' untranslated region of βNGF mRNA due to incomplete synthesis of the second DNA strand or to extensive S1 nuclease treatment; however, Northern blot analysis (Fig. 2) indicated that the poly(A) sequence could not be far beyond the sequences we cloned ($\approx$150 bp).

2.1. Characteristics of Mouse β-Nerve Growth Factor Complementary DNA

The 1164-nucleotide cDNA sequence (see Fig. 5) contains a long (327-amino-acid) open reading frame that includes the known amino acid sequence of mouse βNGF (Angeletti et al., 1973a,b) (amino acids 1–118 in Fig. 5). The predicted βNGF sequence differs at one position from that determined by amino acid sequencing procedures: we predict an aspartic acid residue, rather than asparagine, at amino acid 30 (Angeletti et al., 1973a,b). The βNGF sequence is flanked at its amino and carboxyl termini by polypeptide extensions that presumably represent portions of pre-pro-βNGF precursor. The boundary sequences include basic amino acids (Lys-Arg, Arg-Arg) at both ends, which are present at precursor processing sites in other systems (Noda et al., 1982; Kakidani et al., 1982; Nakanishi et al., 1979; Amara et al., 1982). Processing at the amino-terminal boundary differs from that at the carboxyl-terminal boundary in that it excludes both basic residues from the mature polypeptide, while cleavage at the carboxyl-terminal end appears to occur between the two arginine residues. The resulting mature βNGF chain possesses a carboxyl-terminal arginine residue, thought to be essential for the stability of the 7 S NGF complex (Smith et al., 1968). The entire carboxyl-terminal propeptide is only 2 amino acids in length, which is unprecedented in other prohormone systems.

While the length of the carboxyl-terminal polypeptide extension is readily determined by the occurrence of an in-frame termination signal, determination of the size of the amino-terminal region is more complex, since no amino-terminal polypeptide sequencing information for *in vitro* translated pre-pro-βNGF or pro-βNGF is yet available. Evidence has been presented for the existence of a 22,000-dalton biosynthetic mouse pro-βNGF precursor (Berger and Shooter, 1977). The nucleotide-sequence-predicted precursor is longer than that previously detected; as described in Section 2.2, the entire pre-pro-βNGF sequence is predicted to have a molecular weight of 27,000, the pro-sequence is predicted to be 25,000 daltons, and considering the presence of pairs of arginine residues, processing intermediates of 21,500 and 18,000 daltons may exist within the cell.

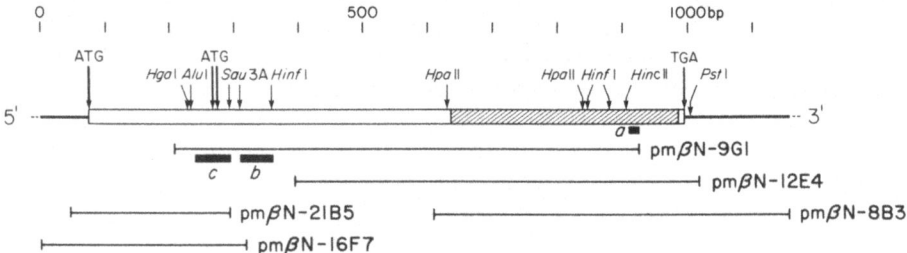

Figure 3. Construction and identification of bacterial clones that contain mouse βNGF cDNA sequences. Primer pool (1) (Fig. 1) was used to specifically prime the synthesis of cDNA on male MSG poly(A)$^+$ RNA. A sample of 220 pmol (1 μg) of each pool of eight primers (440 pmol total, 2 μg) was annealed with 40 μg poly(A)$^+$ RNA in 50 μl 100 mM KCl by incubating for 4 min each at 90, 68, 42, and 37°C. [^{32}P]-cDNA was synthesized in a 100 μl reaction in 50 mM Tris, pH 8.3, 10 mM MgCl$_2$, 10 mM dithiothreitol, and 50 mM KCl. The reaction contained, in addition to the annealed primer–RNA mixture, 500 μM each of dATP, dTTP, and dGTP, 100 μM dCTP, 20 μCi [α-^{32}P]-dCTP [2000 Ci/mmol (Amersham)], 0.5 U/μl ribonuclease inhibitor (Bethesda Research Labs), and 90 U reverse transcriptase. After 60 min at 37°C, the reaction was boiled for 3 min, quenched on ice for 1 min, and centrifuged (60 sec, 12,000g). The supernatant was diluted with an equal volume of H$_2$O, and double-stranded (ds) cDNA was synthesized with 15 U Klenow DNA Polymerase I (Boehringer) for 18 hr at 12°C. After phenol–chloroform extraction and ethanol precipitation, the preparation was digested with 1000 U S1 nuclease (Miles Laboratories) in 150 μl for 1 hr at 37°C. After phenol–chloroform extraction and ethanol precipitation, the cDNA was fractionated by electrophoresis on a 5% polyacrylamide gel. Complementary DNA greater than 200 bp was electroeluted and cloned into pBR322 as described (Wickens *et al.*, 1978). A sample of 10,000 colonies was screened by filter hybridization (Wallace *et al.*, 1981; Grunstein and Hogness, 1975); 1 μg of primer pool (1) was phosphorylated with 200 μCi [γ-^{32}P]-ATP [5000 Ci/mmole (Amersham)] and polynucleotide kinase (P-L Biochemicals). Filters containing the 10,000 clones were incubated with approximately 1 × 10^8 cpm of the ^{32}P-labeled probe at room temperature for 18 hr in a hybridization mix [100 mM Tris HCl, pH 7.5, 0.9 M NaCl, 6 mM EDTA, 1X Denhardt's solution (Wallace *et al.*, 1981), 100 μM ATP, 1 mM sodium pyrophosphate, 0.5% Nonidet P-40, and 0.1 mg/ml yeast RNA (Sigma R-6750)]. Filters were washed three times for 30 min in 6X SSC (0.9 M NaCl, 0.09 M Na$_3$-citrate) at 42°C and exposed to X-ray film for 16 hr at −70°C with an intensifying screen (Dupont Cronex Lightning Plus). Approximately 8% of the colonies (830) gave positive hybridization signals. Three sets of identical filters were prepared from the positive colonies identified in the first round of screening. Filters were hybridized as described above, using radiolabeled pools (1)–(3) (Fig. 1). Nine colonies hybridized with all three oligonucleotide pools. Plasmid DNA was

2.2. Localization of the Initiation Methionine Codon and Signal Sequence

Three methionine residues are candidates for designation as the protein synthesis initiation codon (amino acids -187, -121, and -119); however, several factors strongly implicate amino acid -121 of our sequence as the actual initiation codon employed. The methionine codon closest to the 5' end of mammalian mRNAs is usually used to initiate protein synthesis; however, there are many exceptions to this general scheme (Kozak, 1981). Since βNGF is a secreted protein, the initiation codon is likely to be followed by a signal sequence for cotranslational transfer of this polypeptide into the lumen of the endoplasmic reticulum (Blobel *et al.*, 1979). However, the stretch of amino acids that follows the methionine residue at position -187 (closest to the 5' end) contains a high percentage of polar and charged amino acids and bears no resemblance to any previously described signal sequence (Blobel *et al.*, 1979). On the other hand, amino acids between methionine residue -121 or -119 and residue -104 represent an excellent candidate signal sequence. The 16–18 amino acids that follow residue -121 or -119 include a stretch of 6 completely hydrophobic amino acids (Ala-Phe-Leu-Ile-Gly-Val). Cleavage by signal peptidase could occur between

---

isolated, and the clone with the largest insert was determined by restriction enzyme analysis. The 716-bp insert of clone pmβN-9G1 was completely sequenced by the Maxam–Gilbert method (Maxam and Gilbert, 1977). For the isolation of clones containing complete 5' sequences, we employed two restriction fragments isolated from pmβN-9G1. The ds 58-bp *AluI*-*Sau*3A fragment (*c*) and the 50-bp *Sau*3A–*Hinf*I fragment (*b*) were used separately to prime cDNA synthesis. Approximately 0.2 μg of each fragment (≈6 pmol) was phosphorylated with [γ-^{32}P]-ATP (Amersham) and polynucleotide kinase to a specific activity of approximately 0.5 × 10^8 cpm/μg. Annealing and ds cDNA synthesis procedures were basically as described above. Priming was, however, more specific, resulting in a detectable radioactive band in a polyacrylamide gel for each priming reaction. The bands were electroeluted from gel slices. After phenol–chloroform extraction and ethanol precipitation, the cDNA was cloned into pBR322. A 123-bp *AluI*–*Hinf*I fragment of pmβN-9G1 was used as a radioactive hybridization probe. Clones pmβN-21B5 and pmβN-16F7 had the largest DNA inserts and were completely sequenced by the Maxam–Gilbert method. Clone pmβN-21B5 contained the entire priming-strand sequence (*c*) and a 250-bp cDNA insert, whereas clone pmβN-16F7 contained only the last 6 bases of the priming-strand sequence (*b*) and a 320-bp cDNA insert. Clones pmβN-8B3 and pmβN-12E4 were isolated from 4400 clones by hybridization with a 216-bp mouse βNGF *Hpa*II fragment, which were obtained by oligo(dT)-primed cDNA synthesis on a size fraction of male MSG poly(A)$^+$ RNA (≈800–1500 bases). RNA fractionation on a urea–agarose gel (Lehrach *et al.*, 1977) and cDNA cloning were as described (Wickens *et al.*, 1978). Neither of these cDNA clones contains the entire 3' untranslated sequence of mouse βNGF cDNA. (*a–c*) Positions of DNA fragments used to prime cDNA synthesis (see above). The upper line and boxed region represent the entire region of mouse βNGF cDNA covered by our clones. The base-pair scale begins arbitrarily at zero, which depicts the beginning of our nucleotide sequence. The boxed region encloses the open reading frame, and the hatched area codes for mature βNGF. The three potential sites of translation initiation are indicated (ATG), as is the TGA termination codon. Only those restriction sites that were used to generate specific hybridization probes or primers, and are mentioned in the text, are shown.

the small amino acid alanine -104 and the glutamic acid residue at position -103. Signal peptidase cleaves after an identical alanine–glutamic acid sequence to leave an analogous amino-terminal glutamic acid residue in the case of pre-α-lactalbumin (Lingappa *et al.*, 1978). Because methionine -121 follows Kozak's rule in that it possesses a purine nucleotide 3 residues upstream from the AUG codon (Kozak, 1981), we favor amino acid -121 over methionine -119 as the actual initiator methionine. It seems most likely that methionine -121 is used for translation initiation of mouse pre-pro-βNGF, which would result in a 27,000-dalton preprohormone and a 25,000-dalton pro-βNGF if signal peptide processing occurs at residue -104. The availability of the sequence reported here and the corresponding cDNA and chromosomal gene clones will permit further elucidation of the specificity and order of intracellular processing events for the βNGF precursor protein.

Assignment of methionine -121 as the amino-terminal residue of the initial βNGF translation product leaves an estimated 300-nucleotide 5′ untranslated sequence, which is unusually long but has also been observed for other mRNAs transcribed in the MSG (Gary *et al.*, 1983).

3. ISOLATION AND CHARACTERIZATION OF THE HUMAN CHROMOSOMAL β-NERVE GROWTH FACTOR GENE

A human gene library consisting of 15–20 kilobases (kb), partial *Hae*III–*Alu*I human fetal liver DNA fragments inserted into λCharon 4A (Lawn *et al.*, 1978), was screened using a 665-bp mouse βNGF cloned cDNA fragment (pmβN-9G1 *Hga*I–*Hinc*II in Fig. 3) as a radioactive hybridization probe. A total of 27 hybridization-positive recombinant phages were plaque-purified and partially characterized by *Eco*RI digestion; six different types of restriction patterns were identified. Each pattern category shared restriction fragments and thus appeared to overlap the same genomic region. Clone λhβN8 was further characterized by physical mapping and nucleotide sequencing; Fig. 4A shows a physical map of clone λhβN8 that was derived by multiple restriction digests, DNA sequencing, and Southern blot experiments. An 8700-bp nucleotide sequence derived from subcloned, overlapping *Eco*RI and *Hind*III fragments is shown in Fig. 4B.

Regions within phage λhβN8 that are homologous to our mouse βNGF cDNA sequence were identified by Southern blot hybridization experiments (Southern, 1975), followed by nucleotide sequence analysis (Maxam and Gilbert, 1977) and translation of the sequence into amino acid sequence information. Two regions of extensive homology were identified, 124 bp and 738 bp in length, respectively, which are separated by a 6.5-kb intervening sequence. The 738-bp region encodes a protein that is highly homologous (90%) to mature mouse βNGF (see Section 4 and Fig. 5) and apparently represents the human βNGF protein sequence, which is thus far undetermined. Like mouse βNGF, the putative human βNGF sequence also contains two amino acid codons downstream from the presumed carboxyl terminus of the mature polypeptide, followed by a signal for translational termination (TGA). Farther downstream, one detects multiple AATAAA signals for mRNA polyadenylation (Proudfoot and Brownlee, 1976; Tosi *et al.*, 1981); however, the first one occurs 978 nucleotides beyond the translation-termination site. Unless the 3′ untranslated region of the sequence is interrupted by an intervening sequence, the length of the predicted transcribed region would be substantially longer than the mouse mRNA.

The human pre- and propeptide sequences upstream from the βNGF region are highly homologous to the mouse cDNA and identical in length. The 6.5-kb intervening sequence occurs 12 nucleotides upstream from the methionine residue we have designated as the translation-initiation site (-121). If our initiation-codon assignment is correct, the second homologous region of 124 nucleotides represents part of the 5′ untranslated region and is flanked by potential splice sequences (Sharp, 1981). Interestingly, we were unable to identify a region on phage λhβN8 or any of our other phage isolates that is equally homologous to the 5′ end of the mouse βNGF cDNA. Either the rest of the 5′ end of the human βNGF gene including regulatory sequences is separated from the 124-bp homologous mRNA sequence by an intervening sequence greater than 6 kb and is therefore not present on λhβN8 or human and mouse sequences have diverged more extensively in this region. A Goldberg–Hogness sequence (TATAAA) is present at nucleotide position 1362, with distantly homologous sequences downstream (1500–1560), so it is possible that the 5′-end sequences are more distantly related. However, use of the 5′ mouse cDNA probe on human genomic Southern blots at low hybridization stringency detected a homologous sequence elsewhere in the human genome, supporting the first interpretation (data not shown). Isolation of the human pre-pro-βNGF will be necessary to clarify this point. In addition, Southern blot hybridization experiments using human and mouse chromosomal DNA indicated the presence of a single βNGF gene per haploid genome (data not shown).

4. COMPARISON OF HUMAN AND MOUSE PRE-PRO-β-NERVE GROWTH FACTOR SEQUENCES

Comparison of mouse βNGF nucleotide and amino acid sequences with corresponding human sequences reveals high levels of homology. Figure 5 shows a comparison of 861 bp composed of the small and large exons of the human gene with the mouse cDNA sequence. The nucleotide sequences are highly homologous overall (86%). Although we have tentatively assigned the amino terminus for mouse pre-pro-βNGF to methionine -121, we have included the amino acid sequences of the open reading frame that precede amino acid -121 in both mouse cDNA and human gene for comparison. Most of the sequence upstream from amino acid -121 (40 amino acids) corresponds to the small exon in the human gene and may actually represent 5' untranslated sequences. The level of homology in this region (positions -126 to -166) is significantly lower than that overall (70%). Highly conserved regions within the putative pre-pro-βNGF amino acid sequence are the prepeptide (-121 to -104; 94.5%) and the sequences of mature βNGF (90%). Within the amino- and carboxyl-terminal pro-sequences, regions of more extensive divergence separate several areas of high homology between mouse and human. Interestingly, regions that surround two Arg-Arg dipeptides (positions -41/42 and -72/73) and two potential N-linked glycosylation sites at amino acid -8 and -53 are absolutely conserved, which may reflect the functional importance of these sequences. A third Arg-Arg dipeptide is present at the carboxyl-terminal end of the pro-βNGF sequence and appears to be a protease cleavage site in both human and mouse sequences for the removal of a carboxyl-terminal dipeptide. Within the mature βNGF sequence, the cysteine residues, important for the formation of proper tertiary structure (Angeletti *et al.*, 1973a), are conserved, as are 90% of the other residues. Amino acid differences are mostly conservative with respect to the chemical nature of the residues. Further evidence for the structural homology between mouse and human βNGFs has been obtained by expression of human βNGF sequence in *E. coli* and immunological detection of a protein of the correct size using an anti-mouse βNGF antiserum (Ullrich *et al.*, 1983 (in prep.)).

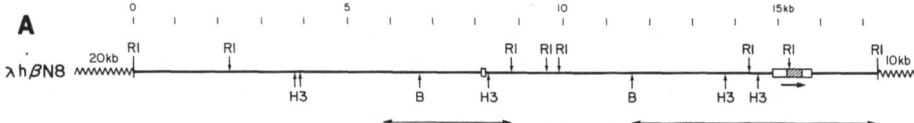

Figure 4. Physical map (A) and partial nucleotide sequence (B) of the human βNGF gene. (A) A human gene library (Lawn *et al.*, 1978) was screened (Grunstein and Hogness, 1975) using a radiolabeled (Taylor *et al.*, 1976) 470-bp *Rsa*I fragment of pmβN-9G1. Clone λhβN8 was analyzed by restriction enzyme cleavage with *Eco*RI (RI), *Hind*III (H3), and *Bam*HI (B); cleavage sites are marked with vertical arrows. Of the 17.3-kb human DNA insert, 15 kb were sequenced; horizontal arrows below the map represent regions for which the sequence is presented here. Zigzag marks represent the λCharon 4A phage arms. The boxes demarcate presumptive mature mRNA regions; the hatched area of the larger box codes for mature βNGF, which would be transcribed from left to right (→). The size scale at the top indicates kilobase pairs, from the 5' end of the phage insert. (B) The nucleotide sequence was determined from pBR322 subcloned *Eco*RI and *Hind*III fragments (Maxam and Gilbert, 1977). The numbers at right denote nucleotide numbers. The regions homologous to the mouse βNGF cDNA clone are translated, and the region that corresponds to mature human βNGF is boxed. Potential initiation methionines are underlined with dashed lines, at positions analogous to the mouse sequence (see Fig. 5). The position of a transcription initiation sequence, TATAA (double underline), is also shown. Three possible AATAAA polyadenylation signals are single-underlined. The nucleotide sequence of 2.8 kb of the intervening sequence is not shown due to space considerations, but was completely determined.

```
TACTTGCTTTAGGAAGTTAAGTATTATGTCTTTATTCTCCTTTGTGTCCCTAGCACCTAACACTTAAAACAGTGGCCAGCACAGGACCTGCAAGTTTAAGT          100
GTTTAATTAATGAAATAAATGAATCCCAATTTTGGGATGAGAGAAAGCACTACTTAAGCATCTAGTAGCAATGCAGCCTGGAAAACATTCAAAGTCACGG          200
AATCTCAGATGATCAGAGCCAAAGGGGACCTTAGCTGTCATCTGTGCCAGCTTCTTATCCTATAGAGGGAAAAGCTCAAAGATGAAATGAATCTCCTTCT          300
ATACAGGAGAAGCTCAGAGTGAACTGAATCAGAATGCGGGTGTGTGGGTTCCAGCCTGCAACCTTTCAGGTTTAGCCAAACACCCAGATGAAGGGTTTGT          400
GGACTAGACGAAACCATCTTCCCATGAGTAATGGGACCCAGATATGCCCACCTCCTTACCCTGGGACACGCATTCTCCCCTCTCCCATGCTAACTCCAAC          500
CCTGGGAGAGCATGAAAATGTTCTTTGTCACAGAATGTAACCTTTAAAGAGTGTCTGAGTATGCATTTCATCACTAGCCTTCAACCCCAATTGAGTATTG          600
AAAGGTTTTTCTGGTACTTTCTGGAGCAAGAAGACTATTTTGAGCAAGATGGGAAAGGAAGAAGAATGGGACATCCCAGGGCTTAATTTCATGATTTCT          700
AGTAACTTGAAGATCACTTTAGAGGTCCTTGCTACCTCCCCATTCTCCAACTCCTCTTCGTGGTTGGCGAATTTGGGGGAGCGATGGTGGCTTTTCTGACA          800
TTTGCTTTCATAGCACAAGCTGAGAGGGAGTTGGATGAAGATATGTGTGGGGGATCCACGCTGGAAAAAGATATTACAGGGAGAAGATTTTTTTGAAGTT          900
GAAGAGAGAATACGGACAGGAAAGTTAAGATGTCATTCTAGAACTTTATTGGGGAGGCATCTCCACCCTACAACAAATTCTGTGATGGACATAATCATTC          1,000
ATTCATTTATCCGTAAATATCACCCTCTTGTTCAAAGCCCTCCACTGCCTTCCTAATATCCTGAAGATAAAACCATAGCTCCTTTGCTGTGTCTCTGTAGA          1,100
CCTGGCTCTTCCTGGCTCTCCAGCTCATTTTCTAGGTCTCGTTACTTCATGCTCAGAACCTTTGTCTTGTTTCTAGCTCAGGGCCTTTGCACTTGTTCTT          1,200
GCTGCCTAGAATGTTCTCTCCCTCATTCCTTCTCATCCTCCAGATCTCAACTTGAAGGCCATCTCCTCAGAGCTCCTCTGCTGAGCGTCCTGTCTCACAGTG          1,300
GCCCCTCGATACATCCCTGCAGTTGCTCTCTATCATCAGACCCTGTAATTGCCTTCATGGCATATAAAGAATCTGGAGTATCTTGCTTATTTACACAACAC          1,400
TGTAAGCTCCATGAGAGCAGAGCCTTGTTTGTCTTGGTTGTTTACTGCTGCTCAGCACCAAAAACAGTGCCTGGCACATAGTCGGTGCCCAGAAAAATTGTG          1,500
AATGAATGAAGTGCCTACATAGATTACATTATAGAAGTGAGAGGAGAATAGAAAACTTCCATTGTTTCTAGAAACTACAGCCTAAAATTGATTTTTTAAA          1,600
ATTGTATCAGCTCCATAGCTTCCAATCCTAAAATCTGCCTTTCAGTGTGGTACTCTGAGATTCCTCTGTGATTCTGTGGAGAGCTCCACATTCTCTCTCAA          1,700
ATGGTCAGTCTGTCTTATTTGTCACCATTACTCATCTGCATTTTATCAAAGCACCAACTTGCTCTGAATTGTCAGGGATTTTGCGTCTGTATAAGGTAT          1,800
TTTAGGCTGGTTCAGAGTTGGATCTGTTATGTCTGCATGTGTAATGTACTGAACAATTTCTATTTGATGCCAGATTAGGGATCTGCTGGGGGCAAGACTT          1,900
TGGCATGTGTTCTAGAAACACCTGCACTAGGTGCAAGATCAGCCATGGACTGTGTCCAGGCTGAAACCAAAAGGTATGGCGCAAGAGTGAGAGGCAGGTGC          2,000
CACCACAGGACCATGAGAGGCCAAGCTCCGGTATAATTTTGTAGACCAAATTCTAGCTCCTTCCTGGGCCTTGATGCTGGTAAAATCCCAGAACTCAAG          2,100

                                                                                              yVa
GAAATGGAATTTGTCCTATTGGCACATGCCTCCCCCACTGTGTAGGGCACAGGGAATGTGGTGAGGTACAGTCTAATGCCAGCTCTCCCCCTCCACAGAGT          2,200
1LeuAlaSerGlyArgAlaVaalValAlnGlyAlaGlyTrpHisAlaGlyProLysLeuSerSerAlaSerGlyProAsnAsnSerPheThrLysGlyAlaAla
TTTGGCCAGTGGTCGTGCAGTCCAAGGGGCTGGATGGCATGCTGGACCCAAGCTCAGCTCAGCGTCCGGACCCAATAACAGTTTTACCAAGGGAGCAGCT          2,300
PheTyrProGlyHisThrGlu ·
TTCTATCCTGGCCACACTGAGGTAAGTGCCTAGGCGCTTGGCCTTGCCAAGGTCCTCCCTCTGCAGCTGCCAGAAGCAGGAGTCCCAAGTGACAGGACCT          2,400

GAGAGGGCAAGTCAGAACCAACTGCTGAGCAGCAGGGGCCTAGAGAAGCTTTCAGTGGTGCAGCAAATGCAGCTCCCCTAGCCCACTCAGGAAAGGAAAT          2,500
CTTTTGTGCATGCAGGTTGGCCCCATGGGCCCCAGGGACATATCAGGAGTCTGAGCTGCATCAAGGGTCTAAAGTTGCCGAAGCACAGGACCTCGAATAG          2,600
ACAGCGTCAGTCAGACCCTCCACACAAACTCTCTGGTCCGATGGGACTTGCAAAATGGAAGGTTCACAGCCACTCTTTGTGCCCTTCACCAGTG          2,700
GAGTAGAGGGGTAGGGAAGAGTTCCAGATCTGCACACAGGAGGTGTGGCTCCGAGTTCCACTTCTGTTTGCTGTGTACCTGTGATGCAATTACTTAATCT          2,800
TACTACATCTCTGGTTTCTCATTGTTAAAGATAATATGGCCAACCTCCCAGGTTTGTGATGGTTAATTAATAAAAATATACCCAATAAAATAATA          2,900
ATTACTATTATTCTCAGCTTCCATTCATGTCAAGGTATGTC----2.8kb----GGCAGAGCAGTCCCTACTGGCTTACAGACAAGGGACTTGGCACC          3,000
AGCTCCTTTGAGCAGGGTGATATGATCAGAGCAGGCTTTAAGAAGCTAAATGGGGTGGCCAGGGATAGGAGGAATTGGAAGAAGAGAGAGATGAGGGTCCAG          3,100
GATATCAGGTGGGAGACTGTTCCCATGACCTGAACTAGGGGCCCAGAGATGAGAATGGAAGATAAAATACAGTTTCAAGATAATGCTACCAAACCAGAAC          3,200
CAAAAGGGCTGGATGGAGGTCAGGGAGGAAAAAGACAGAGGGATAAGAGCAGGTATGGAGCTAGTGGATGAATGATGATGGTTCCAGTAACAGAAAAAGG          3,300
AAACTGTGAAGAAGCGTGTTTAGGGAGATGAGGGAGAGGGGATCAAGAGAGATGGGTAGCTGGTCGTGTGAAAACTATCATCAAATGGTTTCAAAATGGGGCTGTTGC          3,400
TATTGTATGCAGAGAAGGAAAGATGGGGAGAAGATGATCCTAATTATGATTATGGAAGTATGTAAGATTTTTGTTTTGGTAGTCTTGCATTTAAAATGCCAGTA          3,500
TCTATGGAGAATTGCCCTCAGGAAGCAGGAGGGCAGAACTAGCTTCAAAAGACTAGTTGATTGATTAGAGATCCAGGTATGGGAGTCAATTGCACAGAAATTTC          3,600
GATGGGCTCCTTAGAAGAGGAGGAGGCCCACCAAAGGAAGCCAAGAATCCAAGCTTGGAACCTCTGGTAATGGGGAATAGGGAGAGGTATGTCCCCTTATG          3,700
TCCATCGAAACCCTTTTATTTAGTTAATTTCATTTTCCTCTTGAAATTCTAGCAAGTGTTACTAATTTGGATATAAAGGGAATGAATTTCAGTTTTCTTTC          3,800
TTCTCCTTTAATGTATCTGATAAGAAGATATTATGTACCGGAAGAACAAAGAAGTGAAACATGAAAAGGGCAAAACAGAAAAGTAAAAGGTTTAGATTTGCTGCCA          3,900
ACTGAATATTCAGTCCGCTAAGAGTTGACCGGACATTTATCGCAAGACCATTAACAGATCTGTAATGGTCAAAACCCCAGATATGCCATTGCCTCCTTT          4,000
GCATGGTCAGCTGGAGAACTGGAGCAAACAGAAACATCTTCAAAGAAACCAAAGAAGAAAAGAATTTGGTGATTTCCTTCTTGGTATCTCACCAAAGAGC          4,100
TGTCTATTTCCAGCTGAATAAAAGCACAGTATGGAGGAGGAAGAATGTGTTGCCTTAAAACAGATTAGAAAAGCAGTGTGTAGCCATATGCACAAAGGCCA          4,200
AAAGAAACCACACACTGTGATGAAAGAAGGCAGATGGTGAGCTTCTCTTGCCACTCTCATTCATTCTTTCTTTCATTCATTCTCCTTAAATTTTTTGAG          4,300
TACCTACTACATGCCAGACACGGCAACTGGCCCTGGAAATATTGAAATGCAAATTGTGCAGCTTGCAAAGCCCTCTCAGGGAGACAGACAAGTTAATGAT          4,400
TATCCACAACCCTAGGAGAGGTGATAGGGTTATGCACAAACTGCTGTGGAACTCAACAAGGGGAAGTGGGCCAGATACACAAGTGGATGAGACACGGCTCCA          4,500
CCCTCGAGTACCTCCACGTCGGGGTGAGACAATAAAAATGGCTTTGAAGGAGGTCTGAGAGAGGTAAAACCGGCTGAAAGCACAAAAGACAAAGTCCTTG          4,600
ACCACAAGGAGACAGAAATAACTCATGATTTACACAAATTTTGTGAAGAACCTACTTCATCCTAACAACTACTCTGGGATAGTGTTTTTGCCCACCTC          4,700
CGATCAATGGGAACGCTAAGGCTCTGGGAAATTAAATGGGTTTCTTCTGCTAACAAGTAGTGGAGTAGGAATCTCAAACACGATTTTCTAACCCCCAGACC          4,800
CAATGCAGAGTTCTTGTCATTACACTGATTTAGTAGGAGAGAACACACACTCACACTCACATGAACACACACATATGCACATCAATTACGCTATTCT          4,900
GAAGTGCAGAATGGCCAAAAGATCATTTTTGAATTGGATTGATCTGAGAAGGCTTCGAGGACTAGGTCGTATCTGTGTGGAAAACGGAGCAGAATGACCCATA          5,000
AGAGTGGGAAGGGAGAGAGTGAGGAGAAAAGGATAGAATTGGAAGTTTGTAGAGACCTCTTAGCAGGAGATATTTCTGATCCTACCAGGAAGATCTATGGAAGCT          5,100
TTACAGGGGAGCTGACGTTTGCTACACATCTACAAGTATGCATAGGAGCTCCGCGGAGGCCAGTGAGAGGGCCTCCAAGAGCAGAACTAATTCCACAATT          5,200
ACTTGACCAAGTTGGGGGATTATTTGTGGGGTAACTGCAGTGACAGTGTCTTTGGGGGACAGTTAGAGCCATACCATTTTGATCTTACTAGCTACATA          5,300
AGACAAACAATAAAAAGAAAAGACATGCTTAAGAGTGAAGAGAAAGGGAGGAGAAAAGGGAGAGCGGATCATTTGTCATTTTCCTTAATATGTGAAATAA          5,400
TCAACTTTGGATTCTATTTCTGGTTCAGTTTTCATTTGATTTCAGTGCTTTAGTGTTAGTTTCATTTTCTGTAAATCAGTTTCCTTATATGTGAAATAA          5,500
ATATGATAAATCCTAATTGAACTCAATGAACTCATGAGAAGATAGAAGTGAACACATTTTAAAAACATCACACAAAGAGGAACTATTTATGTGGTCCACATT          5,600
TATATATGTGGGGTAGCGTCTGAAGAGGTGCCTGGACTAAGATGGTCCCAGAGCCACAAGGTTTTGCCCAAACATGACGCTTTTGTGAATTCATAACAAGG          5,700
GCTCCAAGTCACCAGATCTTAGAGCTGACCCAGTGCACTGTCTGAAAGGGGGTACCAGTTCTGAGGTCTCCCCAGCAGATCTTCCCCGTG          5,800
CCTTCCCAGAGGATTCAAAACTGTTGAGCAGGACGGCACCATCACATCAAGGCACAAGTGCAGGGAGAGGTGTTAAACTCTCCCCACCAACCTCCCTGG          5,900
TACACACATGGACACTTGCCACCTCCCTCAGCCGCCTTAAGCTTCAGAGAACTCAAAGGACTCTGTAAGTGATGTCTCCAAGCTCATATCGAACTACTGG          6,000
GCAAAATTTCAGGGGCTCTGTCACTTCCCTGGAGAAGCTCGGATGGGGTGACCACACATCCATACTGCCTGAGTCAGCCCCCAGGTTACGCCTGTTGTCCCG          6,100
GTATAACCATTGCTAGCACACCCTTTCCCTCTCAGAAGTGCCCCGGTTTGAATGAAACCTCTTCGTGATCCCCTTGGAGGTCAACTCTGAGGGACCCAGA          6,200

                                                                        ValHisSerValMetSerMetLeuPheTyrThrLeuIle
AACTGCCTTTTGACTGCATTTAGTACTCCATGAAGTCACCCTCATTTCTTTTTCATTCCAGGTGCATAGCGTAATGTCCATGTTGTTCTACACTCTGATC          6,300
ThrAlaPheLeuIleGlyIleGlnAlaGluProHisSerGluSerAsnValProAlaGlyHisThrIleProGlnValHisTrpThrLysLeuGlnHisS
ACAGCTTTTCTGATCGGCATACAGGCGGAACCACACTCAGAGAGCAATGTCCCTGCAGGACACACCATCCCCCCAAGTCCACTGGACTAAACTTCAGCATT          6,400
erLeuAspThrAlaLeuArgArgAlaArgSerAlaProAlaAlaIleAlaAlaArgValAlaGlyGlnThrArgAsnIleThrValAspProArgLe
CCCTTGACACTGCCCTTCGCAGAGCCCGCAGCGCCCCGGCAGCGGCGATAGCTGCACGCGGTGGCGGGGCAGACCCGCAACATTACTGTGGACCCCAGGCT          6,500
uPheLeuLysArgArgLeuArgSerProArgValLeuPheSerThrGlnProProArgGlnValAlaIaAspThrGlnAsnLeuAspPheGluValIGlyGly
GTTTAAAAAGCGGCGACTCCGTTCACCCCGTGTGCTGTTTAGCACCCAGACCTCCCCGTGAAGCTGCAGACACTCAGGATCTGGACTTCGAGGTCGGTGGT          6,600
AIaAIaProPheAsnArgThrHisArgSerLysArgSerSerSerProIlePheSerArgGlyGluPheSerValCysAspSerValSerValTrpV
GCTGCCCCCTTCAACAGGACTCACAGGAGCAAGCGGTCATCATCCCATCCCATCTTCCACAGGGGCGAATTCTCGGTGTGTGACAGTGTCAGCGTGTGGG          6,700
aIaGlyAspLysThrThrAlathrAspIleLysGlyLysGluValMetValLeuGlyGluValAsnIleAsnAsnAsnServalPheLysGlnTyrPhePheGl
TTGGGGATAAGACCACCGCCACAGACATTAAGGGCAAGGAGGTGATGGTGTTGGGAGAGGTGAACATTAACAACAGTGTATTCAAACAGTACTTTTTTGA          6,800
uThrLysCysArgAspProAsnProValAspSerGlyLysCysArgGlyIleAspSerLysSerLysSerHisTrpAsnSerTyrCysThrThrHisThrPheValLys
GACCAAGTGCCGGGACCCAAATCCCGTTGACAGCGGGTGCCGGGGCATTGACTCAAAGCACTGGAACTCATATTGTACCACGACTCACACCTTTGTCAAG          6,900
AIaLeuThrMetAspGlyLysGlnAIaAIaTrpArgPheIeArgIIeAspThrAlaCysValCysLeuSerArgLysSerAlaValArgArgArgAlaEnd
GCGCTGACCATGGATGGCAAGCAGGCTGCCTGGCGGTTTATCCGGATAGATACGGCCTGTGTGTGTGTCTGCAGCAGGAAGGCTGTGAGAGAGAGACCTGAC          7,000

CTGCCGACACGCTCCCTCCCCCTGCCCCTTCTACACTCTCCTGGGCCCCTCCCTACCTCAACCTGTAAATTATTTTAAATTATAAGGACTGCATGGTAAT          7,100
TTATAGTTTTATACAGTTTTAAAGAATCATTATTTATTAAATATTTTGGAAGCATCCTGTGCTGATGCTGGTTATTTTTTTTTGAGTAAAATCATCTGCAA          7,200
GTCTGAGGAAGATGCAGGGGGAATTGTCTGAAGCAACCCCCTGGCTCCTTCTAAGGCCCACCATGTAACCCTTACCCCACCCCCAGCCAGTGCTGCAACTTC          7,300
GGGCATGGGGCTGCCCTGGCCTCAGAGGGCACAGTCTCAGTCCTGGGGGCACTCGATAGACAGGAAGGAGGCAAGCTGTGTACCAAAGATCTCGTAGGCTGCTCATTT          7,400
TACCTAATCATTGGATTACCTGCCCACGGCCTGGAAGCATGGGCCCTGGCTCTCCCTTGGAGAAGTTAGTGGAGCCTCATCAGGAGGGCAGATCACCACT          7,500
GTAGAGGGGAGTGTGCTGTCTGGAGAGTTGTAACAGAGTAACTCATGGCGAAGTAAGATCTAGGGTTTCATGTTCAGCTGGCCTTAGTACTCGT          7,600
TTCAGGGGCCCACCATCAGCCACACGCCAGGGAAGGCAGACAGGGAAGAAGAACTTCTGAAAATATTTTTATTTCTTAAATTTGGATTGATTTTCATAGATGA          7,700
TTCCAGGGGCCCACCATCAGCCACACGCCAGGGAAGGCAGACAGGGAAGAAGAACTTCTGAAAATATTTTTATTTCTTAAATTTGGATTGATTTTCATAGATGA          7,800
AGGTGGGCGGCAGGACCACGGCTTTCTGCTTCTCCCAAACAGGGAAAATGAAACAAGGATGACAAAGAGAAAGGCCAGAGTGGATGTGAAGAAGGGGCA          8,100
GGGGAGGACAGGACGAGGAGGAGGGAGGGTGTATCTCTCATCTTTACCAGCACTCTCCAGCCTCCCAGGAAAGTTAGCAACCTGGGTGGTTCTTGTGGGG          8,200
GCTTTTTTGGATGTGCTAATAAACACTGACTCATCCAGAGCATCTTAGGAGTGAAGGTGACAGAGGGACTCCAGGGTTCAGCCTGCAGCAGGGCTCCACC          8,300
CAGCCTCACACGACCCCTTCCCCTGCAACTTTCATTTTGTATTTTCCTTGAAGCCACAATATTCCCCCAGACAGAGCACTTATATTAAGACAGAAAATT          8,400
TCCCCTCCTCATCCCTGCCTCTCCCCCCGAGGATGAAATGCTACACTGGTTTTCAGAGAACAAGGCCCAAATTGTCCATCCGAGAAAAGACAGCAAATGGT          8,500
CTTATTAGGGTAGAAGTATGATGAAATGTGCCCTGTTTTTTCTATCAGAGCCACTGCAATTGGAATAAGGTTTATGCTCAAGAAACACCAGCTAGAACA          8,600
TTGATTAGGGTAGAAGTATGATGAAATGTGCCCTGTTTTTTCTATCAGAGCCACTGCAATTGGAATAAGGTTTATGCTCAAGAAACACCAGCTAGAACA          8,700
TTCTGTGCTGAGATTCTAAGTGAGATAA-AGGCAT                                                                      8,735
```

Figure 4. Continued.

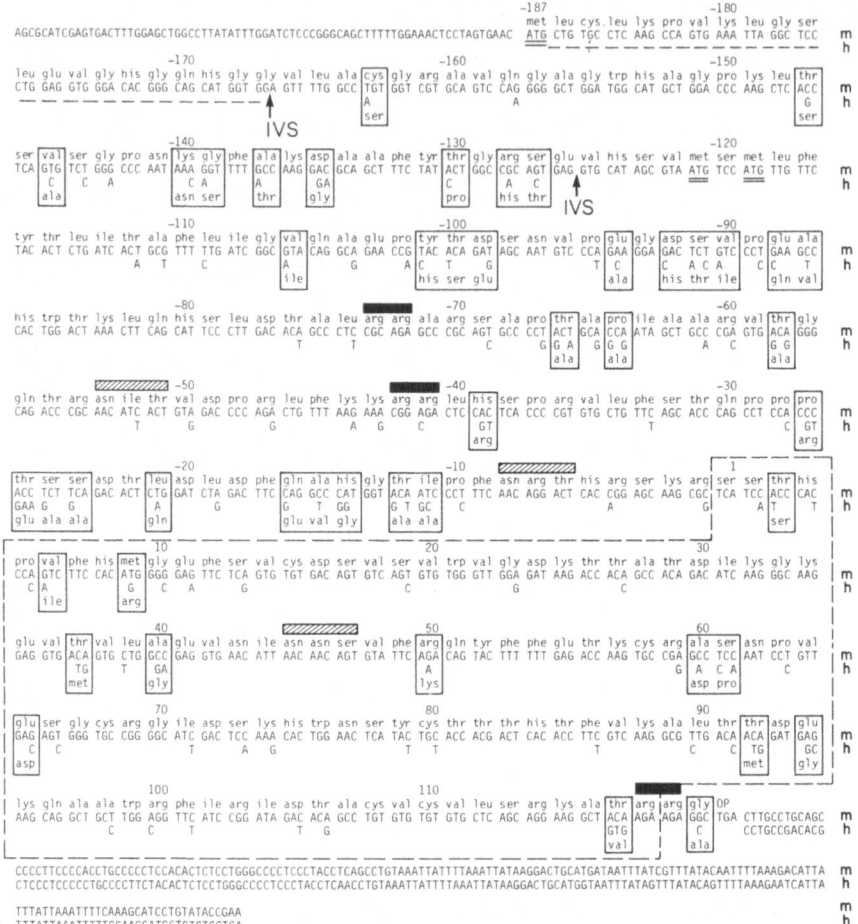

Figure 5. Nucleotide sequence of cloned mouse pre-pro-βNGF cDNA and comparison with human sequences. The mouse cDNA sequence was derived from overlapping clones (Fig. 3) by the procedure of Maxam and Gilbert (1977). Only the coding strand is shown; the longest open reading frame starting with a methionine residue is also presented. A termination codon is present upstream at the 12th nucleotide of the top line. Potential initiation condons are double-underlined; as discussed in the text, we favor methionine − 121. Amino acid positions are numbered, beginning with the first amino-terminal residue of the mature βNGFs, the sequence of which are enclosed by a dashed-line box. The mouse coding-strand nucleotide and amino acid sequences (m) are shown in their entirety; only differences are shown below for the corresponding human (h) sequences. The positions at which a nucleotide difference causes a change in the coded amino acid are enclosed in solid-line boxes. (▨) Potential N-linked glycosylation sites; (■) potential protease cleavage sites; (IVS) position of intervening sequences in the human gene; (OP) opal termination codon (TGA). Sequences upstream from the IVS at position − 166 are known only for the mouse cDNA; single underlines under amino acids − 167 to − 186 denote missing human sequences.

5. IS β-NERVE GROWTH FACTOR A MEMBER OF THE INSULIN GENE FAMILY?

Because of low levels of homology in the amino acid sequences of NGF and the insulin gene family members, insulin, relaxin, and insulinlike growth factors, it has been postulated that βNGF is a member of the insulin gene family (Bradshaw, 1978). The low-level homologies were detected in the B-chain of these proteins. We were therefore interested in comparing the nucleotide sequences of the βNGF and insulin genes (Ullrich *et al.*, 1980; Lomedico *et al.*, 1979). Our analysis revealed no striking additional homologies other than those previously detected; the level of homology is greater than random in the B-chain region, but does not further strengthen the hypothesis that these genes are related. Both genes contain an intervening sequence just upstream from the coding sequences in the 5' untranslated region; however, this is not a particularly unusual gene feature (Breathnach *et al.*, 1978). The absence of an intervening sequence within the βNGF coding region is unlike most of the insulin genes thus far isolated; however, the rat I insulin gene also lacks an intervening sequence in the coding region (Lomedico *et al.*, 1979), so the significance of this βNGF gene feature remains uncertain. Isolation and characterization of other genes in the insulin gene family will be necessary to shed additional light on this interesting hypothesis (Bradshaw, 1978).

ACKNOWLEDGMENTS. We are grateful to Mark Vasser for carrying out oligonucleotide synthesis and to Dr. Peter Seeburg, Joel Hayflick, and Dr. Ellson Chen for their sequencing assistance. We thank Lisa Coussens for laboratory support and acknowledge the contribution of Dr. Stacey Sias in the early phases of this work. We also extend thanks to Drs. Xandra O. Breakefield and Suzanne Pfeffer for their continued interest and useful discussions, to Jeanne Arch for her expert preparation of this manuscript, and to Robert Swanson, President of Genentech, Inc., for his interest and support.

REFERENCES

Amara, S.G., Jonmas, V., Rosenfeld, M.G., Ong, E.S., and Evans, R.M., 1982, Alternative RNA processing in calcitonin gene expression generates mRNAs encoding different polypeptide products, *Nature (London)* **298**:240–244.

Angeletti, R.H., Hermodson, M.A., and Bradshaw, R.A., 1973a, Amino acid sequences of mouse 2.5S nerve growth factor. II. Isolation and characterization of the thermolytic and peptic peptides and the complete covalent structure, *Biochemistry* **12**:100–115.

Angeletti, R.H., Mercanti, D., and Bradshaw, R.A., 1973b, Amino acid sequences of mouse 2.5S nerve growth factor. I. Isolation and characterization of the soluble tryptic and chymotryptic peptides, *Biochemistry* **12**:90–100.

Berger, E.A., and Shooter, E.M., 1977, Evidence for pro-β-nerve growth factor, a biosynthetic precursor to β-nerve growth factor, *Proc. Natl. Acad. Sci. U.S.A.* **74**:3647–3651.

Blobel, G., Walter, P., Chang, C.N., Goldman, B.M., Erickson, A.H., and Lingappa, V.R., 1979, Translocation of proteins across membranes: The signal hypothesis and beyond, in: *Society for Experimental Biology Symposium XXXIII: Secretory Mechanisms* C.R. Hopkins and C.J. Duncan, eds.), Cambridge University Press, pp. 9–36.

Bradshaw, R.A., 1978, Nerve growth factor, *Annu. Rev. Biochem.* **47**:191–216.

Breathnach, R., Benoist, C., O'Hare, K., Gannon, F., and Chambon, P., 1978, Ovalbumin gene: Evidence for a leader sequence in mRNA and DNA sequences at the exon–intron boundaries, *Proc. Natl. Acad. Sci. U.S.A.* **75**:4853–4857.

Crea, R., and Horn, T., 1980, Synthesis of oligonucleotides on cellulose by a phosphotriester method, *Nucleic Acids Res.* **8**:2331–2348.

Gray, A., Dull, T.J., and Ullrich, A., 1983, Nucleotide sequence of epidermal growth factor cDNA predicts a 128,000-molecular weight protein precursor, *Nature* **303**:722–725.

Grunstein, M., and Hogness, D.S., 1975, Colony hybridization: A method for the isolation of cloned DNAs that contain a specific gene, *Proc. Natl. Acad. Sci. U.S.A.* **72**:3961–3965.

Harper, G.P., and Thoenen, H., 1980, The distribution of nerve growth factor in the male sex organs of mammals, *J. Neurochem.* **34**:893–903.

Ishii, D.N., and Shooter, E.M., 1975, Regulation of nerve growth factor synthesis in mouse submaxillary glands by testosterone, *J. Neurochem.* **25**:843–851.

Kakidani, H., Furutani, Y., Takahashi, H., Noda, M., Morimoto, Y., Hirose, T., Asai, M., Inayama, S., Nakanishi, S., and Numa, S., 1982, Cloning and sequence analysis of cDNA for porcine β-neo-endorphin/dynorphin precursor, *Nature (London)* **298**:245–249.

Kozak, M., 1981, Possible role of flanking nucleotides in recognition of the AUG initiator codon by eukaryotic ribosomes, *Nucleic Acids Res.* **9**:5233–5252.

Lawn, R.M., Fritsch, E.F., Parker, R.C., Lake, G.B., and Maniatis, T., 1978, The isolation and characterization of linked δ- and β globin genes from a cloned library of human DNA, *Cell* **15**:1157–1163.

Lehrach, H., Diamond, D., Wozney, J. M., and Boedtker, H., 1977, RNA molecular weight determinations by gel electrophoresis under denaturing conditions, a critical reexamination, *Biochemistry* **16**:4743–4751.

Lingappa, V.R., Lingappa, J.R., Prasad, R., Ebner, K.B., and Blobel, G., 1978, Coupled cell-free synthesis, segregation, and core glycosylation of a secretory protein, *Proc. Natl. Acad. Sci. U.S.A.* **75**:2338–2342.

Lomedico, P., Rosenthal, N., Efstratiadis, A., Gilbert, W., Kolodner, R., and Tizard, R., 1979, The structure and evolution of the two nonallelic rat preproinsulin genes, *Cell* **18**:545–556.

Maxam, A.M., and Gilbert, W., 1977, A new method for sequencing DNA, *Proc. Natl. Acad. Sci. U.S.A.* **74**:560–564.

Nakanishi, S., Inoue, A., Kita, T., Nakamura, M., Chang, A.C.Y., Cohen, S.N., and Numa, S., 1979, Nucleotide sequence of cloned cDNA for bovine corticotropin-β-lipoprotein precursor, *Nature (London)* **278**:423–427.

Noda, M., Furutani, Y., Takahashi, H., Toyosato, M., Hirose, T., Inayama, S., Nakanishi, S., and Numa, S., 1982, Cloning and sequence analysis of cDNA for bovine adrenal preproenkephalin, *Nature (London)* **295**:202–206.

Proudfoot, N.J., and Brownlee, G.G., 1976, 3' Non-coding region sequences in eukaryotic messenger RNA, *Nature (London)* **263**:211–214.

Sharp, P.A., 1981, Speculations on RNA splicing, *Cell* **23**:643–646.

Smith, A.P., Varon, S., and Shooter, E.M., 1968, Multiple forms of the nerve growth factor protein and its subunits, *Biochemistry* **7**:3259–3268.

Southern, E., 1975, Detection of specific sequences among DNA fragments separated by gel electrophoresis, *J. Mol. Biol.* **98**:503–517.

Taylor, J.M., Illmensee, R., and Summers, J., 1976, Efficient transcription of RNA into DNA by avian sarcoma virus polymerase, *Biochim. Biophys. Acta* **4**:324–330.

Tosi, M., Young, R.A., Hagenbuchle, O., and Schibler, U., 1981, Multiple adenylation sites in a mouse α-amylase gene, *Nucleic Acids Res.* **9**:2313–2323.

Ullrich, A., Shine, J., Chirgwin, J., Pictet, R., Tischer, E., Rutter, W.J., and Goodman, H.M., 1977, Rat insulin genes: Construction of plasmids containing the coding sequences, *Science* **196**:1313–1317.

Ullrich, A., Dull, T.J., Gray, A., Brosius, J., and Sures, I., 1980, Genetic variation in the human insulin gene, *Science* **209**:612–615.

Wallace, B., Johnson, M.J., Hirose, T., Miyake, T., Kawashima, E.H., and Itakura, K., 1981, The use of synthetic oligonucleotides as hybridization probes. II. Hybridization of oligonucleotides of mixed sequence to rabbit β-globin DNA, *Nucleic Acids Res.* **9**:879–894.

Wickens, M.P., Buell, G.N., and Schimke, R.T., 1978, Synthesis of double-stranded DNA complementary to lysozyme, ovomucoid, and ovalbumin mRNAs: Optimization for full length second strand synthesis by *Escherichia coli* DNA polymerase I, *J. Biol. Chem.* **253**:2483–2495.

Chapter 20

Linkage Analysis in Familial Dysautonomia Using Variations in DNA Sequence in the β-Nerve Growth Factor Gene Region: A Beginning

Xandra O. Breakefield, Carmela M. Castiglione, Lisa M. Coussens, Felicia B. Axelrod, and Axel Ullrich

1. INTRODUCTION

The availability of cloned human DNA sequences combined with classic methods of genetic linkage analysis allows us to establish whether a mutation in a given gene produces an inherited disease state. In this chapter, we will describe how cloned DNA probes corresponding to the human gene for β-nerve growth factor (βNGF) can be used to determine whether an alteration in the structure or expression of this gene is the primary defect in the inherited neurological disease familial dysautonomia. Our preliminary findings are also presented.

XANDRA O. BREAKEFIELD and CARMELA M. CASTIGLIONE ● Department of Human Genetics, Yale University School of Medicine, New Haven, Connecticut 06510.   LISA M. COUSSENS and AXEL ULLRICH ● Department of Molecular Biology, Genentech, Inc., South San Francisco, California 94080.   FELICIA B. AXELROD ● Department of Pediatrics, New York University Medical Center, New York, New York 10016.

2. ALTERED β-NERVE GROWTH FACTOR IN DYSAUTONOMIA

2.1. Hypothesis and Clinical Findings

In 1974, Pearson et al. (1974) suggested the possibility that βNGF or receptors for this neurotrophic protein might be altered in dysautonomia. Their hypothesis was based on the loss of sensory and sympathetic nerve functions in these patients (for further descriptions of the disease state, see Chapter 21 and 22). The neurons responsible for these functions, which are located in the sensory and sympathetic ganglia, depend on βNGF for their development and survival (for reviews, see Greene and Shooter, 1980; Harper and Thoenen, 1980). Newborn and fetal mice exposed to antibodies to βNGF show substantial loss of neurons in sympathetic and sensory ganglia (Levi-Montalcini, 1972; Gorin and Johnson, 1980a). Neuropathological examination of nervous tissue from dysautonomia patients also shows neuronal loss in sympathetic and sensory ganglia, as well as in some parasympathetic ganglia (Pearson and Pytel, 1978a,b; Pearson et al., 1978). The main argument against a role of βNGF in dysautonomia has been that the variety of neuronal dysfunctions seen in this disease cannot all be explained by the known actions of βNGF in experimental animals. It is not clear in these experiments, however, how completely antibodies to βNGF block its action. By exposing mice and other rodents to βNGF antibodies prior to birth, moreover, Gorin and Johnson (1980a) and others (Aloe et al., 1981) have demonstrated that this neurotrophic protein affects the development of more neural cell types than previously suspected. Without knowing the full range of action of βNGF in development, it is not possible to predict exactly what neurons would be affected if the endogenous βNGF protein or its receptor were defective.

2.2. Biochemical and Immunological Studies of Human β-Nerve Growth Factor

To establish whether patients with dysautonomia make an altered form of βNGF, it is necessary to characterize the human form of this protein. This has been difficult, and although there are a number of reports on preliminary identification of a human βNGF-like molecule (Goldstein et al., 1978; Walker et al., 1980), no one has conclusively demonstrated its presence.

Because so much is known about the structure and function of mouse βNGF (for reviews, see Bradshaw, 1978; Greene and Shooter, 1980; Harper and Thoenen, 1980), the presence of human βNGF-like molecules can be assessed by looking for mouse βNGF-like molecules. Putative human βNGF has been measured by stimulation of the outgrowth of neurites from chick dorsal-root-ganglia neurons in culture (classic bioassay) and by competition of human proteins with [^{125}I]mouse βNGF for binding to βNGF receptors on pheochromocytoma cells (radioreceptor assay) or to antibodies against mouse βNGF (radioimmunoassay). These methods are complicated, however, by a number of factors: our incomplete understanding of other neuronal growth factors, the presence of human proteins that bind [^{125}I]mouse βNGF nonspecifically and hence give artifactual estimates of the presence of human βNGF (Suda *et al.*, 1978; Rosenberg *et al.*, 1983), the necessity of doing assays across species determinants due to a lack of authentic human βNGF, and the apparently minuscule amounts of human βNGF present in all tissues examined to date. Studies showing reduced biological activity of a βNGF-like protein in sera (Siggers *et al.*, 1976) and skin fibroblasts (Schwartz and Breakefield, 1980) from patients with dysautonomia must be viewed in light of these complications.

In attempts to find a source of human βNGF, we have examined a number of fetal (14- to 16-week gestation) and adult tissues, including placenta (unpublished data). Although immunoreactive molecules were demonstrated using antibodies to purified mouse βNGF, none had the appropriate properties of βNGF with respect to molecular weight and net charge, and none could be rigorously proved to have the biological activity and receptor-binding features of βNGF. Rigorous proof of βNGF, as defined by Harper and Thoenen (1980), is biological activity in the classic bioassay that can be blocked with titratable amounts of antibodies to mouse βNGF, and the detection of immunoreactive molecules by antibodies that recognize βNGF at at least two separate antigenic sites.

2.3. Molecular Biological Studies of β-Nerve Growth Factor

The use of recombinant DNA technology presents a dramatic example of how the structure of a human protein and its role in disease can be studied without having access to the protein itself. Ullrich and co-workers (see Chapter 19) have cloned a complementary (cDNA) homologous to the messenger RNA (mRNA) that codes for mouse βNGF from the male submaxillary gland, a rich source of this protein (Greene and Shooter, 1980). Enrichment for these cDNAs was facilitated by using as primers for reverse transcriptase synthetic oligonucleotides constructed to be homologous to sequences of this mRNA. The appropriate synthetic sequences were construed from the known amino acid sequence of mouse βNGF (Bradshaw, 1978) and selected for a small amount of degeneracy in possible triplet codon sequences. The authenticity of the cloned cDNA was verified by determining its complete nucleotide sequence. This cDNA was then used to identify by hybridization homology a cloned sequence of human genomic DNA containing the βNGF gene. There appears to be only one human βNGF gene locus, and this gene has been almost completely sequenced by Ullrich *et al.* (1983). Whether or not there are other related gene loci that arose by gene duplication and have undergone changes in nucleotide sequence during evolution remains to be determined. Through the presence of an apparently functional human βNGF gene, one can predict that a human βNGF protein exists and that it is very similar in amino acid sequence to mouse βNGF (about 90% homology).

3. LINKAGE ANALYSIS USING VARIATIONS IN DNA SEQUENCE

3.1. Theoretical Considerations

The availability of the cloned human βNGF gene allows us to determine whether or not this gene region is altered in dysautonomia. This is most efficiently analyzed in human pedigrees as the coinheritance (linkage) of variations in DNA sequence in or near the βNGF gene with the disease state. The closer two gene loci (or sequences of DNA) are to each other on a chromosome, the less likely it is that a recombinational event (crossover) will occur between them during meiosis and the more likely it is that specific alleles will be transmitted together in the gametes. The simple linear arrangement of genes along a chromosome and the distance constraint to DNA exchange between homologous chromosomes during meiosis (Sturtevant, 1913) provide a powerful means for understanding the distribution and nature of genes in the human genome. Distance between gene loci can be measured in recombinational units where a 1% chance of recombination between two loci is defined as a distance of 1 centiMorgan (cM). This relates approximately to physical distance along the DNA sequence.

The position of a gene locus in the human genome can be mapped by establishing its location relative to other gene loci, through analysis of the coinheritance of identifiable gene products or DNA sequences. Linkage studies have been limited by the small number of commonly occurring variable phenotypes that could be readily measured in humans and were inherited as single gene loci. These phenotypic markers consist mainly of variations in the charge or antigenicity of serum or blood-cell proteins and in the staining pattern of human chromosomes (see McKusick, 1980). To be useful, each marker must have at least two variant forms occurring with a high frequency in the population. The larger the number of variant forms (i.e., the more distinct products of different alleles there are at a given locus) and the more often these variant forms occur in the population, the more useful they are in linkage analysis (Botstein *et al.*, 1980). Using all the phenotypic markers currently available, it is possible to find the position of a variant phenotype (e.g., an inherited disease state) linked to a preexisting marker less than 30% of the time (McKusick, 1980). That is, these phenotypic markers plot out coordinates over only about 30% of the genome, leaving 70% uncharted.

The capability of linkage analysis has been expanded dramatically by the realization that inherited variations in DNA sequence ("genotypic" markers) can be used as markers of location (Botstein *et al.*, 1980) and by the rapidly increasing availability of cloned DNA probes homologous to unique sequences in the human genome (for reviews, see Davies, 1981; Housman *et al.*, 1982; Kurnit and Hoehn, 1979). The applicability of this approach has been affirmed by demonstrating the codominant Mendelian inheritance of variations in DNA sequence (Wyman and White, 1980) and the relatively high frequency of single-base-pair differences among homologous nucleotide sequences of DNA (Jeffreys, 1979; Higgs *et al.*, 1981). For example, information from the β-globin gene region has shown that approximately 1 of every 100 nucleotides [base pairs (bp)] in a given sequence varies once in every 50 people (100 homologous chromosomes) and that these variations are more frequent in noncoding than in coding regions (Jeffreys, 1979).

Since it is impractical to sequence DNA from large numbers of individuals, an alternate approach has been developed to detect discrete differences in sequence. This approach involves cutting double-stranded DNA (dsDNA) with bacterial enzymes (restriction endonucleases) that recognize and cut at specific sequences, usually 4 or 6 nucleotides in length. These dsDNA fragments are then resolved on the basis of molecular weight by agarose gel electrophoresis, denatured to single-stranded (ss) molecules, transferred to nitrocellulose, and hybridized to a labeled, ssDNA probe that recognizes a unique complementary sequence of genomic DNA (Fig. 1). Only those fragments that include part of the unique sequence will be labeled by this procedure; if the unique sequence is cut by the endonuclease, more than one fragment will be labeled. Variations in the size of the hydridizing fragments among individuals indicate differences in DNA sequence in or near the sequence homologous to the probe (Fig. 2).

There are about 200 restriction endonucleases known, and many of them are commercially available. If we assume that a DNA sequence (site) recognized by a restriction endonuclease occurs at random intervals throughout the genome, the chance of a site varying in or around a given gene locus can be calculated (Housman *et al.*, 1982). A specific 6-bp sequence would occur randomly about once every 4000 bp [4 kilobases (4kb)]. Assuming 1 variation per 100 nucleotides, about 1 of every 17 sites would have a variation in nucleotide sequence that would render it uncuttable. For any given gene locus, the chance that a variation in an endonuclease site will occur in or near it is 1/17 times the number of sites that are cut to generate detectable fragments. Further, other changes in DNA sequence besides simple nucleotide substitutions can occur that will also give rise to fragments of different length (Wyman and White, 1980; Ullrich *et al.*, 1982; Bell *et al.*, 1981); these include insertions, deletions, and inversions (Fig. 2). One can make a rough prediction that to find useful variations in restriction-fragment length in and around a particular gene, one would have to cut with about 5 different endonucleases, which recognize nonoverlapping 6-bp

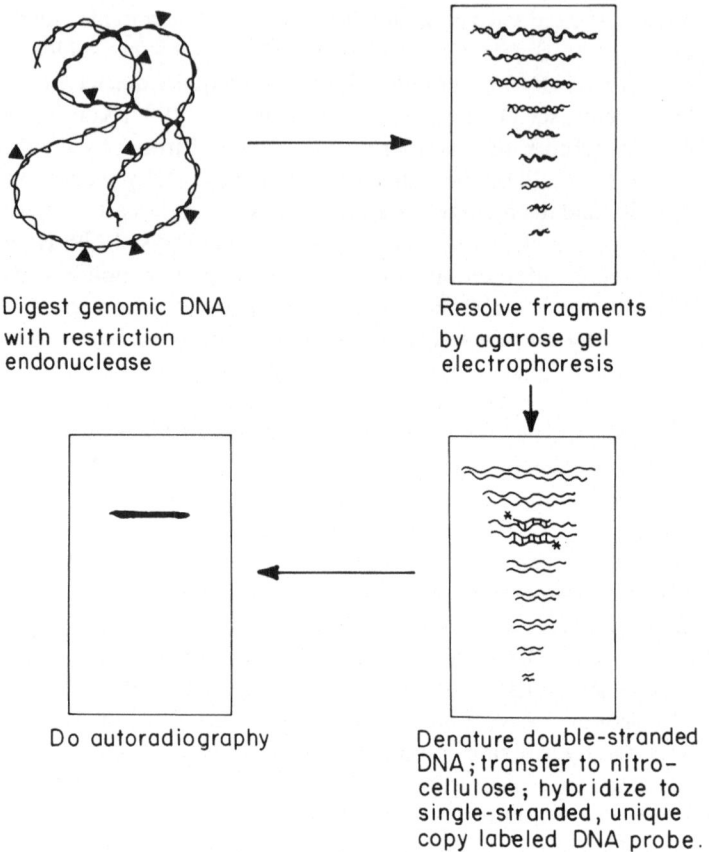

Digest genomic DNA
with restriction
endonuclease

Resolve fragments
by agarose gel
electrophoresis

Do autoradiography

Denature double-stranded
DNA; transfer to nitro-
cellulose; hybridize to
single-stranded, unique
copy labeled DNA probe.

Figure 1. Schematic diagram of techniques used to resolve restriction fragments homologous to a DNA sequence of interest.

sites, using DNA from 50 people (100 alleles), or with about 20 endonucleases using DNA from 10 people.

Several considerations relating to the nature of DNA and the techniques being used should be kept in mind when interpreting variations in restriction-fragment lengths. Human genomic DNA has been calculated to consist of approximately 70% moderately to highly repetitive sequences and 30% "single-copy" sequences (Schmid and Jelinek, 1982). For linkage studies, DNA probes are constructed to hybridize only with the latter category, although distance

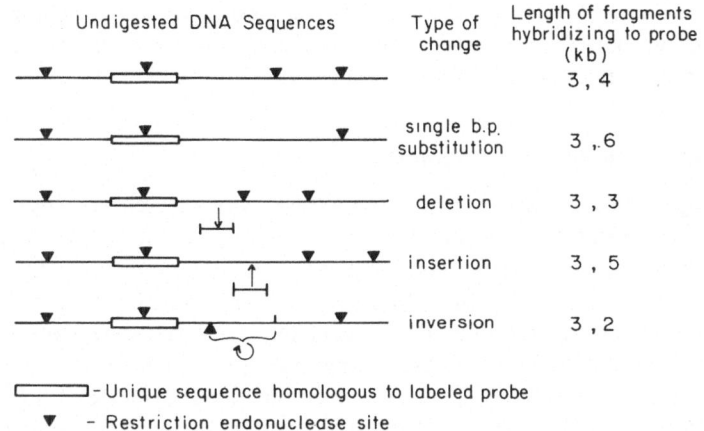

Figure 2. Hypothetical changes in DNA sequence that cause variations in restriction-fragment lengths.

measurements include the entire DNA sequence irrespective of the extent of repetition. Because multiple copies of a gene "locus" can be present in the genome, including expressed and nonexpressed (pseudogene) loci, the presence of a number of fragments that hybridize to a single-copy probe may indicate either that the probe spans one or more endonuclease sites or that it recognizes similar related DNA sequences that may be very far apart in the genome. Some secondary modifications of DNA, such as methylation, can affect the susceptibility of sites to certain restriction endonucleases, and these may be specific to a tissue or a stage of development. Further, the techniques used to analyze fragment lengths can yield misleading results due to incomplete digestion of the DNA, to difficulties in resolving fragments that differ less than 0.1 kb in length, and to the inability to detect fragments that are smaller than 0.5 kb or have only a short region of homology with the probe. Thus, the molecular basis of fragment-length variants must be examined carefully before conclusions about the nature of the variation and linkage can be inferred.

Much of the current work using DNA polymorphisms in linkage analysis is directed toward saturating the entire human genome with marker polymorphisms at recombinational distances of 10–20 cM (equivalent to about 10,000–20,000 kb). This will ultimately require a set of about 165 properly spaced markers, but to select this set, about 750 candidates will have to be identified and positioned on the genome (Lange and Boehnke, 1982). This represents a major ongoing effort on the part of a large number of investigators. Once this marker set is established, virtually any human gene that causes a distinct phenotype and is inherited in a Mendelian fashion can be positioned on the genome.

3.2. Applications to Dysautonomia

Linkage analysis can also be used to ask very discrete questions about inherited disease states. In dysautonomia, we can ask whether or not the disease is caused by an alteration in the DNA sequence that contains the gene for βNGF. This represents a greatly simplified linkage study, since one need test only DNA probes homologous to the human βNGF gene region. If this gene is responsible, variations in endonuclease sites detected by restriction-fragment-length analysis that lie within the gene (equivalent to a recombinational distance of 0.0 cM) or very near it (e.g., 0.001 cM away from it, equivalent to about 1 kb) can be used to follow the inheritance of normal and defective copies of the gene in families with one or more affected individuals. The variant site does not need to be the sequence that causes the defect, nor does it need to lie within the coding region of the gene; it simply needs to coinherit with specific alleles at that gene locus and thus mark them.

Several aspects of dysautonomia make it amenable to this type of linkage analysis, while several others make analysis difficult. In the first category, the early age at which the disease can be diagnosed (shortly after birth) and the expression of the disease in all individuals homozygous for the mutant gene (complete penetrance) means that all affected individuals can be identified. The low prevalence of the disease, the relatively high frequency of the mutant gene in a particular ethnic group (Ashkenazic Jewish), and the lack of heterogeneity in the mode of inheritance of the disease state suggest that the same mutation is responsible for all cases of the disease. In the second category, the recessive mode of inheritance means that unaffected siblings cannot be identified with respect to homozygous-normal or heterozygous-carrier status. The small number of affected individuals and the small family sizes limit the number of families informative for linkage studies. Further, the ethnic similarity of individuals may reduce the number of variations in DNA sequence among them. These latter considerations would discourage the use of linkage analysis to find the position of an unknown gene responsible for dysautonomia. Given the availability of the cloned human βNGF gene, however, it is relatively easy to ask whether or not this region of DNA is involved and thus to provide a simple yes or no answer to a question that has eluded biochemists for several years. A theoretical example of the possible cosegregation of a disease inherited in an autosomal recessive manner and DNA restriction-fragment-length polymorphisms detectable with a gene probe is illustrated in Fig. 3.

A. FAMILY WITH AN AUTOSOMAL RECESSIVE DISEASE

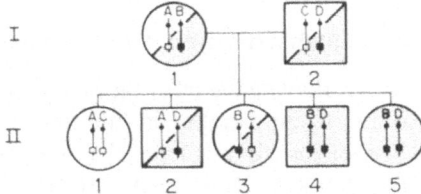

B. POSSIBLE RESTRICTION ENDONUCLEASE SITES (↓) AROUND RESPONSIBLE GENE (▭■)

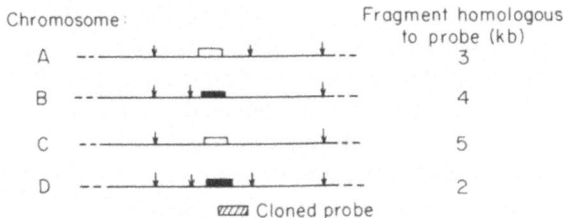

C. AUTORADIOGRAM OF RADIOACTIVE PROBE HYBRIDIZED TO RESTRICTION ENDONUCLEASE–DIGESTED DNA FROM FAMILY MEMBERS

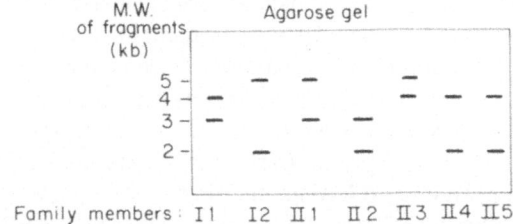

Figure 3. Theoretical diagram of the inheritance of a mutant allele that causes an autosomal recessively inherited disease and the use of restriction-fragment-length variations near the mutant allele to identify its presence in family members. (A) A family in which two children (II4 and II5) have the disease and both parents (I1 and I2) and two children (II2 and II3) are carriers of the mutant allele. (A–D) Chromosomes; (●) centromere; responsible gene: normal (□), mutant (■). (B) Theoretical positions of specific restriction endonuclease cleavage sites in the DNA from different chromosomes around the region of the gene and sizes of DNA fragments generated by cleavage and hybridization to the cloned gene probe (□). (C) DNA isolated from cells of family members and digested with this endonuclease. DNA fragments are separated on the basis of molecular weight in agarose gels. The position of the fragments that contain the responsible gene is determined by hybridization to a ^{32}P-labeled probe, followed by autoradiography. The gene contribution from each chromosome of each parent can be distinguished on the basis of the size fragment within which it is contained. In this ideal example, both affected children inherit the same fragment sizes, and none of the unaffected children has both these same fragments.

4. PRELIMINARY STUDIES LOOKING FOR VARIATIONS IN DNA SEQUENCE NEAR THE β-NERVE GROWTH FACTOR GENE

4.1. Selection of Families and Establishment of Lymphoblast Lines

Linkage analysis depends on following the inheritance of two markers, each of which occurs in several forms. In the families in this study, one marker is expression or lack of expression of the disease state; the other is variations in DNA sequence in or near the βNGF gene. With respect to the disease marker, two genotypes can be clearly defined: (1) affected individuals who are homozygous for the mutant allele that causes the disease and (2) their parents, who, if normal, are obligate heterozygotes carrying a mutant allele and a normal allele at the disease locus. Siblings of the affected individuals are less informative, since it is usually not possible to establish whether they are heterozygous for mutant and normal alleles or homozygous for normal alleles at this locus.

In this study, families were preferentially selected that had two or three affected children and for which samples could be obtained from parents. Four families in North America that met these criteria were identified (Fig. 4). Two other families were chosen on the basis of a relatively large number of siblings (Family 3) and availability of a grandparent (Family 4; father's parent is living). Lymphoblast cell lines have been established from all living members. In taking blood samples, 20 ml of blood were withdrawn into heparinized tubes (B-D Vacutainer green-capped tubes), maintained at room temperature, and shipped within 48 hr to tissue-culture facilities. Using a protocol of Anderson and Gusella (personal communication), serum was separated by centrifugation at $450g$ for 10 min. Lymphocytes were isolated by centrifugation through a gradient of Percoll [colloidal PVP-coated silica (Pharmacia)] at $800g$ for 45 min and transformed with the B95-8 strain of Epstein–Barr virus (Robinson and Smith, 1981) in 2 μg/ml cyclosporin (Sandoz) (Gordon and Singer, 1979). Cells were maintained in suspension culture in roller bottles [850 cm^2 (Falcon)] in RPMI medium (Flow 10-601-240) containing penicillin, streptomycin (each 1000 U/ml), 2 mM glutamine, and 15% fetal calf serum (Flow). Transformation was judged successful when cell clumps were visible. Viable frozen cell stocks are maintained in the vapor phase of liquid nitrogen.

4.2. Screening for Variations in DNA Sequence

In this study, the genotypic markers used in linkage analysis are variations in DNA sequence in the βNGF gene region. Studies are under way to identify such variations using a variety of restriction endonucleases. Several approaches are being taken to increase the chances of finding appropriate variations. Since the parents in these families possess all the potentially useful variations in se-

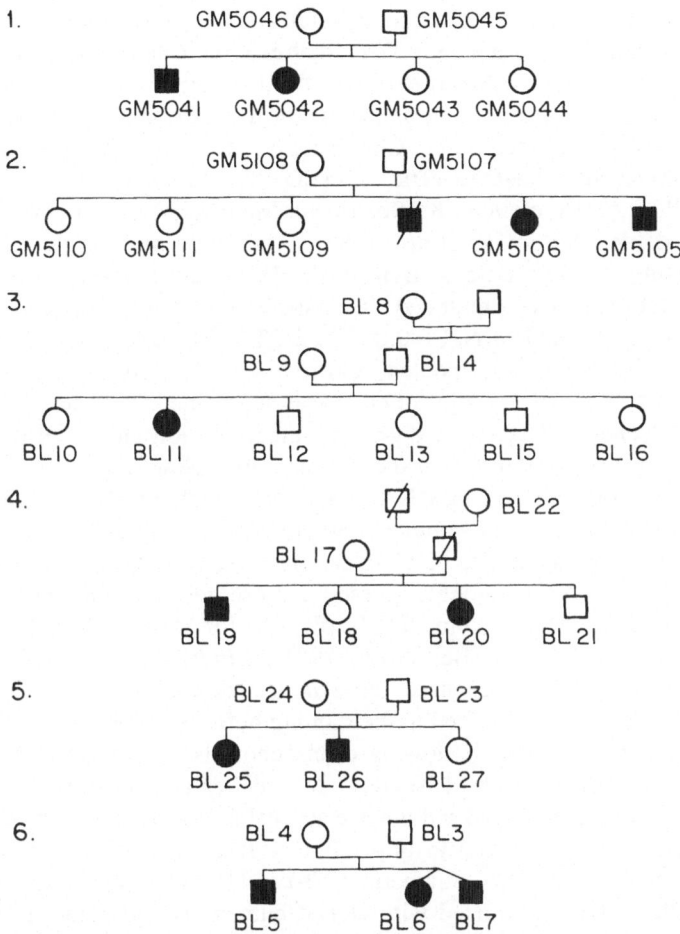

Figure 4. Dysautonomia pedigrees of families used in these studies. (○) Unaffected female; (●) affected female; (□) unaffected male; (■) affected male; (/) dead. In Family 6, BL6 and BL7 are dizygotic twins. In the designations of lymphoblast lines derived from these individuals, the prefix GM denotes lines generated at the Institute for Medical Research (Camden, New Jersey); the prefix BL, lines generated at Yale.

quence, they are being tested first. In the case in which one parent is no longer living (Family 4), blood samples have been obtained from the grandparent. Because most of the coding and noncoding sequences of the normal human βNGF gene region are known (see Chapter 19), endonuclease sites known to exist in this region are being tested. Also, several probes that cover different regions of the gene are being used for hybridization to maximize the number of homologous fragments of genomic DNA that can be visualized (Fig. 5).

For these studies, DNA was extracted from lymphoblasts as follows: Cells were collected by centrifugation at 200 g for 10 min and rinsed twice with isotonic phosphate-buffered saline. Cells were lysed in 0.075 M NaCl containing 0.025 M ethylenediamine tetraacetic acid (EDTA), 1% sodium dodecyl sulfate, and 100–200 μg/ml Proteinase K (Boehringer Mannheim) at 37°C for 15–18 hr as described (de Martinville *et al.*, 1982). High-molecular-weight DNA was prepared using the procedure of Bell *et al.* (1981) and Kunkel *et al.* (1977), except that the phenol was hydrated successively with distilled water, 200 mM Tris, pH 8.0, and 0.075 M NaCl–0.025 M EDTA. The DNA was precipitated with 2.5 vol.100% ethanol, spooled, and redissolved in 10 mM Tris–0.01 M EDTA, pH 7.5. The concentration of DNA was determined by spectrophotometry at 260 nm. Genomic DNA was digested with restriction endonucleases using assay conditions recommended by the manufacturer (New England Biolabs). A typical sample containing 10 ng DNA was digested with 30–40 U of one or two restriction endonucleases in a volume of 30–50 μl for 3–5 hr at 37°C. Digestions were terminated by heating to 65°C. Electrophoresis was carried out on a horizontal 0.8% (or 1.0% for double digests) agarose gel in 40 mM Tris–20 mM sodium acetate–2 mM disodium EDTA, pH 7.8, for 15 hr at 80 mA and then for 3–4 hr at 100–120 mA (Bell *et al.*, 1981). *Hind*III-digested lambda DNA (Boehringer Mannheim) was used for a molecular-weight marker. Ethidium bromide, 0.25 μg/ml, was added to the running buffer and the gel to view DNA fragments under UV light. Following electrophoresis, gels were immersed in 0.25 N HCl and then in 0.5 M NaOH–1 M NaCl as described by Wahl *et al.* (1979) and neutralized in 3 M sodium acetate, pH 5.5, as described by Southern (1979). DNA was transferred to nitrocellulose filters (Schleicher and Schuell BA85) (Huttner *et al.*, 1979; Southern, 1979) and baked at 80°C under vacuum for 2 hr. Blots were hybridized with cloned human genomic sequences corresponding to coding regions of the βNGF gene (see Fig. 5 and Chapter 19) labeled with ^{32}P using avian sarcoma polymerase (Taylor *et al.*, 1976).

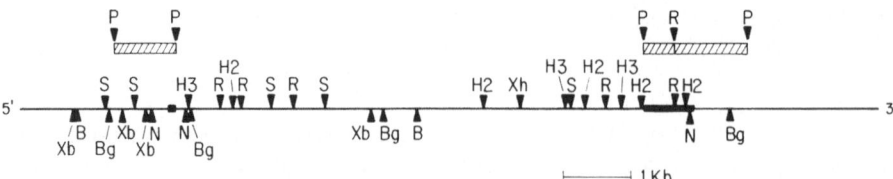

Figure 5.   Region of human genomic DNA that contains βNGF gene and cloned probes used in this study. (■) Coding regions for most of pro βNGF (including βNGF); (□) three cloned human genomic probes used to detect fragments; (▼, ▲) restriction endonuclease sites: (B) *Bam*HI; (Bg) *Bgl*II; (H2) *Hinc*II; (H3) *Hind*III; (N) *Nco*I; (P) *Pst*I; (R) *Eco*RI; (S) *Sac*I; (Xb) *Xba*I; (Xh) *Xho*I. From Ullrich *et al.* (1983).

Table I. Restriction Endonucleases Examined

Restriction endonuclease digestion[a]	Recognition site 5'–3'	Fragments observed (number)	Sites tested (number)	Number of individuals examined (families)	Polymorphism
SacI	GAGCŤC	3	5	13 (1,2)	−
HindIII	ǍAGCTT	2	4	13 (1,2)	−
BglI	GCCNNNNNGGC	2	4	13 (1,2)	−
EcoRI	GATAŤC	2	4	13 (1,2)	−
BamHI and NcoI[b]	ǦGATCC and ČCATGG	3	5	6 (1–3)	−
MspI	ČCGG	2	4	13 (1,2)	−
XbaI and XhoI[b]	ŤCTAGA and ČTCGAG	4	6	6 (1–3)	−
HincII	GTPyↃPuAC	3	5	6 (1–3)	+
BglII	ǍGATCT	3	5	12 (1–6)	+

[a] All blots were probed with the human genomic βNGF probes shown in Fig. 5 except for those generated from BglI, which were probed with a cloned cDNA for the mouse βNGF mRNA.
[b] Double digest.

So far, we have cleaved DNA obtained from individuals in three families with 11 restriction endonucleases and have examined fragments by Southern blot analysis using three probes for the βNGF gene (Table I and Fig. 5). Most of the sites within the human βNGF gene region for the restriction endonucleases tested are shown in Fig. 5, as are the positions of the sequences homologous to the probes tested. The two probes on the 3' side overlap all the βNGF coding region, part of the pro-βNGF coding sequence, and part of the 3' flanking region. The single probe on the 5' side overlaps an additional part of the pro-βNGF coding sequence and part of a 5' intron region. The size and number of the hybridizing fragments of genomic DNA observed were similar to those anticipated from the normal gene sequence, given the limits of resolution of the agarose gels used. So far, in four parents, we have tested about 35 endonuclease sites for which 6 nucleotides must be in the correct sequence for cutting; this corresponds to over 150 nucleotides in noncoding regions and over 20 nucleotides in the coding regions. Two restriction-fragment-length variants have been identified in eight parents using HincII and BglII. A lack of fragment-length variation at three BglI sites (two labeled fragments) is shown for purposes of illustration (Fig. 6). More endonucleases are currently being tested.

4.3. Future Characterization and Identification of Variant Sites

For linkage analysis, it is necessary to be sure that the molecular variations in restriction-fragment lengths that are observed are inherited as codominant alleles. Variations due to incomplete digestion or modified DNA will not prove

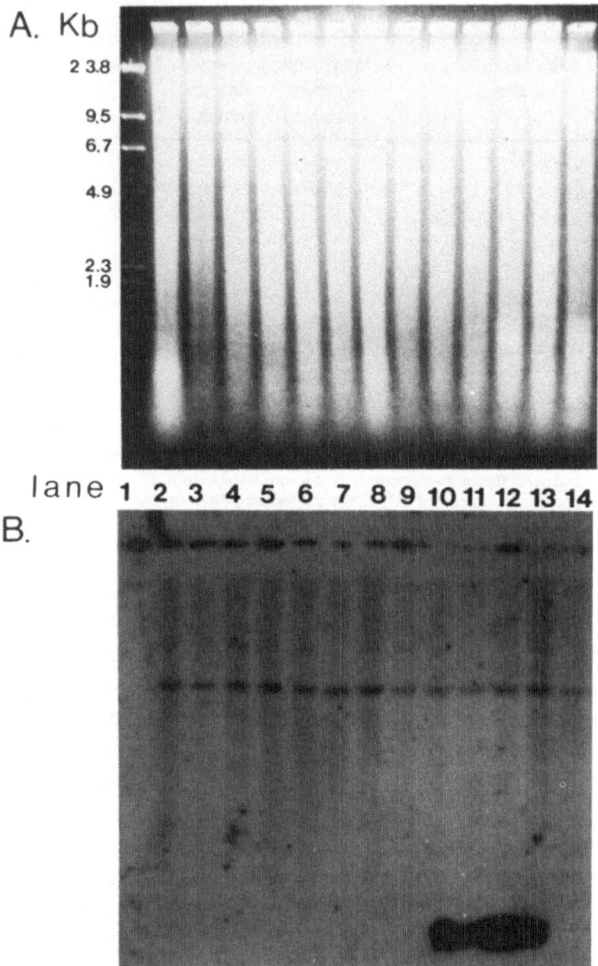

Figure 6. Gel and Southern blot of genomic DNA digested with *Bg*II. (A) Ethidium-bromide-stained gel visualized under UV light. (1) λ DNA digested with *Hind*III; (2–14) genomic DNA from Families 1 and 2 following digestion. (B) Autoradiogram of nitrocellulose filter made from the gel in (A) and hybridized to labeled cDNA for the mouse βNGF mRNA.

to be heritable in families and may show variability in different tissue samples from the same individual (skin fibroblasts, amniotic fluid cells, and untransformed lymphocytes are available from some of these individuals and can also be tested). The issue of whether a variation in fragment length results from a single base-pair change in an endonuclease site or from an insertion, deletion, or inversion of DNA within the fragment (see Fig. 2) can be resolved by cutting genomic DNA with other endonucleases that have sites near the variant one. A variation in length due to events in the latter category will be revealed by several endonucleases that generate fragments containing this variant stretch (for examples, see Wyman and White, 1980; Bell *et al.*, 1981; Ullrich *et al.*, 1982).

Using the strategy described herein, we find that for each individual, a restriction endonuclease generates about three fragments of DNA that hybridize with the probes used. This corresponds to about four endonuclease sites or, for an endonuclease that recognizes 6 bp, to 24 bp per haploid sequence or 48 bp per individual. To be sure to find a variation in one site recognized by this endonuclease, we would have to screen DNA from about 100 individuals. In this study, however, the number of people who will have informative variations is limited to 12 parents. Therefore, it will be necessary to use a large number of restriction endonucleases to be sure of finding enough variations to distinguish the βNGF alleles contributed by each parent in all the families.

In this study, it is easier to exclude involvement of the structural gene for βNGF than to prove it. Thus, a family can be informative even if only one allele for the βNGF gene is marked by a fragment-length variation in one of the parents in a family. In this case, we can predict that if an alteration in the βNGF gene causes the disease, every affected child should have received the same copy of the gene (recognizable by its fragment-length pattern) from that parent—or, to state it another way, that all affected children should have the same labeled fragment lengths. If there is no association between this gene and the disease, about 50% of the time the affected children will receive the same allele and about 50% of the time a different allele from that parent. The demonstration of a single case of two affected children in a family receiving different alleles of the βNGF gene from one parent will essentially exclude a role for this gene in the disease. To prove an association, we will need to follow the coinheritance of marked βNGF genes with the disease state in all six families.

The preliminary nature of the findings described herein makes it impossible to predict the number of endonucleases and families that will need to be tested to exclude or prove linkage of a DNA sequence variation in or near the βNGF gene with dysautonomia. The present state of knowledge about the use of restriction-fragment-length variations in linkage analysis, however, allows us to be confident that this approach will eventually provide an answer to the question of whether alterations in the βNGF gene region are responsible for dysautonomia.

5. USE OF INFORMATION DERIVED FROM LINKAGE ANALYSIS OF THE β-NERVE GROWTH FACTOR GENE AND DYSAUTONOMIA

5.1. If Linkage Does Exist

A positive finding of linkage would open up many opportunities for further understanding of the disease process, genetic counseling, and therapy. It is

important to keep in mind that a variation in DNA sequence identified by restriction endonuclease fragments may lie either within or near the βNGF gene
and would only in an exceptional case have any direct relationship to the alteration
in DNA sequence that actually causes the disease. Further, if the mutation that
causes dysautonomia were in either the coding region of the gene (and hence
caused formation of a structurally altered βNGF precursor or protein) or the
noncoding region of the gene (and hence affected expression of the gene product),
it could be detected equally well by linkage analysis.

If the βNGF gene were implicated in dysautonomia, the next step would
be to prepare libraries of cloned genomic DNA fragments from several patients
and controls. Cloned fragments that contain sequences homologous to the existing
normal βNGF gene would be identified by hybridization to it, and their DNA
sequences would be determined and compared. Given that there may be non-
pathological variation in the structure of this gene, it may be necessary to analyze
DNA from several affected and normal individuals. Sequencing data should
elucidate the position and nature of altered sequences associated with the disease
state, as well as of those that generate restriction-fragment-length variations.

A DNA sequence variation very near to the mutation that causes dysautonomia could be used in heterozygote detection and prenatal diagnosis once the
coupling phase of a particular marker variation and the mutant allele had been
established in a family. This approach is limited, however, by the requirement
for information about other family members. Although in some cases a particular
fragment-length variation has been found to be associated in the population with
a mutant gene more often than with a normal gene, e.g., in sickle-cell disease
(Kan and Dozy, 1978; Panny et al., 1981) and the thalassemias (Orkin et al.,
1982), this association has not been absolutely predictive. The possibility should
not be excluded, however, that a variation in DNA sequence that causes dysautonomia and another one that causes a restriction-length variation near the βNGF
gene are so close together on the genome that they have remained together
through successive generations.

Ideally, it may be possible to identify the mutant lesion directly and thus
be able to classify individuals without family data. This has been possible for
the single base-pair alteration in the βglobin gene that causes sickle-cell disease,
since this variation eliminates a particular restriction endonuclease site (Chang
and Kan, 1982). Further, synthetic nucleotide sequences can distinguish between
DNA fragments that contain the mutant and normal globin alleles by virtue of
their ability to hybridize or not hybridize with them under carefully controlled
conditions (Conner et al., 1983). Heterozygote detection in this latter case becomes a matter of quantitating the extent of hybridization in DNA from heterozygotes as compared to homozygous controls and affected individuals. If the
alteration in DNA sequence that causes dysautonomia affects a larger extent of
DNA through an insertion, deletion, or inversion, it may also be possible to
cleave the DNA in such a way that a particular fragment length becomes pa-
thognomonic of the presence of the mutant gene.

If the βNGF gene is implicated in dysautonomia, it would become important to evaluate the possibility that administration of human βNGF to these patients might be beneficial. The successful cloning of the human βNGF gene allows production of this protein using recombinant DNA technology, as has been done for human insulin (Johnson, 1983). The findings of progressive loss of some neurons and neuronal functions in experimental adult animals exposed to βNGF antibodies (Gorin and Johnson, 1980b; Otten *et al.*, 1979) suggest that βNGF may serve a maintenance and survival function for some neurons after birth and throughout adult life. Patients with dysautonomia also show with age a gradual loss of sympathetic neurons (Pearson *et al.*, 1978) and sensory functions (Axelrod *et al.*, 1981). This suggests the possibility that administration of βNGF to dysautonomic patients after birth might help to maintain some neural functions throughout life. Clearly, in considering clinical trials, it is important to remember that administration of exogenous βNGF may have undesired effects, such as the production of antibodies against it, and thus against any residually active βNGF, or a detrimental action on neurons and other cell types. Further, exogenous βNGF may have no beneficial effects because of lack of access to appropriate neurons or inability of neurons to respond to it at the time of administration.

5.2. If Linkage *Does Not* Exist

A negative finding of linkage provides little information of use in understanding this disease. It would establish definitively that the structural gene for βNGF is not involved and thus end futile research efforts to determine whether it is using biochemical techniques. It would not, however, rule out a primary role for βNGF in this disease state. Defective βNGF function might be caused by alterations in other subunits (α and γ) of the NGF protein storage complex, in proteolytic enzymes necessary for processing of precursor forms of the NGF subunits, in receptors for βNGF, or in other proteins involved in mediating the actions of this hormone. In the case of the α and γ subunits, and in the case of receptors, it is reasonable to assume that cloned DNA probes that correspond to the genes that code for them will become available in the near future and that these probes could also be used in linkage studies as described here.

Although it is theoretically possible to begin a systematic search throughout the genome for the sequence of DNA that causes dysautonomia by linkage analysis, this prospect is unappealing for several reasons. First, it is a very rare disease with only about 500 known affected individuals in the United States, Canada, and Israel; moreover, all the available families are relatively small. Second, the autosomal recessive mode of inheritance means that many individuals in the pedigree are of unknown status with respect to being homozygous normal or heterozygous for the mutant allele and are hence minimally informative. Third, the construction of a marker map consisting of some 165 common polymorphic DNA sequences spaced at 20-cM distances along the human genome is just under way, and the requisite type of linkage analysis will be much easier when it is completed.

5.3. Prospects

Irrespective of the outcome of these studies, they illustrate the tremendous potential provided by techniques of molecular biology for the study of inherited diseases in humans. It is possible to predict the number of human genes that code for a protein, their location in the genome, and the structure of the protein they encode without ever isolating the human form of the protein. Further, it is possible to make substantial amounts of the protein through recombinant DNA technology. Using linkage analysis, the role of this gene in a particular disease can be assessed, and this information can be used to identify individuals who possess one or two copies of the mutant allele. In inherited neurological diseases where we have been limited by the availability of appropriate neural tissue for study, this technology allows us to resolve some long-standing questions about etiology and provides the potential for breakthroughs in genetic counseling and patient therapy.

ACKNOWLEDGMENTS. We are deeply grateful to the dysautonomia families who have provided blood samples for these studies and to the physicians—Dr. Ralph Cash, Dr. Kenneth Keer, Dr. James Selwyn, and Dr. Leonard Pinsky—and their associates who have supplied us with these samples. We would also like to thank Alane Gray, Thomas Dull, and Cara Berman for sharing their unpublished information on the human βNGF gene; Kenneth Kidd and James Gusella for advice on linkage analysis; Bérengère de Martinville, Judith Kidd, James Gusella, and Mary Anne Anderson for helping us (X.O.B. and C.M.C.) master the techniques used; and David Corey, Mark Grossman, and Michael Rosenberg for their thoughtful comments. This work was supported by NIH Grant NS17803 and funds from the Dysautonomia Foundation and the McKnight Foundation to X.O.B.

Note Added in Proof

Subsequent work has excluded a role for the human β-NGF gene in familial dysautonomia (Breakefield et al., 1983a; Breakefield et al., 1983b).

REFERENCES

Aloe, L., Cozzari, C., Callissano, P., and Levi-Montalcini, R., 1981, Somatic and behavioral postnatal effects of fetal injections of nerve growth factor antibodies in the rat, Nature (London) 291:413–415.

Axelrod, F.B., Iyer, K., Fish, I., Pearson, J., Sein, M.E., and Spielholz, N., 1981, Progressive sensory loss in familial dysautonomia, Pediatrics 67:517–522.

Bell, G.J., Karam, J.H., and Rutter, W.J., 1981, Polymorphic DNA region adjacent to the 5' end of the human insulin gene, Proc. Natl. Acad. Sci. U.S.A. 78:5759–5763.

Botstein, D., White, R.L., Skolnick, M., and Davis, R.W., 1980, Construction of a genetic linkage map in man using restriction fragment length polymorphisms, *Am. J. Hum. Genet.* **32**:314–331.

Bradshaw, R.A., 1978, Nerve growth factor, *Annu. Rev. Biochem.* **47**:191–216.

Breakefield, X.O., Orloff, G., Castiglione, C.M., Axelrod, F.B., Coussens, L., and Ullrich, A., 1983a The gene for β-nerve growth factor is not defective in familial dysautonomia, *Proc. Natl. Acad. Sci. U.S.A.* (submitted).

Breakefield, X.O., Orloff, G., Castiglione, C.M., Axelrod, F.B., Coussens, L., and Ullrich, A., 1983b, Genetic linkage analysis in familial dysautonomia using a DNA probe for the β-nerve growth factor gene, in: *Biochemical and Clinical Aspects of Neuropeptides, Synthesis, Processing and Gene Structure* (G. Koch and D. Richter, eds.) Academic Press, New York (in press).

Chang, J.C., and Kan, Y.W., 1982, A sensitive new prenatal test for sickle-cell anemia, *N. Engl. J. Med.* **307**:30–32.

Conner, B.J., Reyes, A., Morin, C., Itakura, K., Teplitz, R.L., and Wallace, R.B., 1983, Detection of sickle cell β^s-globin allele by hybridization with synthetic oligonucleotides, *Proc. Natl. Acad. Sci. U.S.A.* **801**:278–282.

Davies, K.E., 1981, The application of DNA recombinant technology to the analysis of the human genome and genetic disease, *Hum. Genet.* **58**:351–357.

De Martinville, B., Wyman, A.R., White, R., and Francke, U., 1982, Assignment of the first random restriction fragment length polymorphism (RFLP) locus *(D14S1)* to a region of human chromosome 14, *Am. J. Hum. Genet.* **34**:216–226.

Goldstein, L.D., Reynolds, C.P., and Perez-Polo, J.R., 1978, Isolation of human nerve growth factor from placental tissue, *Neurochem. Res.* **3**:175–183.

Gordon, M.Y., and Singer, J.W., 1979, Selective effects of cyclosporin A on colony-forming lymphoid and myeloid cells in man, *Nature (London)* **279**:433–435.

Gorin, P.D., and Johnson, E.M., Jr., 1980a, Effects of exposure to nerve growth factor antibodies on the developing nervous system of the rat: An experimental approach, *Dev. Biol.* **80**:313–323.

Gorin, P.D., and Johnson, E.M., Jr., 1980b, Effects of long-term nerve growth factor deprivation on the nervous system of the adult rat: An experimental autoimmune approach, *Brain Res.* **198**:27–42.

Greene, L.A., and Shooter, E.M., 1980, The nerve growth factor: Biochemistry, synthesis and mechanism of action, *Annu. Rev. Neurosci.* **3**:353–402.

Harper, G.P., and Thoenen, H., 1980, Nerve growth factor: Biological significance, measurement and distribution, *J. Neurochem.* **34**:5–16.

Higgs, D.R., Goodbourn, S.E.Y., Wainscoat, J.S., Clegg, J.B., and Weatherall, D.J., 1981, Highly variable regions of DNA flank the human α globin genes, *Nucleic Acids Res.* **9**:4213–4224.

Housman, D., Kidd, K., and Gusella, J.F., 1982, Recombinant DNA approach to neurogenetic disorders, *Trends Neurosci.* **5**:320–323.

Huttner, K.M., Scangos, G.A., and Ruddle, F.H., 1979, DNA-mediated gene transfer of a circular plasmid into murine cells, *Proc. Natl. Acad. Sci. U.S.A.* **76**:5820–5824.

Jeffreys, A.J., 1979, DNA sequence variants in the ^{G}γ-, ^{A}γ-, δ- and β -globin genes of man, *Cell* **18**:1–10.

Johnson, I.S., 1983, Human insulin from recombinant DNA technology, *Science* **219**:632–637.

Kan, Y.A., and Dozy, A.M., 1978, Polymorphisms of DNA sequence adjacent to human β-globin structural gene: Relationship to sickle mutation, *Proc. Natl. Acad. Sci. U.S.A.* **75**:5631–5635.

Kunkel, L.M., Smith, K.D., Boyer, S.H., Borgaonkar, D.S., Wachtel, S.S., Miller, O.J., Breg, W.R., Jones, H.W., Jr., and Rary, J.M., 1977, Analysis of human Y-chromosome-specific reiterated DNA in chromosome variants, *Proc. Natl. Acad. Sci. U.S.A.* **74**:1245–1249.

Kurnit, D.M., and Hoehn, H., 1979, Prenatal diagnosis of human genome variation, *Annu. Rev. Genet.* **13**:235–258.

Lange, K., and Boehnke, M., 1982, How many polymorphic genes will it take to span the human genome?, *Am. J. Hum. Genet.* **34**:832–845.

Levi-Montalcini, R., 1972, The morphological effects of immunosympathectomy, in: *Immunosympathectomy* (G. Steiner and E. Schönbaum, eds.), Elsevier, New York, pp. 55–78.

McKusick, V.A., 1980, The anatomy of the human genome, *J. Hered.* **71:**370–391.

Orkin, S.H., Kazazian, H.H., Jr., Antonarakis, S.E., Goff, S.C., Boehm, C.D., Sexton, J.P., Waber, P.G., and Giardina, P.J.V., 1982, Linkage of β-thalassaemia mutations and β-globin gene polymorphisms with DNA polymorphisms in human β-globin gene cluster, *Nature (London)* **296:**627–631.

Otten, U., Goldert, M., Schwab, M., and Thibault, J., 1979, Immunization of adult rats against 2.5S NGF: Effects on the peripheral sympathetic nervous system, *Brain Res.* **176:**79–90.

Panny, S.P., Scott, A.F., Smith, K.D., Phillips, J.A., III, Kazazian, H.H., Jr., Talbot, C.C., Jr., and Boehm, C.D., 1981, Population heterogeneity of the *Hpa I* restriction site associated with the *β globin* gene: Implications for prenatal diagnosis, *Am. J. Hum. Genet.* **33:**25–35.

Pearson, J., and Pytel, B., 1978a, Quantitative studies of ciliary and sphenopalatine ganglia in familial dysautonomia, *J. Neurol. Sci.* **39:**123–130.

Pearson, J., and Pytel, B., 1978b, Quantitative studies of sympathetic ganglia and spinal cord intermedio-lateral gray columns in familial dysautonomia, *J. Neurol. Sci.* **39:**47–59.

Pearson, J., Axelrod, F.B., and Dancis, J., 1974, Current concepts of dysautonomia: Neuropathological defects, *Ann. N. Y. Acad. Sci.* **228:**288–300.

Pearson, J., Pytel, B., Grover-Johnson, N, Axelrod, F., and Dancis, J., 1978, Quantitative studies of dorsal root ganglia and neuropathologic observations on spinal cords in familial dysautonomia, *J. Neurol. Sci.* **35:**77–92.

Robinson, J., and Smith, D., 1981, Infection of human B lymphoctyes with high multiplicities of Epstein–Barr virus: Kinetics of EBNA expression, cellular DNA synthesis, and mitosis, *Virology* **109:**336–343.

Rosenberg, M.B., Grossman, M.H., and Breakefield, X.O., 1983, Artifactual presence of β nerve growth factor in adult mouse brain, *Brain Res.* (in press).

Schmid, C.W., and Jelinek, W.R., 1982, The Alu family of dispersed and repetitive sequences, *Science* **216:**1065–1070.

Schwartz, J.P., and Breakefield, X.O., 1980, Altered nerve growth factor in fibroblasts from patients with familial dysautonomia, *Proc. Natl. Acad. Sci. U.S.A.* **77:**1154–1158.

Siggers, D.C., Rogers, J.G., Boyer, S.H., Margolet, L., Dorkin, H., Banerjee, S.P., and Shooter, E.M., 1976, Increased nerve growth factor β-chain cross reacting material in familial dysautonomia, *N. Engl. J. Med.* **295:**629–634.

Southern, E., 1979, Gel electrophoresis of restriction fragments, *Methods Enzymol.* **68:**152–176.

Sturtevant, A., 1913, The linear arrangement of six sex-linked factors in *Drosophila* as shown by their mode of association, *J. Exp. Zool.* **14:**43–59.

Suda, K., Barde, Y.A., and Thoenen, H., 1978, Nerve growth factor in mouse and rat serum: Correlation between bioassay and radioimmunoassay determinations, *Proc. Natl. Acad. Sci. U.S.A.* **75:**4042–4046.

Taylor, J.M., Illmensee, R., and Summers, J., 1976, Efficient transcription of RNA into DNA by avian sarcoma virus polymerase, *Biochim. Biophys. Acta* **442:**324–330.

Ullrich, A., Dull, T.J., Gray, A., Philips, J.A., and Peter, S., 1982, Variation in the sequence and modification state of the human insulin gene flanking regions, *Nucleic Acids Res.* **10:**2225–2240.

Ullrich, A., Gray, A., Berman, C., and Dull, T.J., 1983, The nucleotide sequence of human nerve growth factor β subunit gene is highly homologous to that of mouse, *Nature* **303:**821–825.

Wahl, G.M., Stern, M., and Stark, G.R., 1979, Efficient transfer of large DNA fragments from agarose gels to diazobenzyloxylmethyl-paper and rapid hybridization by using dextran sulfate, *Proc. Natl. Acad. Sci. U.S.A.* **76:**3683–3687.

Walker, P., Weichsel, M.E., Jr., and Fisher, D.A., 1980, Human nerve growth factor: Lack of immunocrossreactivity with mouse nerve growth factor, *Life Sci.* **26:**195–200.

Wyman, A.R., and White, R., 1980, A highly polymorphic locus in human DNA, *Proc. Natl. Acad. Sci. U.S.A.* **77:**6754–6758.

Diseases of Development

Session Chairman: Ira B. Black

Chapter 21

Familial Dysautonomia and Other Congenital Sensory and Autonomic Neuropathies

Felicia B. Axelrod

1. INTRODUCTION

Since we are concerned at this conference with neuronal development, it is appropriate to consider those human disorders that may represent aberrations in development or function. These would be the congenital sensory neuropathies, rather than those that are acquired or of adult onset. Congenital sensory neuropathies are rare and can present as diagnostic problems, since analgesia and autonomic function are difficult to evaluate objectively in the young child. Controversy over terminology further confuses classification of the congenital sensory neuropathies (Axelrod, 1983; Dyck and Ohta, 1975).

Familial dysautonomia is one of six known congenital sensory neuropathies. Since it has been the most extensively studied, it can be used as the prototype against which other disorders can be compared.

2. FAMILIAL DYSAUTONOMIA

2.1. Natural History

Familial dysautonomia is an autosomal recessive disorder with a disease incidence in the Ashkenazi Jewish population of 1 in 10,000–20,000 live births

FELICIA B. AXELROD ● Department of Pediatrics, New York University School of Medicine, New York, New York 10016.

(Brunt and McKusik, 1970; Moses *et al.*, 1967). The estimated carrier rate is 1 in 50. Familial dysautonomia is a neurological disorder, but the clinical manifestations and severity of symptoms will vary among affected individuals. However, there is a generally characteristic clinical history that can be described as follows:

The dysautonomic infant is usually born at term, but is often small for dates (Axelrod and Dancis, 1973). Breech presentation occurs in 25% of patients, an incidence far in excess of the general population (3%) or of the uninvolved siblings (0%) (Axelrod *et al.*, 1974b). Symptoms often arise in the nursery. There may be difficulty in initiating respiration. Feeding problems frequently become evident with poor sucking and swallowing, and aspiration pneumonia may ensue. Hypotonia, hypothermia, and opisthotonus are often seen (Axelrod *et al.*, 1974a).

Poor physical development is accompanied by delayed motor and verbal milestones. Walking is delayed beyond 18 months of age in 68% and intelligible speech beyond $3\frac{1}{2}$ years of age in 25% of children (Axelrod *et al.*, 1974a). Yet as a group, patients have average intelligence with a general tendency for better verbal than motor performance. Many patients have verbal scores in the superior range (Welton *et al.*, 1979).

Difficulty in feeding, often associated with aspiration, is the major complaint throughout childhood. Recurrent misdirection, especially of liquids, results in frequent pneumonias and may lead to bronchiectatic disease, atelectasis, and even lung abcesses (Axelrod and Dancis, 1976). Gastroesophageal reflux further increases the risk of aspiration in some patients (Axelrod *et al.*, 1982). Cinesophagrams have demonstrated delay in cricopharyngeal relaxation, abnormal gastrointestinal motility, and reflux (Margulies *et al.*, 1968).

Vomiting crises associated with irritability, negativistic behavior, hypertension, tachycardia, blotchy erythema, and diaphoresis are observed in 40% of patients (Axelrod *et al.*, 1974a). Crises, which may occur as often as once a week, typically last for 1–3 days, but can last longer. Dehydration and aspiration of vomitus can be fatal. The crises are associated with elevated circulating serum catecholamines.

Compounding problems of aspiration and chronic lung disease are abnormal responses to hypoxic or hypercarbic states (Filler *et al.*, 1965; Edelman *et al.*, 1970). When oxygen tension is lowered (as at high elevations, during air travel, or in tunnels), ventilation does not increase. Ensuing hypoxemia may lead to hypotension or coma. Diving or underwater swimming is especially hazardous. Agitated patients are able to hold their breath without discomfort long enough to produce cyanosis, syncope, decerebrate posturing, and seizures.

Alacrima is the most distinctive of the ocular manifestations (Riley *et al.*, 1949). Baseline eye moisture varies among affected individuals. Since corneal hypoesthesia or anesthesia is a component of the disease, reflex tearing in response to foreign material is absent. Overflow tears are not produced with emotional crying. Corneal de-epitheliazation, ulceration, and scarring often occur. The pupil responds appropriately to light and accommodation, suggesting intact parasympathetic pathways, yet is supersensitive to infused methacholine (Korczyn *et al.*, 1981; Smith *et al.*, 1965c). Abnormal permeability of the damaged cornea might contribute to this sensitivity. Strabismus and myopia also occur with a high frequency (Smith *et al.*, 1965a).

Cardiovascular instability is noted throughout life. Blotchy erythema of the skin occurs during eating or with emotional excitement and may be accompanied by hypertension and peripheral vasoconstriction. Postural hypotension is common, since there is no appropriate sympathetic response to changes in position (Ziegler *et al.*, 1976).

Orthopedic problems are frequent. By 16 years of age, 95% of patients have spinal curvature, scoliosis or kyphosis or both (Axelrod *et al.*, 1974a). Response to trauma is blunted due to decreased pain perception, and unrecognized fractures can heal with deformities.

Progressive diminution of renal function with age is a frequent observation (Pearson *et al.*, 1980). A number of adult patients have died in renal failure. Moderate azotemia is an early sign. Creatinine clearances decrease, and many patients have subnormal renin excretion.

With improved supportive measures, survival and general prognosis have improved (Axelrod and Abularrage, 1982). Mortality has decreased in the early years of life, and the probability that a patient will survive past 30 years is now 50%. Many of these adults are employed, some have married, and women have borne and men have begotten normal offspring.

2.2. Diagnosis

The protean clinical manifestations are all attributed to the basic neurological lesion: a developmental arrest of the sensory and autonomic systems, with sympathetic development more widely affected than parasympathetic. The following description of the consistent neuropathological lesions is by Dr. Pearson.

Diagnosis of familial dysautonomia must be based on recognition of sensory and autonomic dysfunction. Although there is marked variability in expression of the disease, there are specific signs that will occur in more than 90% of patients. These most consistently found signs are listed in Table I.

Sensory involvement is prominent in temperature discrimination, with poor ability to distinguish between hot and cold, especially on the trunk (Axelrod *et al.*, 1981). Hypoalgesia is often first noted by the blunted response to trauma, either a corneal lesion or a bone or skin injury. However, visceral pain sensation appears intact, with the patients having appropriate responses to pleural effusions, esophageal irritation, and menstrual cramping. Touch sensation is well preserved,

Table I. Diagnostic Signs of Familial Dysautonomia

Signs of sensory dysfunction
  1. Decreased response to pain and temperature
  2. Absence of axon flare after intradermal histamine
  3. Decreased or absent deep tendon reflexes
  4. Decreased or absent corneal reflexes
  5. Absence of fungiform papillae on the tongue and decreased taste
Signs of autonomic dysfunction
  1. Absence of overflow tears
  2. Postural hypotension
  3. Miosis following intraocular administration of dilute mecholyl 2.5% or pilocarpine 0.0625%
  4. Blotching
  5. Increased sweating

and dysthesias are often noted with heightened sensitivity in particular areas, e.g., neck, ears, and genitalia (Axelrod *et al.*, 1974a; Riley, 1957). Many patients are extremely ticklish.

Defective afferent conduction probably accounts for the absence of deep tendon reflexes and the lack of an axon flare following intradermal histamine (Smith and Dancis, 1963).

Sensory branches of cranial nerves are also affected, as manifested by decreased taste, gag, and corneal reflexes. Suggesting even more widespread central involvement is the dysfunction of the respiratory center in compensating for hypoxia and hypercapnia, the frequent finding of diffusely abnormal electroencephalograms, and poor spatial sense. In addition, some patients have demonstrated central apnea.

Peripheral abnormalities of *autonomic function* are noted in both sympathetic and parasympathetic systems. Alacrima is present (Smith *et al.*, 1965a). Orthostatic hypotension is explained by sympathetic denervation (Ziegler *et al.*, 1976; Axelrod *et al.*, 1974a). Both dopamine-β-hydroxylase and norepinephrine levels fail to rise normally when patients assume erect posture, yet during emotional crises, dopamine and norepinephrine are markedly elevated. Intermittent blotching and excessive sweating are noted with excitement (Axelrod *et al.*, 1974a; Riley, 1957). Patients are hypersensitive to infused sympathomimetic and parasympathomimetic agents (Smith and Dancis, 1964; Smith *et al.*, 1965c). Small doses of norepinephrine cause hypertension without compensatory bradycardia. Hypersensitivity to methacholine results in profound falls in blood pressure without tachycardia. Sympathetic denervation and unopposed parasympathetic activity increase the possibility of vasovagal reactions.

Further supportive evidence includes feeding difficulties, repeated aspiration, episodes of hypothermia, breath-holding spells, hypotonia and delayed motor development, repeated vomiting, abnormal cinesophagram, spinal curvature, poor somatic growth, Ashkenazi Jewish parentage, and normal intelligence.

3. OTHER CONGENITAL SENSORY NEUROPATHIES

3.1. Diagnostic Differences

There are at least five congenital sensory neuropathies other than familial dysautonomia, some of which also have autonomic dysfunction. They are even rarer than familial dysautonomia, and in most instances the genetics are unknown. Since familial dysautonomia has been the most extensively described, the afore-described clinical diagnostic signs can be used to evaluate the other disorders (Table II). The following description of the neuropathological lesions that can be used to further distinguish the disorders is by Dr. Pearson.

All five disorders presented in this section lack an axon flare following intradermal histamine, indicating peripheral sensory fiber dysfunction. The presence of an axon flare following injection of intradermal histamine phosphate was originally considered to depend on reflexive conduction within the peripheral sensory axons (Smith and Dancis, 1963). Although an abnormal histamine test indicates sensory dysfunction, it is not selective in predicting the neuropathological lesion. Lack of axon flare not only occurs with neuronal depletion, but also has been reported with lesions cephalad to the pons (Cooper, 1950). Neurotransmitter abnormalities can affect the axon flare and may account for abnormal results in atopic dermatitis (Cooper, 1950). Recently, substance P deficiency has been postulated to play a role in neurotransmission of the axon flare (Nicoll *et al.*, 1980). Transient lack of an axon flare has also been reported in a patient with primary deficiency of propionyl coenzyme A carboxylase (Harris *et al.*, 1980). However, the histamine test remains a good method of determining sensory-fiber function, and when an axon flare is absent, sensory function should be considered impaired on the basis of either neuronal depletion or neurotransmitter dysfunction.

Absent or *hypoactive deep tendon reflexes* are characteristic of patients with familial dysautonomia, yet 5% of patients have normal deep tendon reflexes (Axelrod *et al.*, 1981). All the congenital sensory neuropathies, with the exception of congenital sensory neuropathy with anhidrosis, have hypoactive deep tendon reflexes.

Pain sensation as well as cranial sensory nerve involvement are further tested by examining *corneal reflexes* and inspecting the tongue for *fungiform papillae*. Reduced numbers of vascularized lingual papillae appear to be associated with neuronal depletion (Pearson *et al.*, 1970) and result in decreased taste (Smith and Dancis, 1965b). All six congenital sensory neuropathies have diminished corneal reflexes, but only patients with congenital sensory neuropathy with anhidrosis have normal lingual fungiform papillae.

Table II. Affected Neurological Functions and Genetics of Six Congenital Sensory Disorders

Neurological function	Familial dysautonomia (HSN Type III)	Congenital sensory neuropathy with aniydrosis (HSN Type IV)	Congenital sensory neuropathy (HSN Type II)	Congenital sensory neuropathy with skeletal dysplasia	Progressive panneuropathy with hypotonia	Congenital autonomic dysfunction with universal pain loss
Sensory						
Pain	Reduced	Absent	Absent	Absent	Reduced	Absent
Temperature	Reduced	Reduced	Absent	Absent	Reduced	Absent
Touch	Normal	Reduced	Absent	Absent	Normal	Reduced
Position sense	Reduced–normal	Normal	Reduced	?	Reduced	?
Visceral pain	Normal	Reduced	Reduced	Absent	Normal	Absent
Axon flare	Absent	Absent	Absent	Absent	Absent	Absent
Reflexes						
Tendon	Hypoactive or absent	Hypoactive or normal	Hypoactive	Absent	Absent	Absent
Superficial	Hypoactive or normal	Hypoactive or normal	Hypoactive	Absent	?	Absent
Cranial nerves						
Corneals (V)	Normal, diminished, or absent	Normal	Diminished	Absent	Diminished	Absent
Gag (IX, X)	Diminished or norm.	Normal	Diminished	Diminished	Diminished	Diminished
Pain (V)	Reduced	Reduced	Absent	Reduced	?	Reduced
Taste (XII)	Reduced	Normal	Reduced	?	?	?
Autonomic						
Sweating	Increased (with stress)	Absent	Reduced or normal	Reduced	Increased	Increased (with stress)
Tear production	Decreased	Normal	Normal	Normal	Normal, then reduced	Decreased
Postural hypotension	Present	Not present	Not present	Not present	Present	Not present
GI motility	Abnormal	Normal	Variably abnormal	Abnormal	Abnormal	Variably abnormal
Intelligence	Normal	Decreased	Normal	Decreased	Normal	Normal
Genetics	Autosomal recessive	Autosomal recessive	Autosomal recessive	?	Autosomal recessive	?

Patients should then be examined for autonomic dysfunction. Objective tests of parasympathetic function are available.

Alacrima, or impaired tear secretion, is characteristic of decreased cholinergic function. When cholinergic function is impaired on a widespread basis, then sweating and saliva production are decreased as well. In familial dysautonomia, there is a paradox: alacrima coexists with episodic excessive sweating and often copious drooling.

Absence of overflow tears should be carefully assessed. Some patients were thought initially to have alacrima, but had delayed tearing. Lack of overflow tearing is considered normal up to 6 months of age, but may be delayed further in a neurologically immature infant or one who is dehydrated. In addition to clinical history of alacrima, baseline eye moisture can be tested with the Schirmer test.

The *intraocular mecholyl test* is based on the presumption that there is effector hypersensitivity with parasympathetic denervation. We did not find pupillary hypersensitivity to parasympathetic agents of aid in differentiating among the various disorders. Abnormal responses may also be explained by enhanced penetration of the drug via corneal lesions or by lack of dilution by reflex tearing in an anesthetic eye (Korczyn *et al.,* 1981).

Cinesophagrams and *manometric studies* assess gastrointestinal motility. Abnormalities of peristalsis as well as incompetence of the lower esophageal sphincter may be noted (Margulies *et al.,* 1968; Axelrod *et al.,* 1981).

Sympathetic dysfunction is characterized by *orthostatic hypotension* and constancy of heart rate with hypotension. Blood pressure should be taken with the patient supine and then erect. In a young child who cannot stand, a tilt table can be used. *Infusion of sympathetic agents* such as norepinephrine and epinephrine results in exaggerated responses. Serum dopamine-β-hydroxylase levels and urine levels of catecholamine metabolites can be done, but inconsistent results are noted even with familial dysautonomia (Goodall *et al.,* 1971; Weinshilbaum and Axelrod, 1971; Goldstein *et al.,* 1972).

3.2. Summary

Congenital sensory neuropathy with anhidrosis (HSN Type IV) can be easily distinguished clinically by the presence of lingual fungiform papillae, overflow tearing, normal deep tendon reflexes, and absence of truncal sweating. In this disorder, the sensory deficit is extensive and self-mutilation is prominent, as distinct from familial dysautonomia, in which the sensory deficit is prominent but limited and self-mutilation rare. Autonomic dysfunction appears to be confined to the parasympathetic system. Neither blotching nor hypotension, anomalies attributable to sympathetic dysfunction, is present; esophageal motility and serum catecholamines are normal (Swanson, 1963; Lee *et al.,* 1976; Pinsky and DiGeorge, 1966).

There are at least three other congenital sensory neuropathies that share with familial dysautonomia absence of axon flare, diminished corneal reflexes, decreased tendon jerks, and absent fungiform papillae but are clinically distinct in being associated with normal formation of overflow tears.

Patients with *congenital sensory neuropathy* (HSN Type II) have universal pain loss, including loss of touch, and profound hypotonia (Barry *et al.*, 1974; Winklemann *et al.*, 1962; Ogden *et al.*, 1959). In familial dysautonomia, pain loss is not universal and hypotonia is mild or absent. Apnea and poor suck may be due to loss of peripheral afferent components of reflexes, but could also reflect intrinsic brainstem dysfunction in these patients and in familial dysautonomia. Peripheral autonomic function is normal in congenital sensory neuropathy, but episodic fevers occur. Consistent abnormalities in serum catecholamines are not present.

Congenital sensory neuropathy with skeletal dysplasia is a disorder that appears to have abnormalities in mesodermal as well as neurological development (Axelrod *et al.*, 1983b). There appears to be a peripheral distribution of the skeletal and neurological abnormalities. There is no evidence of peripheral autonomic problems, since there is neither blotching nor blood pressure lability. Hypohidrosis may account for the episodic fevers. Central nervous system dysfunction is indicated by abnormal electroencephalogram.

The term *progressive panneuropathy with hypotonia* is used to describe a neuropathy that occurred in two Norwegian brothers (Ørbeck and Oftedal, 1977). Both sensory and autonomic dysfunction are apparent clinically. Hypotonia is much more severe than in familial dysautonomia. Serum catecholamines are normal. The disorder rapidly worsens with age as more clinical anomalies become apparent. A degenerative disease is speculated. Autosomal recessive transmission is postulated.

Patients with *congenital autonomic dysfunction with universal insensitivity to pain* fulfill all the clinical criteria used to diagnose familial dysautonomia (Axelrod *et al.*, 1983a). Use of additional clinical findings can aid in distinguishing these patients from those with familial dysautonomia. In familial dysautonomia, insensitivity to pain is variable in severity and distribution. In this group of patients, loss of pain sensation is severe and involves the entire body. Hypotonia is also more severe. Prominence of frontal bones gives rise to a triangular facial form in these patients, whereas in the young child with familial dysautonomia, the face is normal in shape. A tendency to self-mutilate the eyes and skin is distinct from familial dysautonomia.

4. CONCLUSION

Congenital sensory neuropathies are difficult to diagnose and classify with certainty. We have found Table II helpful in approaching differential diagnosis.

The use of intradermal histamine appears to be a reliable method of clinically ascertaining a sensory neuropathy and predicting peripheral pathology. Clinical information regarding sensory and autonomic function is directive, but pathological examination, such as sural nerve biopsies, aids in classification of more complex cases.

It is unlikely that the other disorders described herein merely reflect different functional penetrance of the same genetic anomaly as familial dysautonomia because distinctive morphological differences are observed in the sural nerve. As yet, we believe that familial dysautonomia is still confined to individuals of Ashkenazi Jewish extraction and that cases of non-Jewish familial dysautonomia are examples of the other congenital sensory neuropathies. Accurate diagnosis and classification are necessary in order to provide insight into developmental differences and to allow for appropriate genetic counseling.

REFERENCES

Axelrod, F.B., 1983, Autonomic and sensory disorders, in: *Principles and Practice of Medical Genetics* (A.E.H. Emery and D.L. Rimoin, eds.), Churchill Livingstone, Edinburgh pp. 284–295.

Axelrod, F.B., and Abularrage, J.J., 1982, Familial dysautonomia: A prospective study of survival, *J. Pediatr.* **101**:234–236.

Axelrod, F.B., and Dancis, J., 1973, Intrauterine growth retardation in familial dysautonomia, *Am. J. Dis. Child.* **125**:379–380.

Axelrod, F.B., and Dancis, J., 1977, Familial dysautonomia, in: *Disorders of the Respiratory Tract in Children*, Vol. 2 (E. Kendig, ed.), W.B. Saunders, Philadelphia, pp. 785–789.

Axelrod, F.B., Nachtigal, R., and Dancis, J., 1974a, Familial dysautonomia: Diagnosis, pathogenesis and management, *Adv. Pediatr.* **21**:75–96.

Axelrod, F.B., Leistner, H., and Porges, R.F., 1974b, Breech presentation in familial dysautonomia, *J. Pediatr.* **84**:107–108.

Axelrod, F.B., Iyer, K., Fish, I., Pearson, J., Sein, M.E., and Spielholz, N., 1981, Progressive sensory loss in familial dysautonomia, *Pediatrics* **67**:517–522.

Axelrod, F.B., Schneider, K.M., Ament, M.E., Kutin, N.D., and Fonkalsrud, E.W., 1982, Gastroesophageal fundoplication and gastrostomy in familial dysautonomia, *Ann. Surg.* **195**:253–258.

Axelrod, F.B., Cash, R., and Pearson, J., 1983a, Congenital autonomic dysfunction with universal pain loss, *J. Pediatr.* **103**:60–64.

Axelrod, F.B., Pearson, J., Tepperberg, G., and Ackerman, B., 1983b, Congenital sensory neuropathy with skeletal dysplasia, *J. Pediatr.* **102**:727–730.

Barry, J.E., Hopkins, I.J., and Neal, B.W., 1974, Congenital sensory neuropathy, *Arch. Dis. Child.* **49**:128–132.

Brunt, P.W., and McKusick, V.A., 1970, Familial dysautonomia, a report of genetic and clinical studies, with a review of the literature, *Medicine* **49**:343–374.

Cooper, I.S., 1950, A neurologic evaluation of the cutaneous histamine reaction, *J. Clin. Invest.* **29**:465–469.

Dyck, P., and Ohta, M., 1975, Neuronal atrophy and degeneration predominantly affecting peripheral sensory neurons, in: *Peripheral Neuropathy*, Vol. 2 (P.J. Dyck, P.K. Thomas, and E.H. Lambert, eds.), W.B. Saunders, Philadelphia, pp. 791–824.

Edelman, N.H., Cherniack, N.S., Lahiri, S., Richards, E., and Fishman, A.P., 1970, The effects of abnormal sympathetic nervous function upon the ventilatory response to hypoxia, *J. Clin. Invest.* **41**:1153–1165.

Filler, J., Smith, A.A., Stone, S., and Dancis, J., 1965, Respiratory control in familial dysautonomia, *J. Pediatr.* **81**:509–516.

Goldstein, M., Dancis, J., Freedman, L., and Axelrod, F., 1972, Serum dopamine-beta-hydroxylase in normal subjects and patients with familial dysautonomia, *Nature (London)* **236**:310–311.

Goodall, G., Gitlow, S.E., and Alton, H., 1971, Decreased noradrenaline synthesis in familial dysautonomia, *J. Clin. Invest.* **50**:2734–2740.

Harris, D.J., Yang, B.I.Y., Wolf, B., and Snodgrass, P.J., 1980, Dysautonomia in an infant with secondary hyperammonemia due to propionyl coenzyme A carboxylase deficiency, *Pediatrics* **65**:107–110.

Korczyn, A., Rubenstein, A., Yahr, M., and Axelrod, F., 1981, The pupil in familial dysautonomia, *Neurology* **31**:628–629.

Lee, E.L., Oh, G.C., Lam, K.L., and Parameswaran, N., 1976, Congenital sensory neuropathy with anhydrosis—a case report, *Pediatrics* **57**:259–262.

Margulies, S.I., Brunt, P., Donner, M., and Silbinger, M., 1968, Familial dysautonomia: A cineradiographic study of the swallowing mechanism, *Radiology* **90**:107–112.

Moses, S.W., Rotem, Y., Jogoda, N., Talmor, N., Eichhorn, F., and Levin, S., 1967, A clinical genetic and biochemical study of familial dysautonomia in Israel, *Isr. J. Med. Sci.* **3**:358–371.

Nicoll, R.A., Schenker, C., and Leeman, S.E., 1980, Substance P as a transmitter candidate, *Annu. Rev. Neurosci.* **3**:227–268.

Ogden, T.E., Robert, F., and Carmichael, E.A., 1959, Some sensory syndromes in children: Indifference to pain and sensory neuropathy, *J. Neurol. Neurosurg. Psychiatry* **22**:267–276.

Ørbeck, H., and Oftedal, G., 1977, Familial dysautonomia in a non-Jewish child, *Acta Paediatr. Scand.* **66**:777–781.

Pearson, J., Finegold, M., and Buzilovich, G., 1970, The tongue and taste in familial dysautonomia, *Pediatrics* **45**:739–745.

Pearson, J., Gallo, G., Gluck, M., and Axelrod, F., 1980, Renal disease in familial dysautonomia, *Kidney Int.* **17**:102–112.

Pinsky, L., and DiGeorge, A.M., 1966, Congenital familial sensory neuropathy with anhydrosis, *J. Pediatr.* **68**:1–13.

Riley, C.M., 1957, Familial dysautonomia, *Adv. Pediatr.* **9**:157–190.

Riley, C.M., Day, R.L., Greely, D.McL., and Langford, W.S., 1949, Central autonomic dysfunction with defective lacrimation, *Pediatrics* **3**:468–477.

Smith, A.A., and Dancis, J., 1963, Response to intradermal histamine in familial dysautonomia— a diagnostic test, *J. Pediatr.* **63**:889–894.

Smith, A.A., and Dancis, J., 1964, Exaggerated response to infused norepinephrine in familial dysautonomia, *N. Engl. J. Med.* **270**:704–707.

Smith, A.A., Dancis, J., and Breinin, G., 1965a, Ocular responses to autonomic drugs in familial dysautonomia, *Invest. Ophthalmol.* **4**:358–361.

Smith, A.A., Farbman, A., and Dancis, J., 1965b, Tongue in familial dysautonomia: A diagnostic sign, *Am. J. Dis. Child.* **110**:152.

Smith, A.A., Hirsch, J.I., and Dancis, J., 1965c, Responses to infused methacholine in familial dysautonomia, *Pediatrics* **36**:225–230.

Swanson, A.G., 1963, Congenital insensitivity to pain with anhidrosis—a unique syndrome in two male siblings, *Arch. Neurol.* **8**:299.

Winshilbaum, R.M., and Axelrod, J., 1971, Reduced plasma dopamine β -hydroxylase activity in familial dysautonomia, *N. Engl. J. Med.* **285**:938–942.

Welton, W., Clayson, D., Axelrod, F., and Levine, D.B., 1979, Intellectual development in familial dysautonomia, *Pediatrics* **63**:708–712.

Winkelmann, R.K., Lambert, E.H., and Hayles, A.B., 1962, Congenital absence of pain, *Arch. Dermatol.* **85**:325–339.

Ziegler, M.G., Lake, R.C., and Kopin, I.J., 1976, Deficient sympathetic nervous response in familial dysautonomia, *N. Engl. J. Med.* **294**:630–633.

Chapter 22

Developmental Neurobiology of Human Disease: Familial Dysautonomia and Related Disorders

John Pearson

To experimental developmental neurobiologists, the human diseases of most interest are those in which an intrinsic error distorts the developing architecture of the nervous system. A very large number of crippling neurological deficits would remain even if it were possible to repair all such intrinsic errors. These result when unpredictable accidents such as infection, ischemia, or trauma distort what would otherwise have been a perfect organism. It is worthwhile to classify disorders of development of the human nervous system as shown in Table I, to highlight those that do have their basis in intrinsic anomalies and to determine those aspects of developmental neurobiology that are likely to contribute to the understanding of familial dysautonomia and related conditions.

This permits the luxury of setting aside a great many problems for others to struggle with. Gone are hemorrhages into the germinal matrix layer occurring in the perinatal period or damage to the same region by viruses. Nor need one consider hydranencephaly and porencephalic cysts resulting from infarction of previously normal tissue, status marmoratus in the basal ganglia resulting from ischemia, damage to multiple nuclei by bilirubin in the syndrome of kernicterus, and many forms of hydrocephalus that result from secondary constriction of well-formed ventricular systems and foramina. In many conditions, the nervous system appears to be the secondary victim of failure in mesenchymal development, as, for example, when occipital, or more rarely cerebellar, tissues protrude through a posterior skull defect in an encephalocele. Similarly, there is a high possibility

JOHN PEARSON ● New York University Medical Center, New York, New York 10016.

Table I. Human Developmental Neuropathology

A.	Secondary lesions of previously normal brain	
	1. Traumatic	Meninges (hemorrhage)
		Lacerations of brain or spinal-cord tissue with hemorrhage
	2. Ischemic	Periventricular infarction ($\pm$ hemorrhage, secondary ependymitis, and leptomeningeal fibrosis)
		Some forms of leukomalacia
		Status marmoratus of basal ganglia (abnormal myelination and gliosis)
		Diffuse cortical atrophy
		Ulegyria (scarred gyri; often at limits of arterial territories)
		Unilateral cerebral atrophy ($\pm$ crossed cerebellar hemiatrophy)
		Porencephaly
		Hydranencephaly
	3. Hemorrhagic	Hemorrhage may play a role in several of the "ischemic" processes; blood pigments appear to be cleared very efficiently, leaving few clues.
	4. Hypoglycemic	Diffuse neuronal necrosis
	5. "Toxic"	Kernicterus (globus pallidus, subthalamic nucleus, brainsten nuclei)
		Cytotoxic drugs (microcephaly)
		Radiation
	6. Infections	Viral (especially rubella, cytomegalovirus: destruction of developing nuclei or germinal matrix or both)
		(Bacterial and fungal rare *in utero*.)
		Protozoal (toxoplasmosis)
	7. Postependymitis	(Precipitating agent may be infection or blood in the ventricular system.)
		Adhesions of ventricular surfaces.
		Aqueductal stenosis with hydrocephalus
		Leptomeningeal fibrosis with hydrocephalus
	8. Mesenchymal defects	Anencephaly Spina bifida } (? Neurotube closure fails primarily.)
		Arnold-Chiari malformation (small posterior fossa)
		Encephalocoele
		Arachnoid cysts
B.	Primary lesions of the developing brain	
	1. Metabolic	(Many diseases may be lethal *in utero*.)
		Abnormal myelination (e.g., globoid-cell leukodystrophy)
		Storage diseases (e.g., Tay-Sachs)
		Diseases of intermediary metabolism
		(?Diseases of neurotransmitters—yet to be discovered ?Receptor disorders)[a]
	2. Cytological differentiation of germinal matrix	(Failures to form oligodendrocytes, astrocytes, and neurons presumably lethal, if they occur.)
		Megalencephaly (?overproduction of cells or lack of normal death)

(Continued)

Table I. *(continued)*

3.	Development of major structures	Reduplication of spinal cord
		Holoprosencephaly ("arhinencephaly": failure of formation of telencephalic vesicles: trisomy 13–15)
		Agenesis of corpus callosum (failure of commissural plate)
4.	Neuronal migration	
	i. Cerebral	Agyria (no gyri)
		Pachygyria (reduced number of thickened gyri)
		(In the two conditions above, migration of cerebral neuroblasts appears delayed.)
		Tuberous sclerosis (chaotic migration, focal in nature; abnormal astrocytes)
		Polymicrogyria (terminal neuronal malpositioning)
		Abnormal columnar and/or laminar organization of cortex within grossly normal gyri
	ii. Cerebellar	Dandy–Walker syndrome
		Lhermitte–Duclos syndrome
		Hypoplasia
	iii. Neural crest[a]	Hirschprung's disease (aganglionic segment of gut; most migration is intact)
5.	Anomalies of individual neurons[a]	Down's syndrome (trisomy 21)
		Various forms of "mental retardation"
		Werdnig–Hoffmann disease (spinal motoneuron death)
6.	Neuron–target interaction	
	i. Neuron–neuron	?Failure to establish normal circuiting underlies many dysfunctional states in which no gross or light-microscopic changes are seen.
	ii. Neuron–effector tissue or sensory transducer	Facial paralysis and ophthalmoplegia (?primary neuronal dysgenesis)
		Familial dysautonomia

[a] Processes that may be involved in familial dysautonomia and related disorders.

that in the Arnold–Chiari malformation, the impaction of a folded medulla into the upper cervical spinal canal in such a way as to prevent cerebrospinal fluid drainage from the fourth ventricle will be the outcome of overcrowding in the abnormally small mesenchymal lined space of the posterior cranial fossa.

An abundance of fascinating problems remain. Why, with neurulation complete and with the spinal cord, cerebellum, and brainstem intact, does the prosencephalon fail, partially or completely, to form telecephalic vesicles in the various forms of holoprosencephaly? Perhaps this too is a mesenchymal disease, a failure of mysterious inducers somehow linked to the abnormality of trisomy of chromosomes 13–15 known to underlie the neural and other somatic structural changes that occur (Friede, 1975). In contrast to the chaotic disorders of disposal of abnormal astrocytes and of the guidance of neuronal placement in the affected lumps of thickened cerebral cortex in tuberous sclerosis, more subtle defects appear to be involved in the genesis of agyria and pachygyria. Within the small brain, a fairly uniform cortex develops with a molecular layer, a gray zone of often enlarged neurons, and at least one inner limiting zone of myelinated axons. Deep to this, radially migrating columns of immature neurons appear bogged down en route from the germinal matrix layer, giving the white matter an abnormal gray coloration. Some parts of the cortex may be normal and the cerebellum frequently is so, although granular-cell heterotopias do occur. The dentate gyrus and the inferior olive may be deformed. On rare occasions, the cerebral cortex forms its normal six layers and may function adequately, but laminae of neurons are left behind in the white matter. More commonly, there are nodular heterotopias of gray matter within the white matter of both cerebrum and cerebellum. In these conditions, a diffuse primary abnormality in migrational guidance, persistence, or timing seems to be likely (see Chapter 2).

Such mechanisms may occur in some cases of polymicrogyria, a condition in which distorted folds of fused cortex pile one on the other and have malplaced neurons with a generally reduced number of layers. The transition with underlying white matter is sharp and continues across zones where regions of polymicrogyria abrubtly give way to normal cortex. In polymicrogyria, neuroblasts complete their migration to the cortex but do not arrange themselves appropriately on arrival. The defect may be primary in some cases, but frequent association of polymicrogyria with other anomalies such as porencephaly and some forms of hydrocephalus that are considered to be secondary in nature suggests that the abnormality in control of terminal placement may be acquired.

In the cerebellum, neuroblasts migrating centrifugally through the walls of the neural tube as in the cerebrum are joined by a layer of cells migrating over the surface to form the external granular layer. The Dandy–Walker syndrome is the most dramatic form of cerebellar dysgenesis. In this syndrome, a cystic expanded roof of the fourth ventricle is associated with upward and anterior displacement of the tentorium and the vermis; the latter is more or less rudimentary. Varying degress of hydrocephalus are associated, but the exit foramina of the fourth ventricle are open in the majority of cases (Hart et al., 1972). Cerebellar cortical heterotopias, malformation of the olives, and agenesis of the corpus callosum are often seen in association with the syndrome, suggesting involvement of the rhombic lip and commissural plate in a form of developmental arrest that occurs at about the 3rd fetal month. As with all such arrests, the occurrence of a critically timed accident such as a fetal viral infection must be considered, but occasional reports of the Dandy–Walker syndrome in siblings support the idea that it is the result of an intrinsic dysgenetic process. A failure of cerebellar organization is also seen in the Lhermitte–Duclos syndrome. The white matter is attenuated and Purkinje cells lie in random disarray within a hypocellular but normally positioned granular cortex. The defect may be partial and remain clinically undetected (I have seen a case associated with a well-differentiated cerebellar astrocytoma). More frequent than this syndrome are dense heterotopic neuronal masses resembling a hybrid parody of dentate nucleus and cortex lying within otherwise normal cerebellar white matter. Diffuse neuronal heterotopias also occur. Varying degrees of cerebellar hypoplasia are seen in which cortical neurons are drastically reduced and the dentate nucleus is fragmented into small islands. Olivary, pontine, and vestibular nuclei may be abnormal, and rarely the dorsal column of the spinal cord is absent, thus forming a link with the various forms of familial and sporadic spinocerebellar syndromes that are generally classified as degenerative rather than developmental anomalies.

The histologically obvious deficits in appropriate migration and persistence of neurons in the cerebrum, cerebellum, brainstem, and spinal cord may be like the tip of an iceberg. It is certainly possible that finely timed guidance, say of one type of cortical neurons, could be abnormal and cause brain dysfunction without there being a light-microscopically recognizable defect. Development of dendritic spines has been demonstrated to be abnormal in some cases of brain dysfunction (Purpura, 1974). Even if all structures are correctly aligned, the possibility exists for abnormal expression of transmitters and receptors (cf. Chapter 18). Such subtle changes might well underlie central nervous system (CNS) dysfunction in familial dysautonomia.

Patients with dysautonomia demonstrate clinical signs and symptoms that suggest brain involvement. These include vomiting crises that respond to drugs that block dopaminergic systems, poor suck reflex, abnormal coordination of respiratory responses to alterations in blood gases, grimacing and tongue-thrusting, and episodic anomalies in social behavior (see Chapter 21). The electro-encephalogram is occasionally abnormal. While some of these manifestations could be attributed to damage secondary to hypoxic episodes that occur during breath-holding (air hunger is not perceived) or during bouts of pneumonia associated with aspiration of regurgitated food, it seems unlikely that they can all be so explained. So far, no anomalies have been seen in the brain, but it seems probable that the developmental abnormalities in this disease will eventually be found to involve more than just sensory, sympathetic, and selected parasympathetic ganglia. Some apparent CNS dysfunctions may be secondary to established anomalies in peripheral structures. Reversal of cycling of sleeping and wakefulness may be associated with defective autonomic innervation of the pineal gland. Abnormal responses to blood gases may be due to attenuated sensory input from the carotid body.

In congenital diseases of motoneurons, the interrelationship of target and neuron must be considered. In syndromes of ophthalmoplegia and facial paralysis, cranial-nerve nuclei show normal positions, but there is diminution of motoneurons and peripheral tissues are hypoplastic. Such hypoplasia may be secondary to lack of innervation. Alternatively, peripheral tissues may not develop, resulting in neurons failing to make contacts with muscle and then dying in an extension of the normal process of "morphogenetic necrobiosis." In Werdnig–Hoffmann disease, spinal motoneurons are depleted at birth. Muscle fibers are present in great number. Many are hypoplastic, but heterogeneity of histochemical fiber type indicates that innervation did at one time exist during development and was later lost. Thus, this disease could be a result of primary failure of motoneurons to achieve a normal life-span. Currently, it is not possible to be sure. The muscle fibers present may be failing to produce a trophic factor or other environmental condition essential for motoneuron survival, and thus the lesion may again lie at the periphery.

The importance of matrix material laid down by mesenchyme is a subject of intense study, especially in relationship to the early migration of neural-crest cells (see Chapters 3, 14, and 17). Such studies are certainly relevant to the understanding of familial dysautonomia, in which populations of neurons in dorsal-root sensory ganglia, sympathetic ganglia, and sphenopalatine ganglia are diminished (Table II). It seems an unlikely possibility, however, that basic failure of migration accounts for these anomalies. The adrenal medulla, formed from the same neural crest and at the periphery of a migration pathway in which the affected ganglia lie closer to the source, is normal in familial dysautonomia. Nevertheless, there are aspects of migration that may be involved. What factors induce some traveling neural-crest cells to stop when they reach the dorsal-root ganglion and other more persistent wanderers to settle in the sympathetic chain while some press on even further to the gut wall and adrenal medulla? Do early arrivals overcrowd optimal environments, forcing later immigrants to move on? Are there signpost pioneer cells that sort traffic in a manner analogous to those that lay down the rules for nerve formation in insects (see Chapter 4)? Do the migrating neuroblasts represent an initially chemically heterogeneous group of morphological look-alikes that recognize preferred microenvironments when they arrive at ganglionic sites or that are selectively capable of proceeding down paths that change their character at each way station? Are the members of the initially migrating population uniform in all respects but genetically predetermined to move different distances by inbuilt clocking mechanisms that will change their character at appropriate times (see Chapters 1 and 3)? Do neuroblasts migrate normally but then fail to extend neurites to contact the periphery (see Chapter 9)? Defects of such mechanisms could account for selective loss of neural-crest cells along their migration pathways.

Even if all migrating cells reach their appropriate destinations with their distinctive microenvironments, other processes exist that may selectively deplete them. Defects in such processes seem to be important in dysautonomia, in which it can be demonstrated that sensory-neuronal populations, already small in infancy, continue to diminish with age (Pearson, 1979). The adrenal medullary neural-crest derivatives evidently have adequate support in their local environ-

Table II. Lesions Observed in Familial Dysautonomia[a]

Neuron populations	Findings[b]	Controls
Sphenopalatine-ganglion neuron populations	1,150	(Controls: 56,500)
Ciliary-ganglion neuron populations	2,900	(Controls: 3,670)
Superior cervical sympathetic-ganglion neuron populations	120,000 (Tyrosine hydroxylase altered; occasional inflammatory foci; spinal sympathetic ganglia small)	(Controls: 1,060,000)
C8 dorsal-root-ganglion neuron populations	5,810 (Tend to decrease with age)	(Controls: 49,500)
Other sensory ganglia known to be involved	Trigeminal, nodose, and all spinal sensory ganglia	
Sensory-ganglion morphology	Capsular proliferation: residual nodules up to 49%, tending to increase with age	(Controls: residual nodules up to 4% maximum)
Substantia gelatinosa	Substance-P-depleted	
Spinal dorsal columns	Slight degeneration initially; severe by age 30	
Spinal intermediolateral gray neuron populations	6,260	(Controls: 1,100)
Tongue	Absence of fungiform papillae, severe diminution of taste buds and submucosal neurons	—
Geniculate-ganglion neuron populations	Reduced (Tokita *et al.*, 1978)	—
Spinal-cord γ-motoneuron populations	Reduced (Kawamura *et al.*, 1978)	—
Kidney	Absent autonomic nerves; progressive glomerulosclerosis (Pearson *et al.*, 1980)	—
Tissues in which no morphological changes have been detected to date (?methods insensitive)	CNS, gut neurons, adrenal medulla	—

[a] See Pearson (1979).
[b] For data on which mean neuron counts are based, see papers referenced in Pearson (1979).

ment. The sensory neurons need something more. Why such neurons continue to die is one puzzle that makes it difficult to fit dysautonomia into the idea of a simple anomaly in nerve growth factor (NGF) dysfunction. Dr. Le Douarin reports in a personal communication that when spinal cord and notohcord are removed after neural-crest migration is complete, the dorsal-root ganglia and sympathetic ganglia die, but more peripherally placed components persist in a pattern similar to that of familial dysautonomia.

It has long been known that developing sensory ganglia require NGF for survival but that mature ganglia apparently do not (Levi-Montalcini and Angeletti, 1968). This observation is strengthened by studying the effects of antibodies to NGF during intrauterine development and adulthood in mammals (Johnson et al., 1980; Schwartz et al., 1982). It is perhaps possible that dysautonomic sensory neuroblasts that survive NGF privation in utero are constitutionally less able to survive a normal life-span in the adult. Environmental changes may cause adult neurons to revert to NGF dependency (see Chapters 11 and 12). Certainly it is possible that yet unknown trophic factors are required for maintenance of sensory and sympathetic neurons during development and adulthood (see Chapter 16) and that it is one of these factors that fails in familial dysautonomia. Alternatively, surface markers and trophic factors may be incorrectly genetically reprogrammed during development.

If trophic-factor failure does underlie the disease, then no intrinsic neural anomaly need be postulated. A trophic chemical defect is most likely to lie in peripheral sources of such a factor. If it were centrally derived, then it would have to be manufactured in neurons of the salivatory nuclei, intermediolateral gray columns, the spinal and trigeminal substantia gelatinosa, and the nuclei gracilis and cuneatus, since these are the CNS connections of various neuronal systems known to be abnormal. While there are several reasons that adrenal medullary cells could survive lack of such a centrally produced trophic agent, it is worthwhile to remember that they share innervation from the intermediolateral gray column with the sympathetic ganglia that are affected in familial dysautonomia. It is possible that the trophic support that Varon (Chapter 17) suggests is derived from glial cells could be abnormal in dysautonomia.

Even if dysautonomia is related to failure of a trophic factor, there are several possible sites at which dysfunction could occur. The trophic molecule itself may be abnormal (Schwartz and Breakefield, 1980), there may be anomalies in receptors for the molecule on the neuronal surface, axonal transport (see Chapter 10) may selectively malfunction, or the translation mechanisms required to effect a response to the message may be imperfect.

Doubt regarding simple involvement of NGF in dysautonomia arises from observing the effects of exposing the mammalian fetus to antibodies to this factor. The cranial neural-crest derivatives that give rise to sensory ganglia are damaged in a manner similar to their spinal counterparts, but parasympathetic neurons that share the initial parts of their migratory pathways with the trigeminal ganglion are unscathed (Table III) (Pearson *et al.*, 1983). In familial dysautonomia, the trigeminal ganglion is involved, but so also, and to a very marked degree, is the parasympathetic sphenopalatine ganglion, which is spared in the animal model. Developmental immunosympathectomy does not affect the placodally derived nodose ganglion, but this structure is abnormal in familial dysautonomia. It should be remembered that in the model, antibodies can reach the fetus only when the placenta becomes permeable to them in the latter part of gestation, and it is possible that parasympathetic cranial ganglia pass through a transient phase of NGF dependence early in gestation, although there is no evidence to support the idea that these ganglia develop precociously with respect to those of sensory function. Studies on trophic factors for cholinergic parasympathetic neurons currently concentrate on ciliary ganglia (see Chapter 15); more work is needed on other parasympathetic ganglia.

Patients with familial dysautonomia have marked depletion of substance P (SP)-immunoreactive primary sensory terminals in the substantia gelatinosa (Pearson *et al.*, 1982). This may account for blunting of pain sensation and absence of axon flare in the patients, since both these functions may normally depend on the presence of SP (Lembeck and Gamse, 1982). Adult animals, as stated, do not suffer sensory-neuron depletion when exposed to antibodies to NGF, but the content of SP in sensory ganglia does fall (Schwartz *et al.*, 1982). Lability in neurotransmitter content and even type within sympathetic ganglia has also been demonstrated in adult animals (see Chapters 5 and 8). It is possible that similar

Table III. Effects of Intrauterine Exposure of Guinea Pigs to Antibodies to Nerve Growth Factor[a]

Neuron populations[b]	Test	Control
Sphenopalatine ganglion	9,700 ± 1300	9,800 ± 635
Ciliary ganglion	127 ± 29	134 ± 40
Otic ganglion	5,310 ± 975	4,540 ± 500
C8 dorsal-root ganglion	3,120 ± 96	15,500 ± 180
Trigeminal ganglion	9,400 ± 400	51,000 ± 4400
Root ganglion of the 9th cranial nerve	1,610	14,500
Nodose ganglion	15,500 ± 1530	15,900 ± 980
Spiral ganglion	17,200 ± 3200	18,100 ± 1450

[a] From Johnson *et al.* (1980) and Pearson *et al.* (1983).
[b] The number of ganglia in each case is 3, except for the 9th cranial nerve root ganglion, of which there was 1.

alterations induced by environmental changes may ultimately be lethal to the neurons of humans, which tend to live longer than do cell cultures or rodents. Degenerative changes that are observed in sensory neurons in familial dysautonomia are very slow; there may be insufficient time for their counterparts to develop in experimental models. Evidence for sensory degeneration in the disease consists of clinically progressive loss of touch and vibration senses, diminution of spinal dorsal column size with age, increasing prominence of residual nodules of Nageotte, and a tendency for older patients to have fewer ganglion cells.

Dysautonomia is certainly not the only partially degenerative disorder of interest to developmental neurobiologists. In Werdnig–Hoffmann disease, spinal motoneurons begin to die *in utero* and continue to do so postnatally; the spinal sensory ganglia are also abnormal (Marshall and Duchen, 1975). In Kugelberg–Wellander disease, spinal motoneuron death begins in early childhood. A spectrum of adult-onset diseases show similar spinal motoneuron loss; some are familial, such as peroneal muscular atrophy, in which autonomic (James, 1972) and sensory neurons may be involved (Trevor-Hughes and Brownell, 1972). In one family known to me, both peroneal atrophy and familial dysautonomia are observed. In Down's syndrome, there are abnormalities in development of the CNS; adults with the disease show a precocious tendency to have the pathological stigmata of senile dementia of the Alzheimer type, a destructive syndrome of great socioeconomic importance. Again, there is a suggestion of a link between abnormal developemnt and failure of adult maintenance of neuronal structures.

Even within one small part of the human peripheral nervous system (PNS), the sural nerve, the spectrum of human developmental anomalies is surprising. In Chapter 21, Dr. Axelrod describes the clinical aspects of several clinical syndromes that partially mimic familial dysautonomia but that can be distinguished by minute analysis. Observations on the fiber population of the sural nerve tend to confirm these distinctions (Table IV). In control sural nerves, we find mean total populations of 8500 myelinated and 23,000 nonmyelinated axons.

In congenital sensory neuropathy with anhidrosis, diminution of small myelinated axons has been reported (Swanson *et al.*, 1965; Appenzeller and Kornfeld, 1972). In a sural nerve from one of Dr. Axelrod's patients with this condition, we found 3700 myelinated fibers and 290 nonmyelinated fibers.

In congenital sensory neuropathy with normal sweating, myelinated fibers are reported to be absent from nerve biopsies and nonmyelinated fibers are reduced (Comings and Amromin, 1974; Lambert and Hayley, 1962; Miller *et al.*, 1976; Ohta *et al.*, 1973). Such patients have also been reported to lack Meissner and Paccinian corpuscles in the skin.

In one patient classified by Dr. Axelrod as sensory neuropathy with skeletal dysplasia, we found 4700 myelinated axons and 20,000 nonmlyelinated axons. There was ballooning and abnormal granularity of myelin.

In one patient classified as congenital autonomic dysfunction with universal

Table IV.

Sural nerve	Familial dysautonomia (HSN III)	Congenital sensory neuropathy with anhidyosis (HSN IV)	Congenital sensory neuropathy (HSN II)	Sensory neuropathy with skeletal dysplasia	Congenital autonomic dysfunction with universal insensitivity to pain
Myelinated axons per nerve (normal 8500)	1,900	3,700	Severe reduction	4,700	4,100
Nonmyelinated axons per nerve (normal 23,000)	1,300	290	Normal	20,000	16,000
Morphological features	Small nerve Redundant Schwann-cell folds Myelin morphology normal	Small nerve No redundant Schwann-cell folds Myelin morphology normal	Nerve extremely small in diameter — Myelin morphology normal	Nerve normal in diameter Vesicles in one tenth of axons One fifth of myelin sheaths abnormal	Small nerve — Myelin morphology normal

insensitivity to pain, we found 4100 myelinated axons and 16,000 nonmyelinated axons.

In two siblings with progressive panneuropathy and hypotonia, total absence of dorsal-root-ganglion neurons with replacement by residual nodules of Nageotte indicated massive degeneration (Orback and Oftedal, 1977). No axons can have persisted in peripheral sensory nerves.

It must be emphasized that the aforementioned conditions are extremely rare and descriptions of sural-nerve biopsies even scarcer; future studies may show phenotypic variability in peripheral-nerve morphology.

With familial dysautonomia, we are on surer ground, having studied a sufficient number of cases to state with confidence that myelinated fibers are markedly diminished, especially those of smaller diameter, and that nonmyelinated fibers are severely depleted. Clinical and pathological observations suggest that larger axons are lost more slowly than those of smaller diameter.

Reviewing the sural-nerve changes seen in this group of diseases, it can be seen that in some of them, myelinated or nonmyelinated fibers are selectively involved, and that in some, both types of fibers are abnormal. Thus far, there appear to be no developmental neurobiology studies that can shed light on this pathological variability. It is possible that in those diseases in which myelinated fibers are severely affected, a metabolic anomaly causes the majority of axons to be naked. If this were so, the total number of nonmyelinated axons would be expected to be increased, since their number would be augmented by those fibers that would normally have been myelinated. Since no such augmentation has yet been observed, the hypothesis is rejected. Selective loss of dorsal-root-ganglion neurons might be due to an intrinsic failure of the type suggested for programmed cell death in lower organisms or to a selective lack of an essential type -specific trophic factor from the periphery. In the latter context, it is of interest to recall the lack of cutaneous corpuscles in the skin of patients with congenital sensory neuropathy and normal sweating. Is it possible that such corpuscles secrete trophic factors necessary for survival of some sensory axons? Here we are caught in a chicken-and-egg quandary. It is well established experimentally that sensory axons are essential for the trophic maintenance of some peripheral tranducers, notably the lingual taste buds (Zalewski, 1974). Lack of geniculate-ganglion neurons in familial dysautonomia is associated with marked diminution of taste buds. Information regarding possible trophic input from sensory transducer organelles to sensory axons is lacking.

The conference on which this volume is based was convened to discover how far basic developmental neurobiology has come in explaining the clinical and pathological anomalies observed in familial dysautonomia and other primary congenital neurological human diseases. There are some areas in which great progress is being made. Perhaps the most striking is in the field of orientation of neuronal migration in terms of both the guiding structures and the basic mechanisms and biochemistry of the cell–cell recognition, adhesion, and differential locomotion that are essential to this process. It appears that the means by which the general organization of nervous system anatomy is established will soon be known. While this comes close to the critical process by which active neuronal circuits are initiated, the latter remains a largely mysterious process in complex organisms. Great promise comes from progress being made in understanding mechanisms of visual system development (Trisler *et al.*, 1981) and the thrust toward unraveling the morphogenesis of dendrites and interneuronal synapses (Hume and Purves, 1981). The histological techniques by which such circuits can be examined have improved markedly in the past decade. The subtle interactions that govern development of microcircuits in the human brain and the chemical transmitters involved have yet to be unraveled to the point where they may yield clues as to the CNS dysfunction in familial dysautonomia. Indeed, we do not yet have sufficiently refined clinical or pathological tools to point the developmental neurobiologist toward the most likely site of brain lesions, other than to say that catecholaminergic mechanisms are prime suspects.

In the PNS, well-defined trophic mechanisms, particularly those related to NGF, add to knowledge concerning mechanisms of cell migration and to information regarding control of neuronal health. We are still unable to explain how functionally and chemically heterogeneous neuron populations, even in relatively simple structures like a dorsal root ganglion, can be selectively involved in different diseases.

Familial dysautonomia is a complex syndrome. Like the lethal and relatively simple Tay–Sachs disease, this autosomal recessive disorder is likely to be attributable to a single genetic mutation that causes malfunction of a metabolic pathway or a chemical messenger. Even when that biochemical lesion is precisely identified, studies designed to explain the deficits in familial dysautonomia will yield information of widespread value regarding the means whereby the nervous system organizes and regulates its internal interactions and its external interplay with effector and sensor organs.

REFERENCES

Apenzeller, O., and Kornfeld, M., 1972, Indifference to pain: A chronic peripheral neuropathy with mosaic Schwann cells, *Arch. Neurol.* **27**:322–339.

Comings, D.E., and Amromin, G.D., 1974, Autosomal dominant insensitivity to pain with hyperplastic myelinopathy, *Neurology* **24**:838–848.

Friede, R.L.L., 1975, *Developmental Neuropathology*, Springer-Verlag, New York.

Hart, M.N., Malamud, N., and Ellis, W.C., 1972, The Dandy–Walker syndrome, *Neurology* **22**:771–780.

Hume, R.I., and Purves, D., 1981, Geometry of neonatal neurones and the regulation of synapse elimination, *Nature (London)* **293**:469–471.

James, J.L., 1972, The autonomic nervous system in peroneal muscular atrophy, *Arch. Neurol.* **27**:213–220.

Johnson, E.M., Gorin, P.D., Brandeis, L.D., and Pearson, J., 1980, Dorsal root ganglion neurons are destroyed by exposure *in utero* to maternal antibody to nerve growth factor, *Science* **210**:916–918.

Kawamura, Y., Dyck, P.J., Low, P.A., and Shimono, M., 1978, The number and sizes of reconstructed peripheral autonomic, sensory and motoneurons in a case of dysautonomia, *J. Neuropathol. Exp. Neurol.* **37**:741–755.

Lambert, E.H., and Hayley, A.B., 1962, Congenital absence of pain, *R. K. Winkleman Arch. Dermatol.* **85**:325–339.

Lembeck, F., and Gamse, R., 1982, Substance-P in peripheral sensory processes, in: *Substance-P in the Nervous System,* Ciba Foundation, Pitman.

Levi-Montalcini, R., and Angletti, P.U., 1968, Nerve growth factor, *Phsyiol. Rev.* **48**:534–569.

Marshall, A., and Duchen, L.W., 1975, Sensory system involvement in infantile spinal muscular atrophy, *J. Neurol. Sci.* **26**:349–359.

Miller, R.G., Nielsen, S.L., and Summer, A., 1976, Hereditary sensory neuropathy with tonic pupils, *Neurology* **26**:931–935.

Ohta, M., Ellefson, R.D., Lambert, E.H., and Dyck, P.J., 1973, Hereditary sensory neuropathy, type II, *Arch. Neurol.* **29**:23–37.

Orbeck, H., and Oftedal, G., 1977, Familial dysautonomia in a non-Jewish child, *Acta Paediatr. Scand.* **66**:777–781.

Pearson, J., 1979, Familial dysautonomia (a brief review), *J. Autonom. Nerv. Syst.* **1**:119–126.

Pearson, J., Gallo, G., Gluck, M., and Axelrod, F., 1980, Renal disease in familial dysautonomia, *Kidney Int.* **17**:102–112.

Pearson, J., Brandeis, L., and Cuello, A.C., 1982, Depletion of substance-P-containing axons in substantia gelatinosa of patients with diminished pain sensitivity, *Nature* **295**:61–63.

Pearson, J., Johnson, E.M., and Brandeis, L., 1983, Effects of antibodies to nerve growth factor on intrauterine development of derivatives of cranial neural crest and placode in the guinea pig, *Dev. Biol.* **96**:32–37.

Purpura, D.P., 1974, Dendritic spine "dysgenesis" and mental retardation, *Science* **186**:1126–1128.

Schwartz, J.P., and Breakefield, X.O., 1980, Altered nerve growth factor in fibroblasts from patients with familial dysautonomia, *Proc. Natl. Acad. Sci. U.S.A.* **77**:1154–1158.

Schwartz, J.P., Pearson, J., and Johnson, E.M., 1982, Effects of exposure to anti-NGF on sensory neurons of adult rats and guinea pigs, *Brain Res.* **244**:378–381.

Swanson, A.G., Buchan, G.C., and Alvord, E.C., 1965, Anatomic changes in congenital insensitivitiy of pain, *Arch. Neurol.* **12**:18.

Tokita, N., Sekhar, H.K.C., Sachs, M., and Daly, J., 1978, Familial dysautonomia (Riley–Day syndrome): Temporal bone findings and otolaryngological manifestations, *Ann. Otol. Rhinol. Laryngol.* **87**(Suppl. 46):1.

Trevor-Hughes, J., and Brownell, B., 1972, Pathology of peroneal muscular atrophy (Charcot–Marie–Tooth disease), *J. Neurol. Neurosurg. Psychiatry* **35**:648–657.

Trisler, G.D., Schneider, M.D., and Nirenberg, M., 1981, A topographic gradient of molecules in retina can be used to identify neuron position, *Proc. Natl. Acad. Sci. U.S.A.* **78**:2145–2149.

Zalewski, A.A., 1974, Neuronal and tissue specifications involved in taste bud formation, *Ann. N. Y. Acad. Sci.* **228**:344–349.

Index